Primate Locomotion

Contributors

DICK M. BADOUX
JOHN V. BASMAJIAN
MATT CARTMILL
RICHARD LEE DECKER
J. P. GASC
FARISH A. JENKINS, JR.
F. K. JOUFFROY
O. J. LEWIS
DAVID ROBERTS
M. D. ROSE
FREDERICK S. SZALAY
RUSSELL H. TUTTLE
ALAN WALKER

PRIMATE LOCOMOTION

Edited by

FARISH A. JENKINS, JR.

Department of Biology and
The Museum of Comparative Zoology
Harvard University
Cambridge, Massachusetts

ACADEMIC PRESS New York and London 1974

A Subsidiary of Harcourt Brace Jovanovich, Publishers

ACADEMIC PRESS, INC.
111 Fifth Avenue, New York, New York 10003

United Kingdom Edition published by
ACADEMIC PRESS, INC. (LONDON) LTD.
24/28 Oval Road, London NW1

LIBRARY OF CONGRESS CATALOG CARD NUMBER: 73-5305

PRINTED IN THE UNITED STATES OF AMERICA

Contents

3 Tree Shrew Locomotion and the Origins of Primate Arborealism

Farish A. Jenkins, Jr.

4 A Cineradiographical Analysis of Leaping in an African Prosimian (*Galago alleni*)

F. K. Jouffroy and J. P. Gasc

5 The Wrist Articulations of the Anthropoidea

O. J. Lewis

6 Structure and Function of the Primate Scapula

David Roberts

List of Contributors

Numbers in parentheses indicate the pages on which the authors' contributions begin.

DICK M. BADOUX (1), Institute of Veterinary Anatomy, State University of Utrecht, Utrecht, Holland

JOHN V. BASMAJIAN (293), Departments of Anatomy, Physical Medicine, and Psychiatry, Emory University, Atlanta, Georgia

MATT CARTMILL (45), Departments of Anatomy and Anthropology, Duke University, Durham, North Carolina

RICHARD LEE DECKER (223, 261), Department of Anthropology, Hunter College, New York, New York

J. P. GASC (117), Laboratoire de Biologie–Vertébré, Université Paris XI, Paris, France

FARISH A. JENKINS, JR. (85), Department of Biology and The Museum of Comparative Zoology, Harvard University, Cambridge, Massachusetts

F. K. JOUFFROY (117), Centre National de la Recherche Scientifique, Paris, France

O. J. LEWIS (143), Department of Anatomy, St. Bartholomew's Hospital Medical College, Charterhouse Square, London, England

DAVID ROBERTS* (171), Peabody Museum of Natural History, Yale University, New Haven, Connecticut

M. D. ROSE (201), Department of Human Anatomy, University of Nairobi, Nairobi, Kenya

* Present address: The School of Medicine and The School of Dental Medicine, University of Pennsylvania, Philadelphia, Pennsylvania.

FREDERICK S. SZALAY (223, 261), Department of Anthropology, Hunter College, and Department of Vertebrate Paleontology, The American Museum of Natural History, New York, New York

RUSSELL H. TUTTLE (293), Department of Anthropology and Committee on Evolutionary Biology, and the Biology Collegiate Division, University of Chicago, Chicago, Illinois; and Yerkes Regional Primate Research Center, Emory University, Atlanta, Georgia

ALAN WALKER (349), Department of Human Anatomy, University of Nairobi, Nairobi, Kenya

Preface

This book reviews aspects of current work on primate locomotion, and was assembled in an attempt to bring together selected approaches and topics in this field of study which are highly diverse but which at the same time offer important insights into traditional disciplines—anatomy, anthropology, paleontology, and zoology. Therefore, the book is intended not only for researchers dealing with primate locomotion, but equally for students and others who share an interest in mammals and locomotor adaptations.

The first major studies of primate locomotion were based on field observations and emphasized classifications, or so-called locomotor categories, which grouped primate species according to major behavioral and anatomic similarities. However, the concept of "primate locomotion" as a subject of study is now rapidly changing. Although locomotor classifications are still useful as a conceptual framework, current workers are exploring new techniques and concepts that distinguish behavioral and anatomic differences among primate species that traditionally have been assigned to the same locomotor category. The contributors to this book describe in detail some of these new trends as well as provide guidance to areas in which further investigation might be particularly productive. Biomechanical principles, which are at the foundation of our understanding of locomotor adaptations, are outlined by Badoux in the first chapter. The critical importance of understanding primate locomotion in terms of habitat and naturalistic behavior is emphasized in papers by Rose and Walker, whereas the chapters by Tuttle and Basmajian, Jenkins, and Jouffroy and Gasc discuss data from instrumentation presently restricted to laboratory settings. Paleontology brings an increasingly greater perspective to our view of primate adaptations as the chapters by Szalay, Decker, and Walker show. Furthermore, the degree to which paleontological and neontological interpretations may be closely interrelated is repeatedly demonstrated in discussions by Cartmill, Lewis, Tuttle and Basmajian, and others. The anatomy of various aspects of the appendicular musculoskeletal system, still a major basis for comparative study of primate locomotion, is the subject of different styles of functional analysis in reviews by Cartmill, Lewis, and Roberts.

If a single conclusion may be drawn on the basis of these diverse studies, it is that different approaches and techniques are essential to understand the complex phenomena of primate locomotion. Furthermore, the need for further field studies is clearly presaged; the wealth of detailed information derived from captive specimens and experimental situations must be reexamined in the context of each species' natural habitat and behavior. Thus, primate locomotion continues to offer a spectrum of biological problems to challenge both field and laboratory research.

To the staff of Academic Press I am grateful for assistance in every phase of this undertaking. Special recognition is due Professor Charles R. Noback, my friend and former colleague in The Department of Anatomy at The College of Physicians and Surgeons of Columbia University, with whom the concept of this book was initially conceived.

FARISH A. JENKINS, JR.

1

An Introduction to Biomechanical Principles in Primate Locomotion and Structure

DICK M. BADOUX

General Principles

At the present time, a detailed survey of primate biomechanics cannot be made since not enough is known about the various forms of locomotion in primates, in general, and their joint movements, in particular; detailed biomechanical information on specific parts of the locomotor apparatus of nonhuman primates is scarce. However, the present contribution is based upon the supposition that biomechanical principles based on mammalian morphology and locomotion in general apply also to primates. Since my principal field of interest lies in the analysis of parallels and analogs between mammalian and mechanical structures, some of the ideas expressed in this chapter will be personal and perhaps provocative.

Biomechanics is an old science and its practical implications date as far back as man's dawn as a toolmaker; however, it obtained its firm theoretical basis from technical advances made principally in the last three decades. Generally speaking, biomechanics deals with forces, accelerations, and movements acting in and on living bodies, the analysis of which is made possible by a wide variety of specialized and sophisticated techniques. Since living matter is subjected to the same physical laws and rules as inanimate bodies, the subdivision of biomechanics is analogous to that in physical mechanics and distinguishes two subdisciplines: "biodynamics" and "biostatics." Biodynamics is subdivided into "kinematics" and "kinetics." Kinematics analyzes motion in space and time without taking into consideration the forces which cause these motions. Kinetics studies changes in motion caused by an unbalanced system of forces and determines the force required to produce any desired change of motion.

Despite the great variation in primate conformation and the versatility of their movements, the basic mechanical principles underlying primate locomotion are relatively limited and derive from Newton's three laws of motion. The first law states that a body continues in its state of rest or uniform motion in a straight line except insofar as it is compelled by applied forces to change that state. Although the first part of the law with regard to bodies at rest is self-evident, the second part needs further consideration. Consider a monkey jumping from one branch to another. The body leaves the branch with a velocity of V m/sec and would perpetually continue its course in a straight line if there were no gravitation and air resistance. Under the influence of gravitation and air resistance it follows a parabolic curve and comes to rest on the "target" branch. The second law states that change of momentum per unit time is proportional to the applied force and takes place in the direction of the force. The biological implication is that when an animal exerts a propulsive force F (expressed in newtons), the velocity V (in meters) imparted to the animal is proportional to the magnitude of the force and the period of time T (in seconds) during which it acts and is inversely proportional to the mass M (in kilograms) of the body: hence

$$V = FT/M \tag{1}$$

The velocity of the body is increased by the amount F/M for every second for which the force is applied so that the acceleration A (in meters per second per second) is:

$$A = F/M \tag{2}$$

If an animal suddenly comes to rest under the influence of a constant restraining force F (friction between the ground and the soles of the hands and feet of the fixed legs), the whole of the animal's kinetic energy is dissipated when the force moves the animal distance S (in meters) so that

$$FS = \tfrac{1}{2} MV^2 \tag{3}$$

or

$$S = MV^2/2F \tag{4}$$

The negative acceleration follows from Eqs. (2) and (4):

$$F = MA$$

hence

$$S = MV^2/2MA = V^2/2A$$

thus

$$A = V^2/2S \tag{5}$$

The third law states that forces always occur in pairs, each pair consisting of two equal opposites; in other words, for every action there must be a reaction. The implication of this well-known principle is the most important concept in the analysis of animal locomotion: when an animal subjects its body to a forward propulsive force, the environment exerts an equal but opposite background force against the animal. The animal moves forward because the environment resists the movements of the limbs relative to the body.

"Biostatics" deals with forces and their equilibrium acting upon animals and their organs in a state of rest or in motion at uniform velocity in a straight line; it draws upon analogies between the components of the locomotor apparatus—bones, cartilage, muscles, ligaments—and mechanical structures. Furthermore, biostatics includes the study of load effects on locomotor organs and their individual members and is closely related to the subject of "material strength." Biostatic investigations therefore cover such subjects as strength of bone and cartilage, shape and position of trabeculae in spongy bone, lubrication and wear in joints, and many others. The most common concepts in this type of biomechanical research are "stress" and "strain." A stress is constituted by the equal and opposite action and reaction which take place between two bodies or two parts of the same body transmitting forces. A simple example of material subjected to stress is that of a tendon attached to a bone (Fig. 1). Let the force exerted by the muscle be F kg. Consider a planar section XX through the tendon perpendicular to its axis, with an area measuring P cm^2 and dividing the tendon into the portions A and B. The portion A exerts a longitudinal pull on B and vice versa. The average force per cm^2 at the section XX is

$$\sigma_t = F/P \tag{6}$$

and this is the mean intensity of "tensile stress" at the section. If the angle of insertion of the tendon be α, then $F_1 = F \cos \alpha$ is the component of F acting perpendicularly to the axis of the bone. At the point of insertion of the tendon, consider a planar section YY of P cm^2 perpendicular to the long axis of the bone, and dividing the bone into the portions C and D. The force F_1 has the tendency to slide C to the right side relative to D

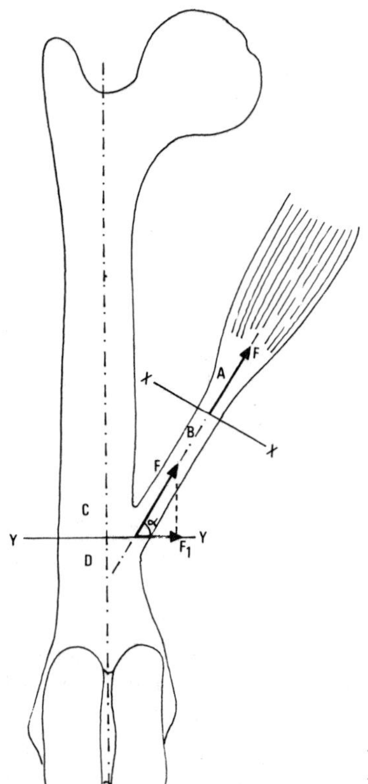

Fig. 1. An example of a femur subjected to stress by muscular force (F) of an adductor inserted at angle α. For further explanation, see text.

and this is reacted upon by a shear that balances F_1. The "shear stress" is given by

$$\sigma_s = F_1/P \tag{7}$$

The third type of stress, "compression stress," exists between two portions of a body when each pushes the other from it. Consider a bone such as the tibia in which the mechanical and morphological axes coincide, and place it under a vertical load of F kg (Fig. 2). At a transverse section XX of P cm^2 dividing the bone into two parts A and B, the material is under compressive stress. The portion A exerts a push on B equal and opposite to $-F$ exerted by B. The average force per cm^2 of the section is

$$\sigma_c = F/P \tag{8}$$

and this is the mean intensity of compressive stress at section XX. If a bone is either acted upon by an eccentric load or by a set of two parallel but

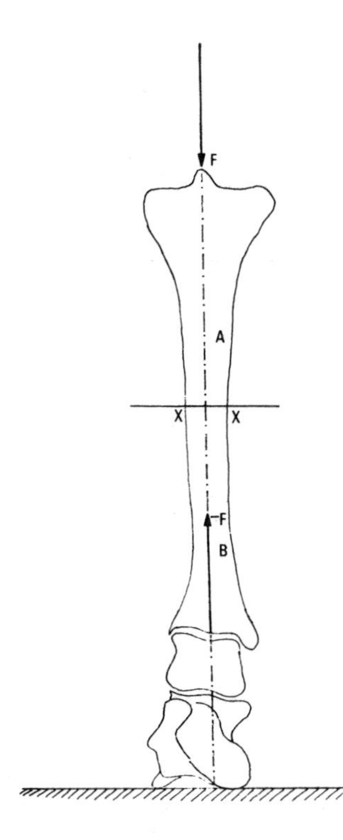

Fig. 2. An example of compression stressing of a tibia with a force of F kg. For further explanation, see text.

opposite forces (a couple), it is subjected to "bending." The induced stress can be calculated from specific formulas (11 through 15) discussed below.

Strain is the alteration of shape or dimension resulting from stress. If a ligament measuring b cm in length is stretched to a length of $b + \Delta b$ cm, the "tensile strain" is

$$\epsilon = \Delta b / b \tag{9}$$

and if a piece of cartilage of a height of b cm is compressed to a height of $b - \Delta b$ cm, the "compressive strain" is again

$$\epsilon = \Delta b / b \tag{10}$$

Tensile stress causes a contraction perpendicular to its own direction and compressive stress produces an elongation perpendicular to its own direction (Fig. 3). A third type, "shear strain," is the angular displacement produced by shear stress. In Fig. 4 an intervertebral disc (D) is subjected to shear stress in a certain plane due to a twist (T) of the vertebral column; the displacement $ZZ' = \alpha$ (measured in radians) is the numerical measure of the resulting shear strain. For further information, see Den Hartog (1949).

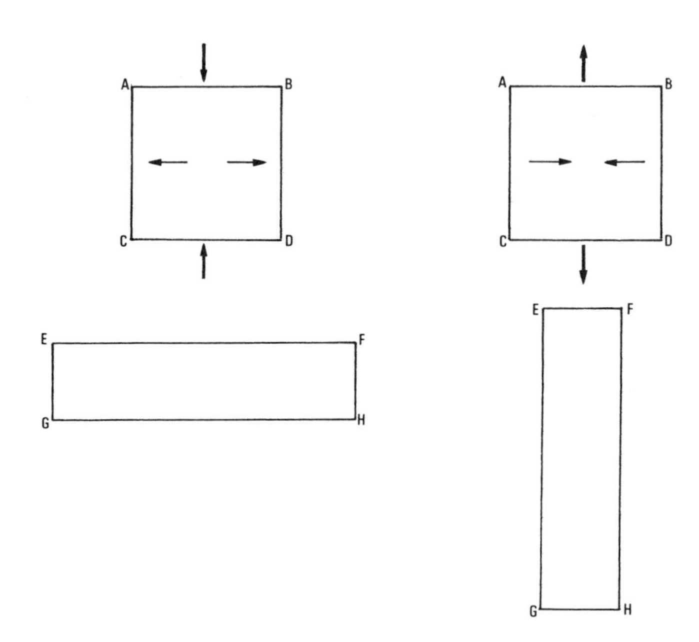

Fig. 3. Deformation of material. The left square ABCD represents a section under compression which will tend to alter its shape to EFGH with material extension in a direction perpendicular to the compressive force. The right square ABCD is under tension and will tend to extend to EFGH in the direction of the tensile force.

Although biomechanics is a precise tool for analyzing biological structure and movement, used alone it is insufficient to explain the great diversity of morphological adaptations. Survival depends upon a variety of habitat adaptations, only a few of which are of a purely mechanical nature. The agile and dexterous brachiation of the gibbon, for example, is accomplished by a specialized locomotor apparatus; yet structural adaptation to climbing and arm swinging is not sufficient per se to ensure the survival of an ape in its natural environment. Special demands are made upon sensory, masticatory, digestive, circulatory, respiratory, excretory, and neurohormonal systems, all of which must be accommodated within a structural framework suitable for arboreal locomotion. Although brachiation as a locomotor mode requires certain biomechanical adaptations, these play only a subordinate role in the totality of structural characteristics acquired during evolution. Indeed, *any animal is a compromise structure;* this statement is valid for the creature as a whole as well as for its organ systems. Examination of the pelvic region of a quadrupedal macaque, for example, will illustrate this point. In the standing animal (Fig. 5), the most favorable position of the bony pelvis is

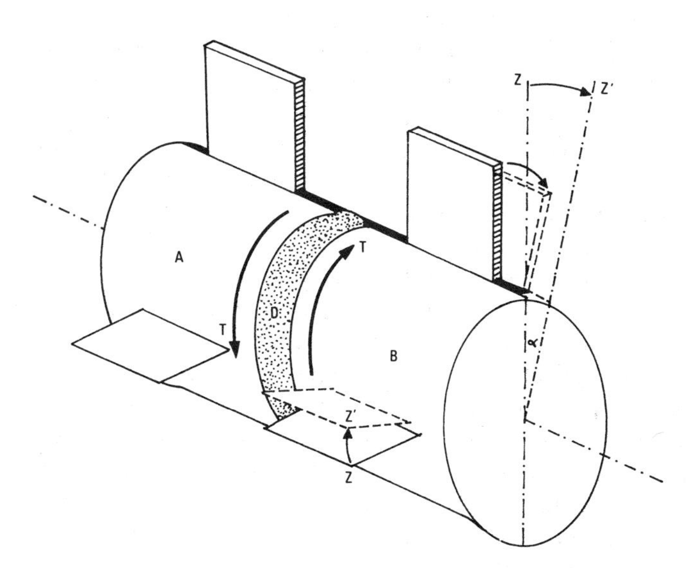

Fig. 4. Angular displacement resulting from shear stress. An intervertebral disc (*D*) is distorted by a twist (*T*) over an angle α from Z to Z'.

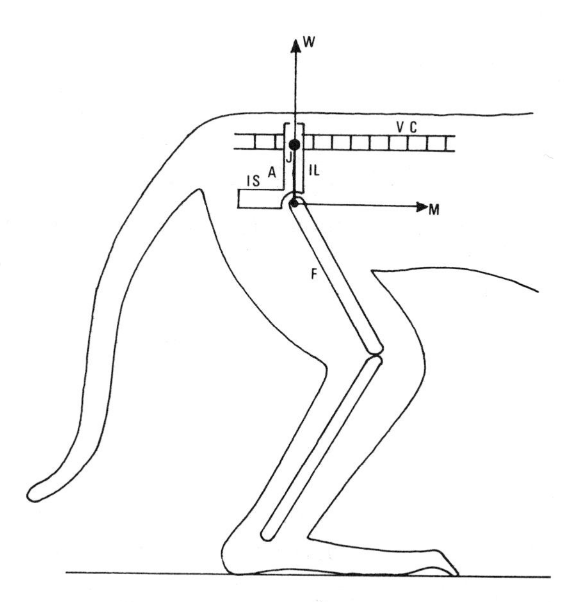

Fig. 5. Diagrammatic illustration of an iliosacral angle of 90° which theoretically is the most favorable orientation for a standing quadruped. *W*, reaction of the femur (*F*) against the acetabulum (*A*) and passing through the iliosacral joint (*J*), its moment is zero. *IL*, ilium; *IS*, ischium; *VC*, vertebral column; *M*, propulsive force.

such that the long axis of the ilium (IL) is perpendicular to the longitudinal axis of the vertebral column (VC): hence, the iliosacral angle measures 90°. In this case, the moment of femoral reaction *W* against the dorsal roof of the acetabulum about the iliosacral joint (J) is zero since its working line passes through the joint. In the running animal, however, this position is inexpedient. The propulsive force *M* is mainly generated by the contraction of the hamstring group (biceps femoris, semitendinosus and semimembranosus muscles) and the moment of *M* about the iliosacral joint (J) is the product of the force pushing forward against the cranial side of the acetabulum and the perpendicular distance between its working line and joint. When the iliosacral angle is 0°, i.e., when the axes of ilium and vertebral column are concurrent, the moment is zero so that no distortion occurs in the joint (Fig. 6). Apparently, static and dynamic requirements are conflicting (iliosacral angle 90° versus 0°). An additional factor must be considered: the iliosacral angle partly determines the space between the sacrum and the pelvic brim through which the young must pass at birth. When this angle is 90°, the passage has its maximum dimension. If, therefore, biostatic, biodynamic, and obstetric factors influence one and the same pelvic structure, it is not surprising to find an iliosacral

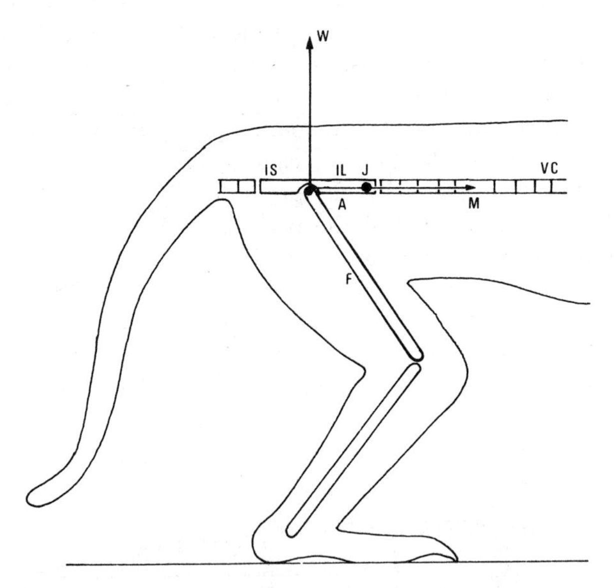

Fig. 6. Diagrammatic illustration of an iliosacral angle of zero, theoretically the most efficient arrangement for a running quadruped in which the propulsive force (*M*) has a zero moment about iliosacral joint (*J*). For further details, see text; other abbreviations as in Fig. 5.

angle of about 35° in the macaque, which represents a compromise between the extremes.

In the following sections I shall discuss some biomechanical principles relevant to mammalian structure, in general, and primate structure, in particular.

The Body Axis

The vertebral column, the intervertebral ligaments, and the epaxial (mainly iliocostal, longissimus, and spinalis) and hypaxial (longus colli and longus capitis) muscles form the central axis of the body. The components of this axis are intrinsic parts of the locomotor apparatus and it is logical that their structure and mechanical properties should relate to the phylogenetic level of the locomotor stage of the animal.

In aquatic lower vertebrates—fishes and some amphibians—the body is surrounded and supported by water, and locomotion is principally effected by lateral movements of the body axis. The rather homogeneous character of the spine in this group reflects the neutralized influence of gravitation on the vertebral column; the body, wholly or partially immersed in the water, it is acted upon by an upward force equal to the weight of the water it displaces. The prime task of the vertebral column during propulsion is to serve as a flexible rod acting as an antagonist to the serially arranged myomeres.

In semiterrestrial amphibians and some reptiles, body weight is carried by weakly developed girdles and limbs, and the ventral side of the trunk and tail may be in contact with the ground (Fig. 7). Propulsion occurs in part by lateral movements of the body axis. The acropodial elements of the limbs function as rotating points of contact with the ground when the animal moves with a diagonal sequence of limb displacements. The horizontal position of the stylopodia is a mechanically inefficient means to transmit gravitational force from the girdles to the limbs. In amphibians, the joint between shoulder girdle and trunk is syndesmotic or synsarcotic

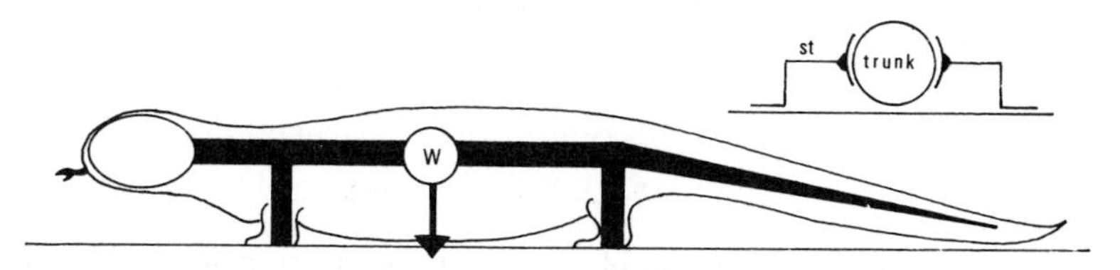

Fig. 7. Diagrammatic illustration of the relationship between body axis limbs in semiterrestrial amphibians and some reptiles. The stylopodia (humeri and femurs) are in a more or less horizontal position. *W*, gravitational weight.

(i.e., joined either by connective or muscular tissue); in reptiles, there is almost an immovable synostotic or syndesmotic connection between clavicle, interclavicle, and sternum.

In quadrupedal and especially cursorial mammals (Fig. 8), the body is lifted up above the ground, the stylopodia have a more upright position,

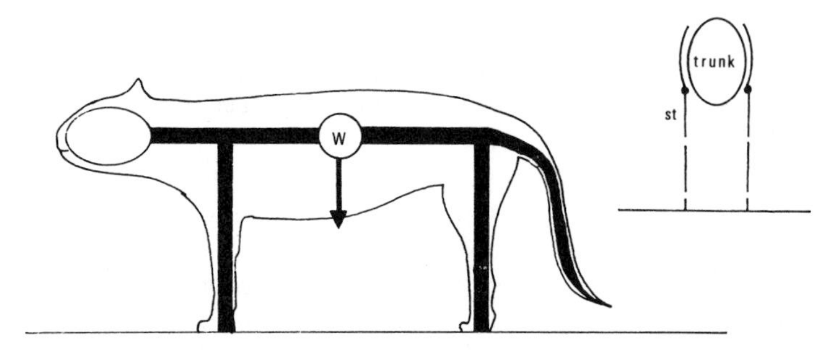

Fig. 8. Diagrammatic illustration of the relationship between body axis and limbs in a quadrupedal, cursorial mammal. Axial flexion is not shown. The stylopodia are more or less vertical in orientation. *W*, gravitational weight.

and the gravitational forces are efficiently transmitted from the trunk to the girdles; locomotion occurs either in the primitive diagonal pattern of footfalls or in other more specialized modes. As far as the primates are concerned, one may assume that the ancestral Cretaceous group moved with a diagonal pattern of limb movements, a mode that is maintained up to the present day by a number of prosimians and simians. From this primitive type of locomotion, more complicated and sometimes highly specialized modes of progression evolved: bipedal jumping in *Tarsius, Galago,* and *Cheirogaleus,* quadrupedal walking-climbing (most of the Cercopithecidae), hanging–climbing or brachiation (*Hylobates, Pongo, Ateles*), semibrachiatory (*Presbytis, Nasalis, Alouatta, Lagothrix*), and finally bipedalism, facultatively adopted by the great apes and permanently by man.

In aquatic lower vertebrates the function of the body axis is primarily dynamic (propulsive); in terrestrial quadrupeds it must bear body weight (gravitation) as well as transmit propulsive forces generated mainly by the hindlimbs. In terrestrial vertebrates, the construction of the body axis represents a compromise between the vectorial directions of gravitational and propulsive forces, and must be able to resist deforming loads as well as absorb impacts in the direction of the resultant of all propulsive forces.

Elevation of the body off the ground promotes a more marked subdivision of the body axis into three segments. The cranial segment (CRS,

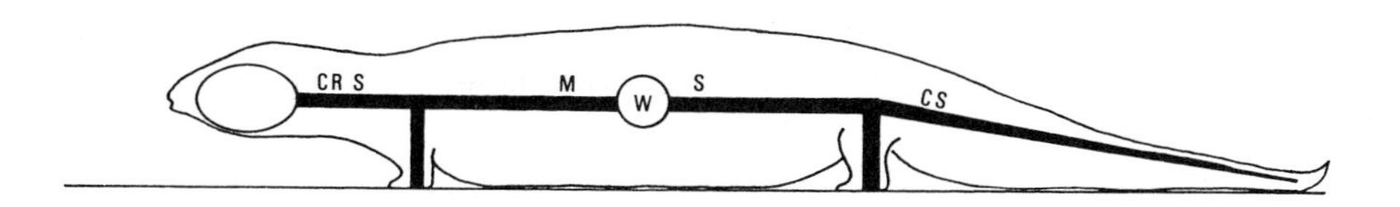

Fig. 9. The body axis of quadrupedal vertebrates is divided into three mechanically different segments: cranial (CRS), middle (MS), and caudal (CS). *W*, gravitational weight.

Fig. 9) comprises head and cervical vertebral column and is statically equivalent to a flexible rod fixed at one end and carrying a weight on the other. A middle segment (MS, Fig. 9), comprising the thoracic, lumbar, and sacral divisions, is in fact a flexible rod fixed on both ends. Finally, the caudal segment (CS, Fig. 9) is represented by the tail, analogous again to a rod fixed at one end. Elevation of the body requires stiffening of these segments, which would otherwise tend to sag under their own weight. Firmness is effected by a bow-and-string construction in which the "bow" consists of alternating solid and elastic elements, vertebrae and ligaments, which form an elastic synarthrodial rod (Fig. 10). The degree of flexion or curvature can be momentarily stabilized by intrinsic elastic

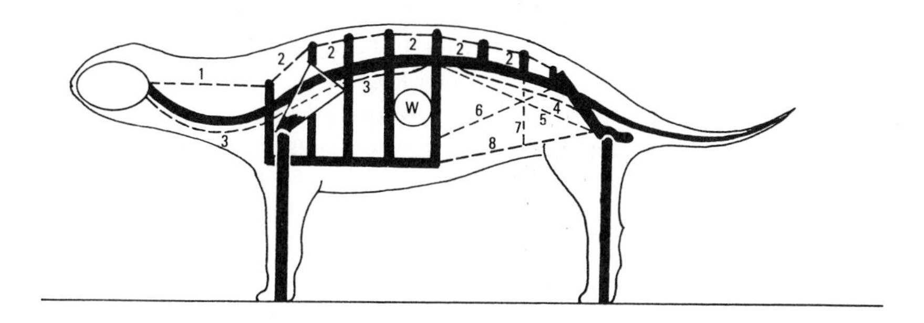

Fig. 10. Diagrammatic representation of the body structure of a quadrupedal mammal. 1, 2, dorsal (epaxial) muscles; 3, ventral (hypaxial) muscles; 4, sublumbar muscles; 5, 6, 7, 8, oblique, transverse, and straight abdominal muscles. *W*, gravitational weight.

elements, i.e., the ligaments and annular parts of the intervertebral discs, and varied by the action of three sets of contractile "strings." The dorsal string—epaxial muscles, mainly the erector spinae—covers the dorsal and dorsolateral aspect of the bow (1–4, Fig. 11) and tends to straighten it.

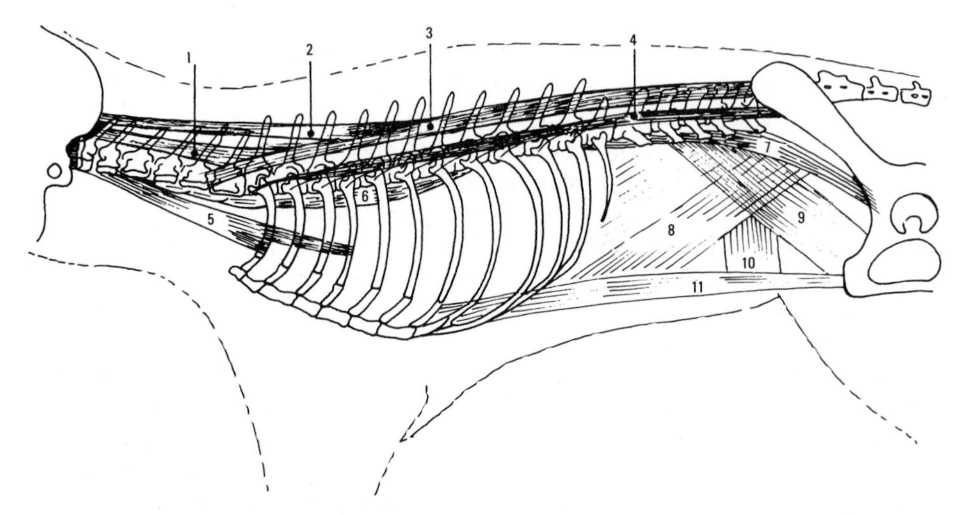

Fig. 11. Semischematic drawing of the important muscles in the bow-and-string construction of the mammalian body axis. 1, Cervical longissimus; 2, spinalis and semispinalis; 3, thoracic and lumbar longissimus; 4, iliocostal; 5, scalenus; 6, longus colli; 7, psoas group; 8, internal oblique 9, external oblique; 10, transverse abdominal; 11, straight abdominal.

The "ventral strings" occur in two groups. The first group (5–7, Fig. 11) is situated closely against the ventral side of the vertebrae in the cervical and cranial thoracic region (m. longus colli and capitis) and in the caudal thoracic, lumbar, and sacral region (psoas group). A short portion in the transitional thoracolumbar region is left deprived of muscles. The second group (8–11, Fig. 11) comprises the oblique, transverse, and the straight abdominal muscles and is indirectly attached to the bow by interposition of ribs and the pelvis. The muscles of the first group flex the respective regions of the bow that they cover; in cooperation with the corresponding divisions of the epaxial muscles they also stabilize the "bow's" curvature. The muscles of the second group are primarily responsible for flexing the bow in the thoracolumbar region and this is seen as the characteristic hump in the back of a sitting monkey; their action is also prominent in spinal flexion and extension during the leaping-gallop. It has been assumed that when a quadrupedal animal stands at ease, the intrinsic resilience of intervertebral discs and ligaments keeps the vertebral column in a state of unstable equilibrium and that only slight muscular activity or perhaps even none is required to maintain this posture. The slightest amount of muscular contraction producing a force that overcomes this intrinsic elasticity is sufficient to break the balance and produces flexion or extension of the "bow."

The functional systems in the bow are the spinal units, each consisting of two adjacent vertebrae, an intercalated intervertebral disc, the intervertebral joints, and the associated ligaments and muscles. At birth the vertebral column consists of a number of rigid bodies—diaphyses (D, Fig. 12A)—and a corresponding number of intercalated "adaptive zones" in the mechanical sense, from which epiphyses, fibrous rings, and nuclei pulposi arise (E, ED, and ID, Fig. 12A). The final shape of these zones depends on the intensity of stresses produced by gravitation and the mode of progression of the animal. Although details of this remodeling are not yet known, there is evidence to suppose that the epiphyseal discs (ED) are set perpendicularly to the force that originates from the contraction of the muscles previously mentioned. The direction and sense of this force can be obtained on the assumptions that (1) the center of rotation between adjoining vertebrae is in the nucleus pulposus and that (2) equilibrium requires that the working line of the resultant force passes through that center. This means that the vertebral bodies are under central compression and this is reflected in the orientation of the trabeculae.

When a vertebral body sustains a centrically applied compressive force, the corresponding stress in any section perpendicular to its longitudinal axis is equally distributed over the area (Fig. 12C) of the section and follows from formula (8). The stressing effect of an eccentric force on a section, however, varies with the distance between the working line of the force and the center of the section. The eccentricity of the force induces bending moments and the calculation of stress is now more cumbersome. In Fig. 13, the rectangle $ABCD$ represents a longitudinal section of a vertebral body. Let L be the working line of the resultants $R_{L\ (\mathrm{eft})}$ and $R_{R\ (\mathrm{ight})}$ of the external compressive forces acting upon the left and right portions of the vertebral body; K is the point of intersection of L and the section XX. The resultant R_L and R_R can be resolved into a force $R'_L = R_L$ and $R'_R = R_R$ and the couples M_L and M_R. The forces R'_L and R'_R induce compression stress and the couples bending stress so that the stress induced in XX can be found from the summation of the stresses caused by R'_L, R'_R, M_L, and M_R. Figure 14 depicts a frontal view of the symmetrical section XX; K lies eccentrically on the axis of symmetry OZ at a distance u from the centroid O. The normal force $R'_L = R'_R = F$ has a bending moment Fu, its axis being perpendicular at OK. The axis OY is

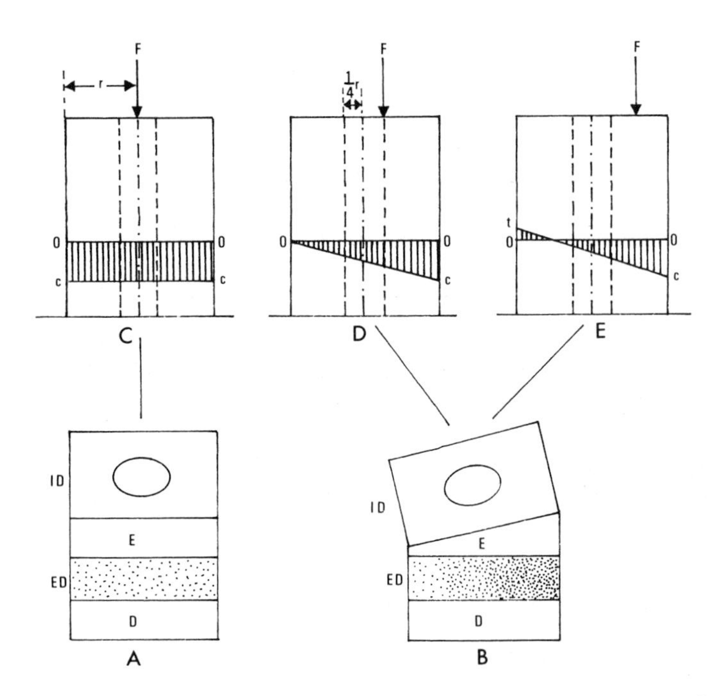

Fig. 12. Diagrammatic sagittal section of the vertebral body in a bipedal mammal to demonstrate the biomechanical factors in growth. (A) Principal components are: D, diaphysis; ED, epiphyseal disc; E, epiphyseal bone; ID, intervertebral disc with nucleus pulposus. (B) With uneven loading, there is intensification of bone deposition at the right side of the epiphyseal disc (heavy dotting) that leads to a more intensive increase of thickness of the epiphyseal disc of bone at the right side of the figure. The result is a wedge-shaped epiphysis that is perpendicularly aligned to the line of action of the force. (C) The load F is centrically applied to a column and the resulting compressive stress c is equally distributed over a plane section 00 of the column. This distribution of stress equals that in the epiphyseal disc in (A) and therefore osteogenesis in that disc is of equal intensity throughout the section. (D) The load is applied at a distance $\frac{1}{4}$ r from the central axis of the column. There is compressive stress c at the right side and zero stress at the left side of the perimeter. (E) The load is applied beyond $\frac{1}{4}$ r from the central axis, and there is compressive stress c at the right and tensile stress t at the left side of the column. The stress distribution in (D) and (E) may lead to the differential growth depicted in (B).

the so-called neutral axis for the stress induced by Fu and the stress at a given point E at distance z from OY follows from

$$\sigma_d = Fuz/I \tag{11}$$

in which I is the moment of inertia of the section. This moment is defined as the integral of the area of each infinite element of the section multiplied

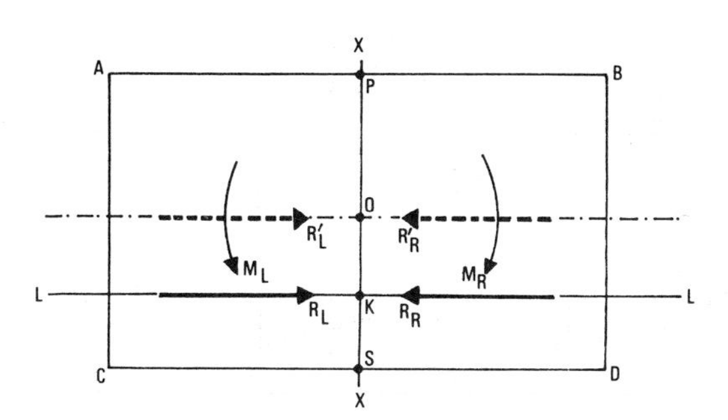

Fig. 13. Longitudinal cross section of a vertebral body *ABCD*. R_L and R_R are compressive forces acting along *L*. For explanation, see text.

by the square of its distance to *O*. Moreover, *XX* is subjected to stress due to the load *F* which is equally distributed over the section, thus:

$$\sigma_n = F/P \text{ (formula 8)}$$

so that the total intensity of the stress in *E* is given by:

$$\sigma = \sigma_d + \sigma_n = Fuz/I + F/P \tag{12}$$

On the axis *OY*, $z = 0$ and $\sigma_d = 0$, so that $\sigma = F/P$, hence, *OY* does not represent the neutral axis for the combined stress, since it has no zero stress. There is compression stress $[(Fue/I) + (F/P)]$ at a point *S* at the perimeter due to both the normal force and the bending couple *Fu*; in *P*, diametrically opposite to *S*, there is tensile stress Fue/I induced by the couple *Fu* which is diminished by the compression stress F/P due to *F*. In a solid circular section of radius *r*, which closely resembles the actual cross section of a cylindrical vertebral body, and with an eccentricity *u*, the stresses in *P* and *S* can be computed from

$$\sigma = F/\pi r^2 + Fur/\tfrac{1}{4}\pi r^4 = F/\pi r^2 \pm Fu/\tfrac{1}{4}\pi r^3 \tag{13}$$

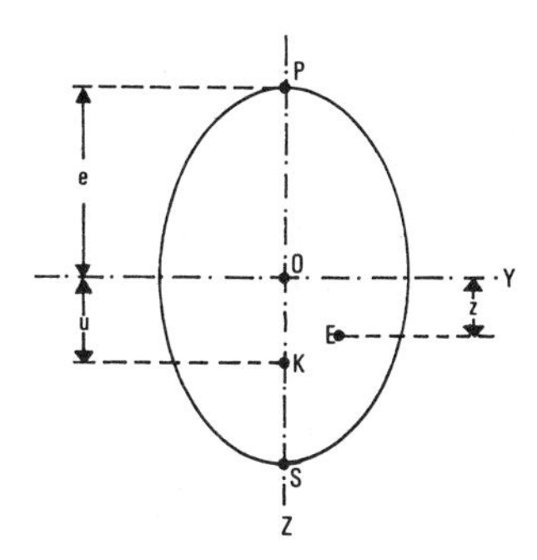

Fig. 14. A frontal view of the section *XX* in Fig. 13. For explanation, see text.

The eccentricity of the force that just produces zero stress in P and doubles the average intensity in S follows from

$$E\pi r^2 = Fu/\tfrac{1}{4}\pi r^3 \tag{14}$$

hence,

$$u = r/4 \tag{15}$$

An eccentric force applied at a distance greater than $r/4$ from the midpoint of the section causes a bending tendency in the vertebral body; that is, induces compressive stress at one side of the section and tensile stress at the other (Fig. 12D and E). The general tendency in musculoskeletal architecture is to reduce bending moments as much as possible. Since the center of the nucleus pulposus is the point of application of the compressive force, the maximum eccentricity of the nucleus pulposus within the intervertebral disc during flexion or extension of the vertebral column is 1/4 of the radius of the body's cross section.

The principles discussed above apply to longitudinal as well as to oblique forces. In the latter case, the force can be resolved into compressive and shearing components, and during vertebral growth the epiphyseal zones are aligned perpendicularly to the line of action of the force. There is marked intensification of bone deposition at the location of the maximum compressive force; osteogenesis is favored on one side of the perimeter of the epiphyseal zone and suppressed on the other (Fig. 12B). The resulting tilt in the position of the epiphyseal zone is responsible for the shape of a sagittal section of the vertebral bodies.

The biomechanical principles outlined above are valid for all quadrupedal primates. The bipedal upright posture has further implications on the mechanics of the body axis. It must be stressed, however, that there are fundamental differences in curvature, conformation, and location of the center of gravity between the great apes, which adopt an "erect" position only incidentally, and man, in which this posture is habitual. Information on the exact biomechanical particulars of the spine, especially in the upright position, is inadequate and I will discuss only some basic principles.

If the bow-and-string model of the quadruped is more and more erected toward an upright position, the weight of the head, shoulders, and forelimbs as well as the partial weight of the trunk is superimposed on the compressive forces already exerted by the intrinsic spinal muscles. The more the bow approaches the vertical, the greater the influence of the load. If this posture becomes permanent, the most favorable solution for the shape of the spine is to develop secondary curves in the neck and lumbar regions. Mechanical analyses of columns under loads (see Den

Hartog, 1949)—which are beyond the scope of this contribution—demonstrate that the curvature of the human spine provides a mechanism to resist deformation (buckling). In the great apes, which only incidently assume an erect or semierect posture, the spine still has only two curves, both concave forward—a condition also found in the newborn human infant. The sinusoidally curved spine, with its cervical and lumbar curvature, is characteristic of adult man only.

The Thorax and the Shoulder Suspension Mechanism

Although it is common practice to deal separately with thorax and shoulder girdle, the mechanical relation between the suspension mechanism of the forelimb and the wall of the thorax is so close that they will be considered together.

The craniocaudal dimension of the thorax is directly related to the number and individual length of the thoracic vertebrae and intervertebral discs; the dorsoventral and transverse diameter, however, are determined by the length and curvature of the ribs. The multitude of functions of the thorax (respiratory movements, protection of lungs and heart, cranial fixation of the muscles in the "bow-and-string" construction, area of insertion of a number of shoulder muscles) leads one to anticipate a compromise structure. In quadrupedal primates, the thorax has a relatively large dorsoventral and a small transverse diameter, and the plates of the shoulder blades lie in planes more or less parallel with the long axis of the vertebral column. A well-developed clavicle acts as a strut between sternum and scapula, resisting laterally directed compressive and tensile forces applied at the shoulder joint (Fig. 15, right side). When comparing this synostotic assemblage with the synsarcotic (aclavicular) type typical of cursorial mammals, one notes that the latter is adapted to sustain tensile forces only in the direction of the fibers of the principal muscles between scapula, humerus, and the trunk.

The biomechanical correlations between shoulder and locomotor types are principally based on the fact that in quadrupedal primates the shoulder sustains mainly compressive forces, in brachiators tensile forces, and in semibrachiators intermittently tensile and compressive forces. Ventrodorsally directed force, i.e., body weight in the standing quadrupedal primate, is resisted by contraction of the serratus magnus muscle (Fig. 16); the dentations of its thoracic part insert on the first eight or nine ribs. The lateral component H_t of the serratus force tends to bend the rib, which has a moment of resistance against bending given by

$$W = I/e \qquad (16)$$

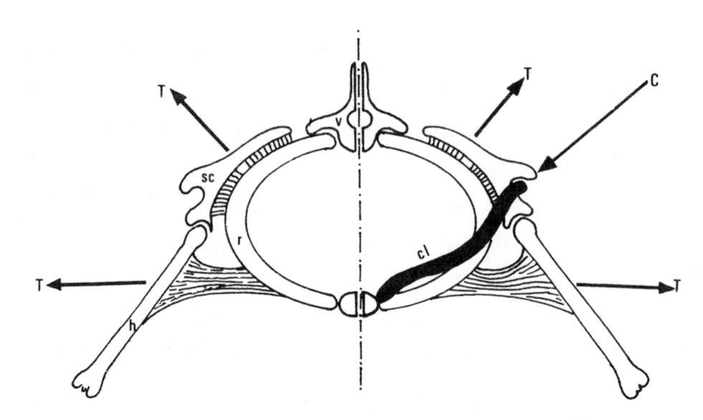

Fig. 15. Right side, schematic cross section of a primate thorax and shoulder girdle. The clavicle (cl) acts as a strut capable of resisting both compressive (*C*) and tensile (*T*) forces. Left side, a synsarcotic (aclavicular) shoulder girdle typical of cursorial mammals; this mechanism is adapted to sustaining tensile forces (*T*) only in the direction of the major muscles suspending the scapula (sc) and humerus (h); r, rib; v, vertebra.

in which *I* is the moment of inertia of the rib section at the serratus insertion, and *e* is the distance from the neutral axis to the perimeter. A computation of *W* for ribs four to eight inclusive in *Macaca mulatta* shows a distinct peak at rib number four. This is, in fact, the rib that lies at the level of insertion of the serratus on the medial side of the shoulder blade, and consequently it sustains the greatest bending stress when the shoulder transmits the body weight. The distinction between load-bearing "true ribs" and the more caudal, freely movable "false ribs" is substantiated by their mechanical qualities.

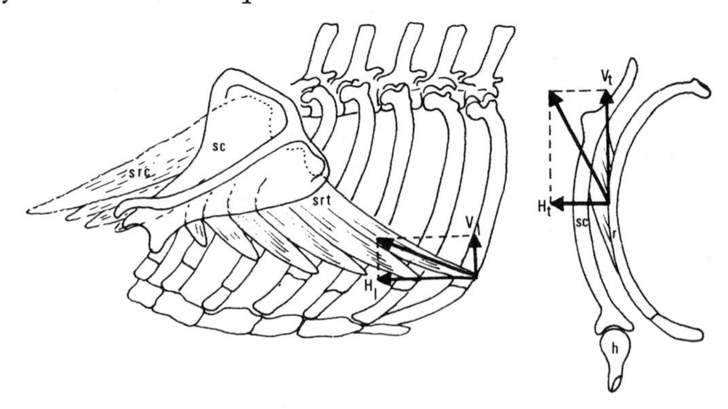

Fig. 16. Left, lateral view; right, transverse section of the serratus suspension mechanism. H_l, V_l, horizontal and vertical components in the lateral plane; H_t, V_t, horizontal and vertical components in the transverse plane; r, rib; sc, scapula; src, cervical serratus; srt, thoracic serratus.

From a comparative point of view, it is interesting to compare the loading of the shoulder blade in quadrupedal and brachiating primates. The muscular arrangement of the shoulder reveals a certain symmetry around the scapular spine. These muscles form a number of functional "couples" that tend to counteract the rotations of the shoulder blade produced by the various manipulations of the forelimb. The most important muscles between shoulder girdle and trunk are the serratus magnus, the pectoral group, the trapezius, and the rhomboid. The fan-shaped serratus diverges from the medial aspect of the shoulder blade to attach to the caudal cervical vertebrae and first eight or nine ribs, a pattern which indicates a diversified action for the various fibers. The same applies generally to the trapezius, rhomboid, pectoral, and latissimus muscles, all of which have broad attachments to the trunk (Fig. 17). In the standing animal, the moment of reaction N (Fig. 18, left) at the shoulder joint tends to turn the glenoidal angle of the scapula craniodorsally. A combined contraction of the levator scapulae and caudal minor pectoral muscles (LS and PM), possibly reinforced by latissimus fibers, forms a "couple" that produces an equal but counteracting moment. In the case of retraction of the upper arm and shoulder joint (R, Fig. 18, right) by the minor pectoral and latissimus, the glenoidal angle of the scapula is drawn ventrally backward; a balancing "couple" can be established by the posterior trapezius (TP) and anterior trapezius (TA). The greater the moments that tend to produce scapular shift along the lateral thoracic wall—such as in the suspensory postures of some heavy apes and in swinging brachiators—the greater the need for equilibrating "couples."

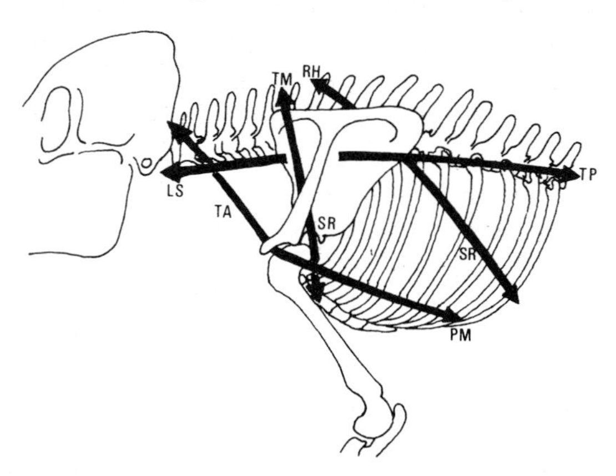

Fig. 17. Schematic diagram of the muscular "couples" acting around the shoulder-blade. LS, levator scapulae; PM, pectoralis minor; RH, romboids; SR, serratus; TA, anterior trapezius; TM, middle trapezius; TP, posterior trapezius.

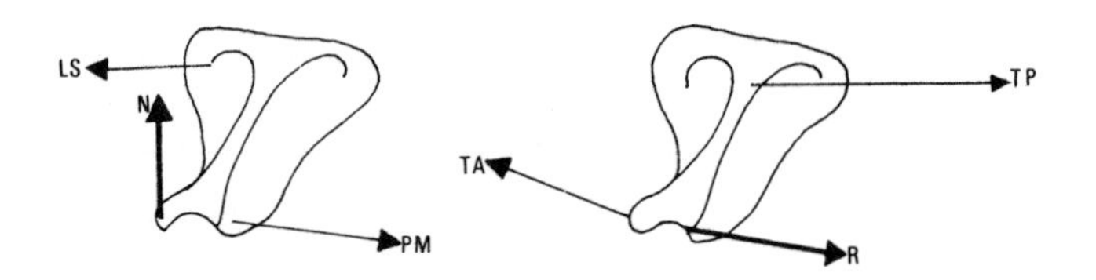

Fig. 18. The principle of scapular rotation and muscular couples. Left, the moment of reaction *N* at the glenoid is counteracted by a muscular couple of levator scapulae (LS) and pectoralis minor (PM). Right, retraction (*R*) of the glenoid may be balanced in part by the posterior (TP) and anterior trapezius (TA).

This requirement can be met both by enlargement of the shoulder muscles as well as by increase of the distance between the working lines of the muscles that attach to the scapula and produce the counteracting "couples"; both arrangements are implemented in brachiators.

The Pelvic Girdle and Limbs

Pelvis

The articulation between pelvis and body axis at the iliosacral joint is commonly syndesmotic in young mammals and synostotic in older ones. Hence it is difficult to say to what extent stability in the joint is affected by properties of intra-articular structures or by the action of surrounding ligaments and muscles. From a phylogenetic viewpoint, the iliosacral joint in early tetrapods was capable of mobility and the question arises how stabilization is achieved in modern forms. In lower terrestrial vertebrates, the iliosacral and hip joints lie on the same vertical line so that an unstable equilibrium exists in both. In quadrupedal mammals, however, the hip joint lies caudoventrally to the ilioscral articulation. In Fig. 19, the reaction at the acetabular roof *N* has a moment *Na* about the iliosacral joint that tends to rotate the pelvic bone in a clockwise direction about the iliosacral axis. Apparently this can be prevented by contraction of the abdominal muscles, especially the rectus abdominus (*M*), provided the moment $Mb = Na$. In addition, the loading of the lumbar vertebral column induces clockwise rotation of the sacrum about the iliosacral joint; this is resisted by tension *L* in the ligaments between sacrum and cranial coccygeal vertebrae, on one side, and the ischium, on the other. The stabilizing effect of ligaments and muscles around the iliosacral joint promotes the development of a synostotic linkage between the two bones by reducing torsional stress in the joint.

In a transverse plane through the iliosacral and hip joints, the pelvis

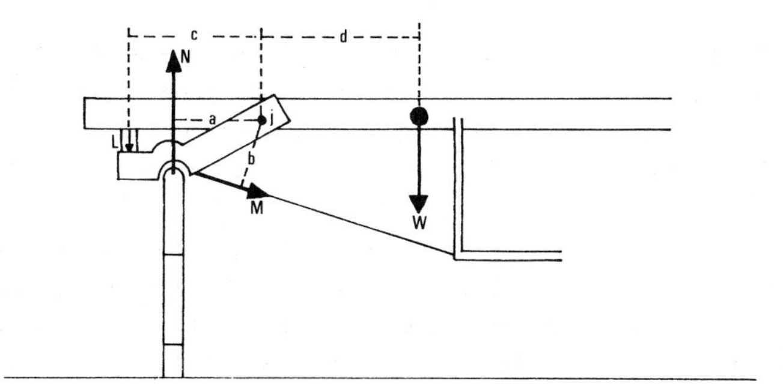

Fig. 19. Vectorial analysis of some moments about the iliosacral joint (j). (a) The lever arm of N at (j), the gravitation vector through the hip joint; (b) the lever arm of M, the rectus abominus muscle; (c) the lever arm of L, the ischiosacral ligaments; (d) the lever arm of W, the loading of the lumbar column. Equilibrium is established when $Na = Mb$ and $Lc = Wd$.

exhibits interesting statical implications. The pelvic halves are in the form of two arcs that attach dorsally to the sacrum and meet ventrally at the pelvic symphysis. The relative position of the iliosacral and hip joints determines the orientation and intensity of stress at the symphysis; in turn their positions depend on the breadth of the sacrum. If the vertical projection of the midpoints of the iliosacral joints falls to the medial side of the acetabula (Fig. 20), the partial body weight W/2 acting at the

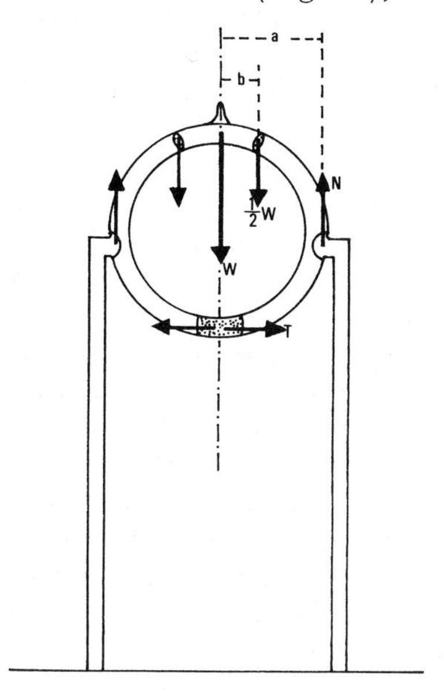

Fig. 20. Schematic transverse section through the iliosacral and hip joints. In the standing animal, the moment Na is greater than $\frac{1}{2}$ Wb, and hence the supra-acetabular part of the right pelvis rotates medially and the pelvic symphysis is under tension (T). For further discussion, see text.

iliosacral joint and the corresponding reaction N at the acetabula constitute a couple that tends to rotate the pelvic bones ventrally outward; hence the pelvic symphysis sustains tensile stress. When the hip joint lies vertically beneath the iliosacral joint, there is zero stress in the symphysis. When the breadth between the sacral wings is greater than that between the hip joints, the symphysis is under compression stress. The first type is common in primates so that their symphysis is under tensile stress.

Femur

The femur has been the subject of numerous biomechanical analyses and there is ample information on its structure and behavior under loads (Frankel and Burstein, 1970). The adaptations of this bone to biostatical requirements can be only understood when considered in relation to equilibrium in the hip and knee joints.

The establishment of static equilibrium in a joint requires that the resultant of all applied forces pass through the momentary point of rotation—the *hypomochlion*—of the joint. Therefore, the hypomochlions of hip and knee joints are the points of application of the resultant loads on the femur. Friction between the surfaces of the joints is zero or nearly zero, so that the joints cannot transmit lateral forces, and loads are applied perpendicularly to the joint surfaces. The working lines of the loads, however, do not coincide with the long axis of the femur, and as a consequence the femoral shaft is subjected to bending moments. Since a principal tendency in skeletal architecture is to minimize bending moments as much as possible, the femoral shaft is curved to such a degree that the working lines of the loads are brought near to the axis of the bone, thus reducing bending moments considerably. The characteristically curved profile of the lateral aspect of the femur is an adaptation to the normal physiological load. This may be demonstrated by an analysis of a model of a hind leg of *Macaca mulatta* in a plantigrade position; the bending moments may be calculated for different postures: quadrupedal, bipedal, and during climbing. The relative dimensions of the model are based on two male skeletons, and the myographical data are based on personal dissections, augmented by data from Uhlmann (1968).

Quadrupedal Posture (Fig. 21)

The (partial) body weight W is transmitted to the pelvis through the iliosacral junction. If it is assumed that about 40% of body weight rests upon the hind limbs, each hind limb bears approximately 20%. The working line of the weight (the length of which in Fig. 21 represents 20 units or 20 % of body weight), passing in front of the acetabulum, has a moment which tends to rotate the pelvic bone around the femoral head in a clockwise direction. The ischiocruralis muscle, a member of the hamstring group, can stabilize the hip joint. The muscle originates from the sciatic tubercle and inserts on the lateral tract of the tensor fasciae latae, the kneecap, the lateral condyle, and the anterior brim of the tibia; its force is represented by M_1 in Fig. 21. The resultant of M_1 and W is R_1 which passes through the hypomochlion H_1. M_1 is 45 units and R_1 is 59 units. A comparable situation is found in the ankle joint; R_2 (78 units)

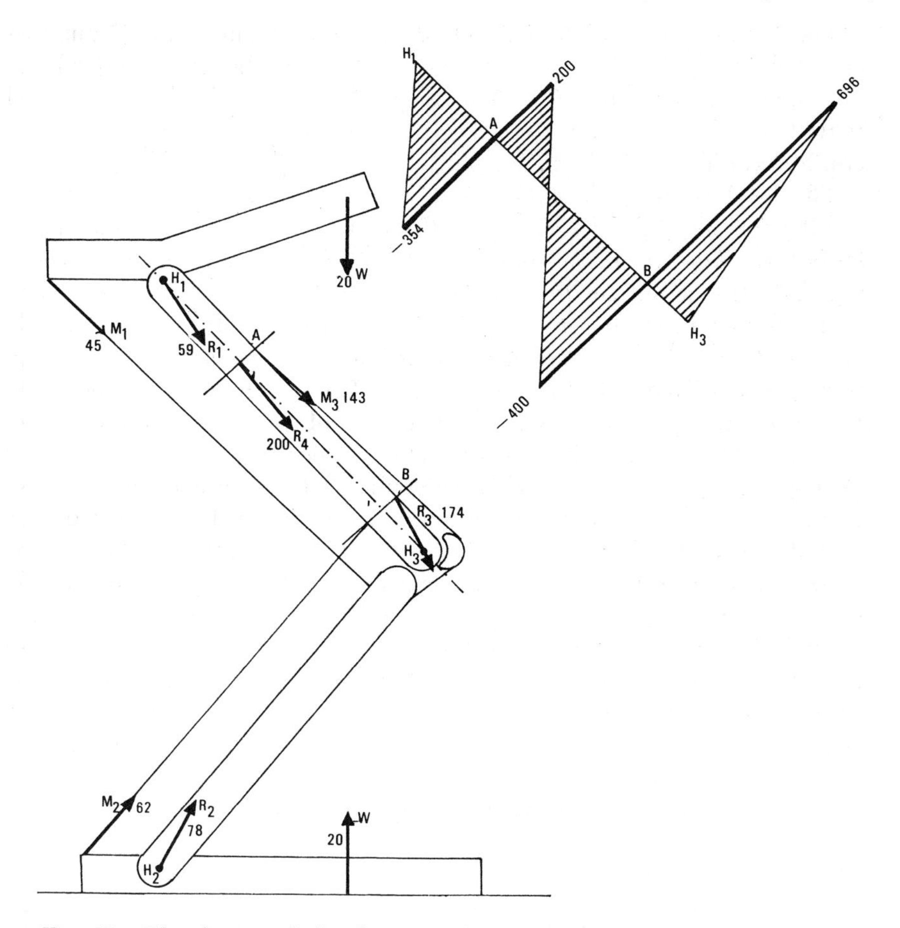

Fig. 21. The theoretical distribution of bending stresses in the femur in a quadrupedal macaque. M_1, M_2, M_3, are, respectively, the ischiocruralis, gastrocnemius, and vastus group of muscles. For other symbols and explanation, see text. At the upper right is a bending moment diagram.

is the resultant of the reaction at the foot is equal but opposite to W (20 units) and the force M_2 (62 units) of the gastrocnemius. The ischio-cruralis and gastrocnemius, being biarticular muscles, have an effect on the knee joint as well as on hip and ankle. Since the ankle is stabilized, the entire lower limb and foot can be mechanically interpreted as a single body, hinged at H_3. There are two forces acting on this body: R_2 (78 units) and the force M_3 (143 units) of the vastus group which originates on the femoral shaft distal to the great trochanter and inserts on the patella and tibial tuberosity. The force R_3 (174 units) is the resultant of R_2 and M_3 and passes through hypomochlion H_3. From the viewpoint of statics, the femur is comparable to a straight, balanced beam divided into three portions. The proximal portion is bounded by H_1 and A, A being the proximal point of origin of the vastus group; the middle portion by A and B, B being the proximal point of origin of the gastrocnemius; and the distal portion by B and H_3. The loads on the three parts differ. The portion H_1A sustains a load R_1, AB is loaded by the resultant R_4 of R_1 and M_3, and BH_3 is loaded by R_3. In none of the cases does the working line of the load coincide with the long axis of the femur, and therefore the bone sustains bending stress. The bending moment diagram (Fig. 21) shows two negative and two positive peaks, with zero bending moments at H_1 and H_3.

Bipedal Posture (Fig. 22)

The entire weight of the bipedal animal rests upon the hind limbs, or 50% per limb; W is 50 units and $2\frac{1}{2}$ times greater than in the quadrupedal posture. Although the magnitude of the forces is higher, the mechanism that maintains equilibrium in the three joints is the same. The bending moment diagram (upper right, Fig. 22) has one negative and two positive peaks. Again, the working lines of the loads do not coincide with the femoral axis of the femur which therefore sustains a bending stress.

Climbing

A climbing macaque is positioned with fully extended fore limbs and moderately bent hind limbs on two branches B_1 and B_2 (Fig. 23). The reactions at these points can be determined from the consideration that the limb vectors form the sides of a force parallelogram which has the weight W (100% = 100 units) as a diagonal; reaction at B_1 is 110 units and at B_2 80 units, or 40% per hind limb. In Fig. 24, the load at the iliosacral joint is $B_2 \cos \alpha = 36$ units. The hip joint is equilibrated by W (36 units) and M_1 (62 units), and their resultant R_1 (90 units) passes

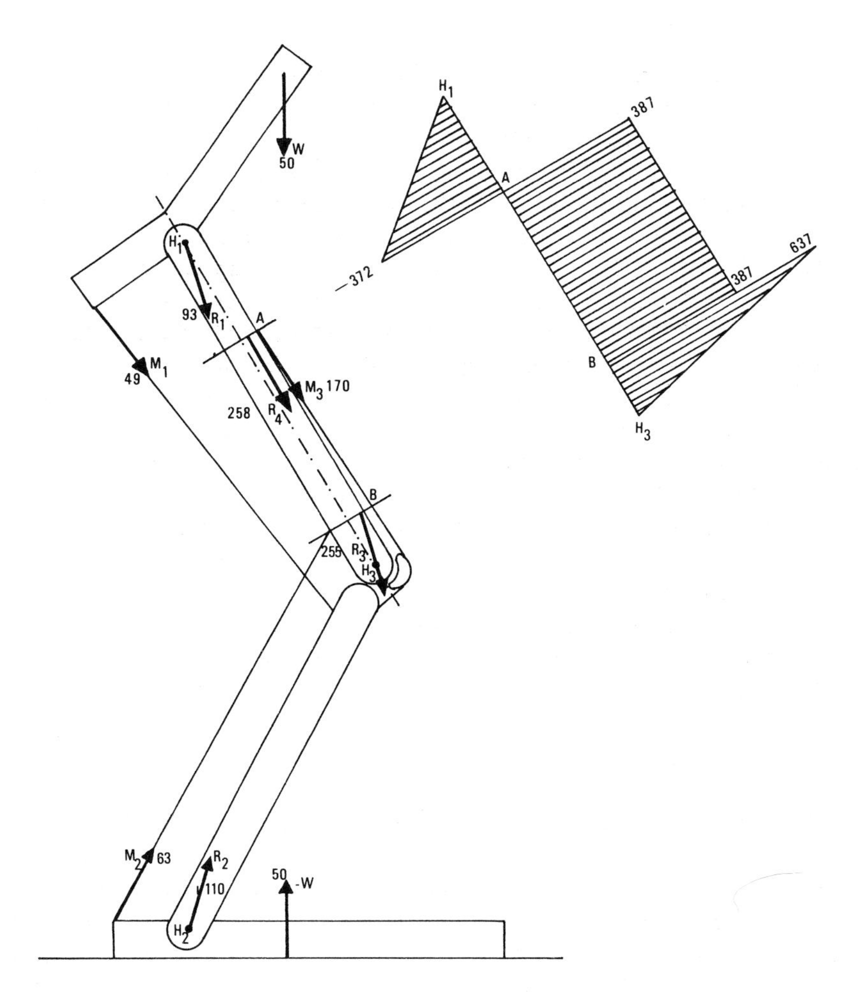

Fig. 22. The theoretical distribution of bending stresses in the femur of a bipedal macaque. Labels as in Fig. 21. For explanation, see text.

through H_1. The ankle is stabilized by B_2 (40 units) and M_2 (185 units), and therefore R_2 is 225 units; R_2 passes a little dorsally to H_3 and consequently tends to extend the knee joint. This means that no contraction of the vastus group is required to balance the knee. Since R_1 passes H_3 at the plantar side and tends to flex the knee, the moment of R_1 about H_3 is equal but opposite to that of R_2 about H_3. The loads on the various portions of the femur differ from those in the preceding case: H_1A is loaded by R_1 and so is AB; BH_3 by the resultant R_3 (290 units) of R_1 and R_2. The bending moment diagram shows one negative and one high positive peak.

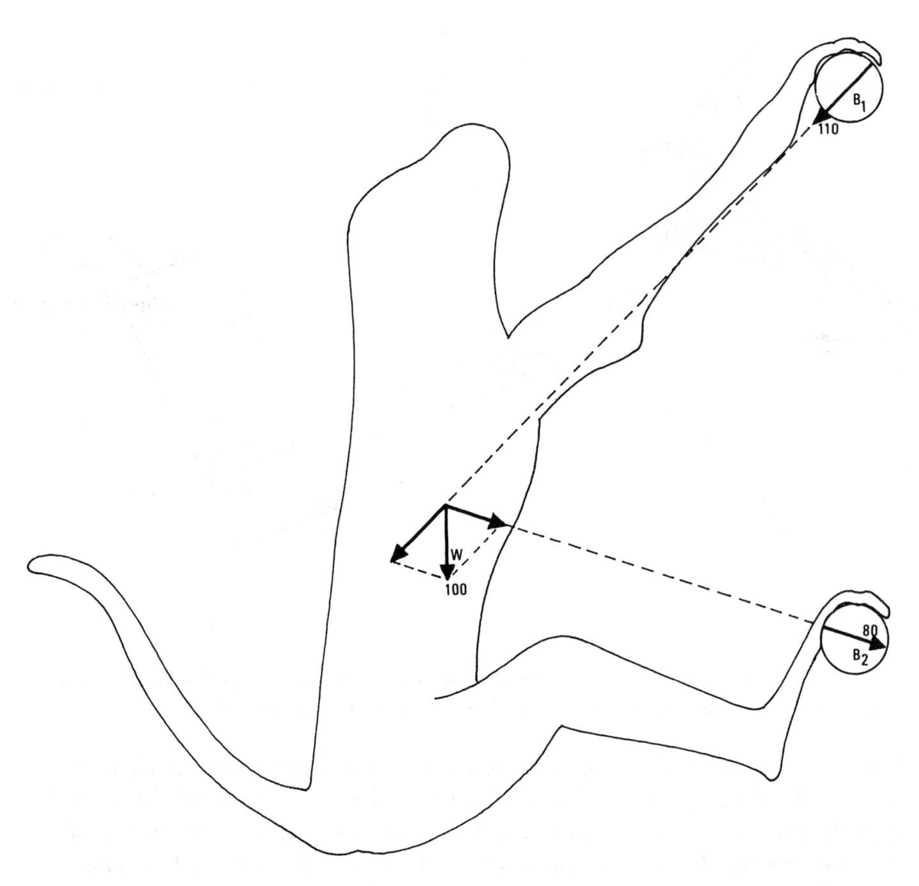

Fig. 23. Schematic diagram of the distribution of forces engendered by the body weight (*W*) of a vertically climbing macaque. For explanation, see text.

The analysis of these very simplified models shows that a straight femur is by no means a favorable construction since in all demonstrated cases, the straight femur implies that the working line of the resultant load on the bone deviates from the central axis. In terms of static loading, it is preferable for the axis of the bone to coincide with the working line of the load. This can be realized in various ways. The first mechanism is similar to that controlling growth in the epiphyseal plates of vertebral bodies. In Fig. 12, it was shown that an eccentric load produces unevenly distributed stress in sections of a bone, with a concomitant concentration of osteogenesis on the side of the maximum stress. The result is that the epiphyseal disc, or part of it, is realigned perpendicularly to the applied

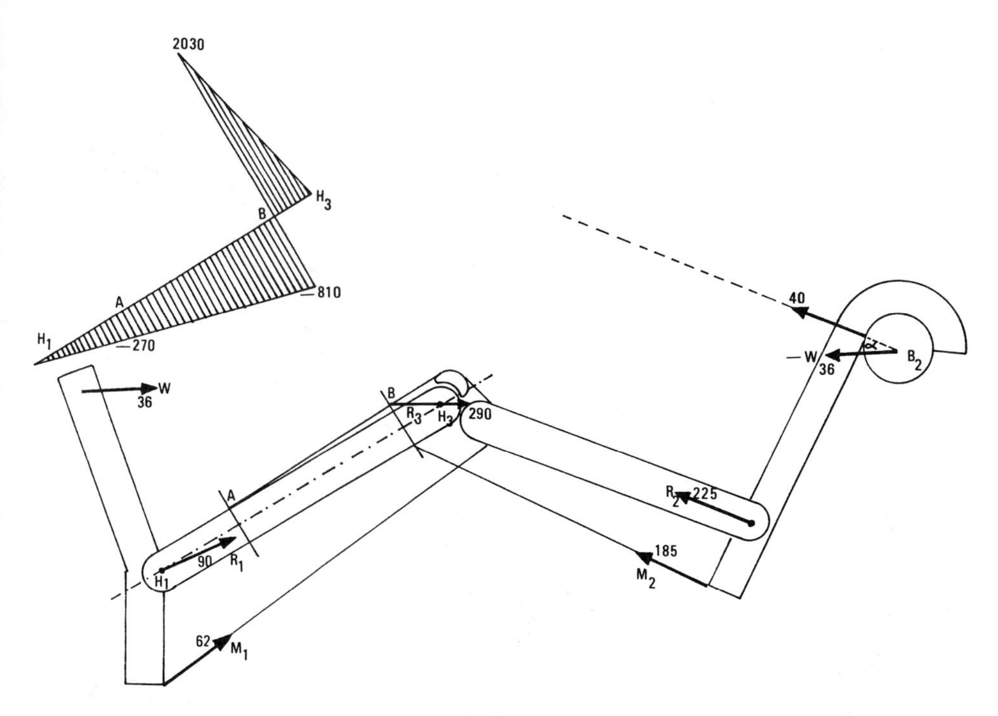

Fig. 24. The theoretical distribution of bending stresses in the femur of a vertically climbing macaque. Labels as in Fig. 21. For explanation, see text.

load. This same mechanism produces the two characteristic bends in the axis of the femur: proximally at the level of the major trochanter, and distally just above the femoral condyles. The second process that tends to approximate the line of action of the load and the axis of the bone is muscular. For example, the contraction of the gluteus minimus, whose fibers pass over the hypomochlion H_1, exerts a force of 63 units and centers the load on the portion H_1A and increases it to 118 units (Fig. 21). However, since the load acts centrically, the computed initial compressive stress of 261 units per cm^2 drops to 78.6 units per cm^2, a reduction of about 70%. The shape and curvature of the femur as seen in a transverse plane are also determined by these mechanisms; body weight and iliotibial tract appear to be the principal factors.

Patella

The patella is a sesamoid in the tendon of the quadriceps femoris tendon and acts as a lever arm. Equilibrium in a knee joint involving a sesamoid is shown in Fig. 25, although it should be noted that the

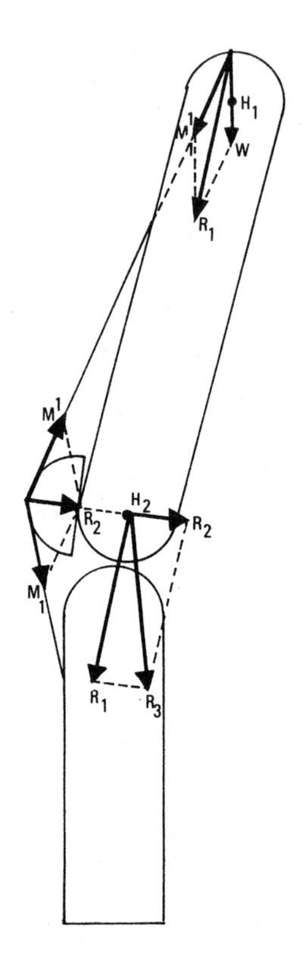

Fig. 25. The lever arm effect of a patella at the knee joint, illustrating joint equilibrium involving a sesamoid. R_1 is the resultant of weight W and muscular force M_1 and passes through H_2. R_2 is the pressure of the patella against the femur, and also passes through H_2. R_3 is the resultant of the system.

hypomochlion varies in position and the simplified figure shows only one position. The patella, resting upon the femoral trochlear ridge, reduces the expenditure of energy during quadriceps contraction by increasing the lever arm of the quadriceps. This is substantiated by the fact that in man congenital absence of the patella leads to moderate hypertrophy of the knee extensors.

Ankle and Foot

The conformation of ankle and foot offers numerous examples of adaptation to the specific modes of posture and progression. In order to demonstrate some basic biomechanical principles I will consider the tarsal assemblage as forming a single mechanical unit to which metatarsal rays are distally attached, and will analyze the forces in a single metatarsophalangeal assemblage. When viewed in a sagittal plane, the position of the elements in a ray is maintained by the action of dorsal extensor and plantar flexor muscles, the latter being assisted by a deep layer of intrinsic pedal muscles and ligaments (e.g., plantar aponeurosis); obviously we have here the problem of the equilibration of a series of consecutive joints (first and second interphalangeal, and metatarsophalangeal).

Whether in the standing position and in various locomotory postures, the plantar side of the ray is always more intensively loaded than the dorsal. The obvious presence of dermal callosities on the former points to the fact that this side sustains heavier loads than the dorsal side. Thus the moments producing plantar flexion of the metatarsophalangeal and interphalangeal joints are greater and occur more frequently than those initiating extension of these joints, and concomitantly the flexor group is better developed than the extensor muscles.

Let us suppose that the animal is walking on branches so that part of its body weight is intermittently supported by the hind feet. The foot sustains forces in two main directions: dorsoplantar and mediolateral. Equilibration of the distal interphalangeal joint in a sagittal (dorsoplantar) plane is depicted in Fig. 26, where R is the resultant of the reaction $(N) = -W$ at the plantar side of the foot and the force M exerted by the deep flexor. The joint can transmit forces acting only perpendicularly to its surface. The semicircular shape of the distal articular surface makes it possible that R intersects the surface at a right angle even when the direction of the vector varies considerably. This condition is not fulfilled when dealing with forces in a mediolateral direction; they induce a shearing component that tends to shift joint surfaces sideways and this would eventually lead to ruptures in the surrounding tissues. Dislocation of the components can be prevented by strong collateral ligaments, a central condylar crest (verticillus), or a curvature of the joint's surface that is coplanar with the vector of the shearing force (e.g., a saddle joint). The actual shape of the distal interphalangeal joint indicates that the magnitude of mediolateral forces can only be moderate; this is clearly demonstrated in climbing where the opposable great toe is always placed on the other side of the branch to

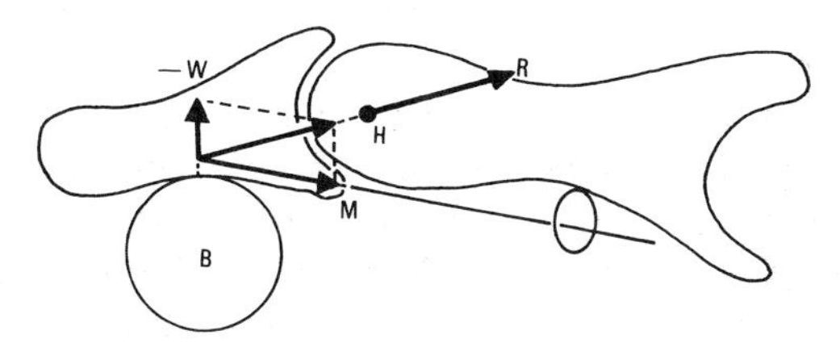

Fig. 26. Medial section through two distal phalanges, illustrating an equilibrium condition in the distal interphalangeal joint. *B*, branch; *H*, axis of flexion-extension; *M*, force exerted by deep flexor muscle; *R*, resultant of *M* and $-W$, weight loading.

the load-bearing toes in order to avoid loading of its distal interphalangeal joint, which, in this case, is not supporting weight.

The curvature of the central axis of the distal phalanx approximates the resultant R and the axis of the bone, and is very conspicuous in species that have flat nails and thick plantar callosities, such as, gorilla, chimpanzee, and man. In *Alouatta* and in individual cases of gibbon, orangutan, and chimpanzee, the point of application of a load may be further shifted distally to a curved and laterally compressed strong nail; the third phalanx and the nail then act as a single body (Fig. 27; see Preuschoft, 1969).

The load R on the third phalanx is transmitted to the second through the distal interphalangeal joint. The tendon of the deep flexor passes through a flexor tendon sheath that is attached to the plantar side of the second phalanx. When the long axis of the third phalanx is aligned with that of the second, the flexor tendon sheath does not exert any force on the bone induced by restraining the flexor tendon. In a flexed joint, however, the reorientation of the tendon at the flexor tendon sheath induces a force $R_2 = 2\ M_1 \cos \alpha$. Equilibration of the proximal interphalangeal joint H_2 may then involve the action of the superficial flexor M_2. The second phalanx is equivalent to a beam fixed at two ends; its shaft is slightly curved to bring the working line of the load near to its axis.

The type of loading of the proximal phalanx resembles that of the second very closely. Since both the tendons of the superficial and the deep flexor are restrained by the flexor tendon sheath, the forces exerted by the flexor tendon sheath on the bone are superimposed. Balance of the metatarsophalangeal joint is effected by the action of the interosseous and lumbrical muscles. The same consideration applies to the metatarsals, especially when the toes perform a firm grasp around a branch (Fig. 28). The resultant R_4 intersects the hypomochlion H_3 of the corresponding metatarsophalangeal joint when this joint is in equilibrium. In most cases, however, the working line of the resultant will pass dorsally

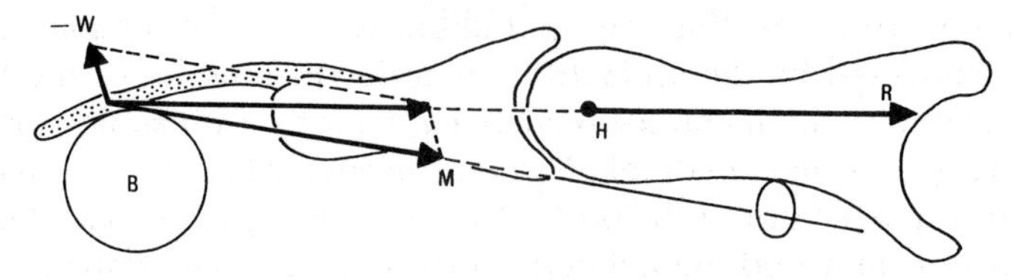

Fig. 27. Medial section through two distal phalanges, illustrating an equilibrium in cases when a strong, laterally compressed nail is the principal point of loading. Abbreviations as in Fig. 26.

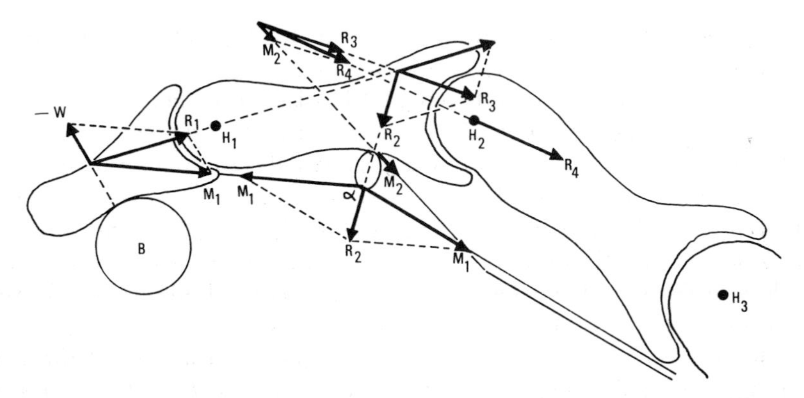

Fig. 28. Median section through a moderately flexed and loaded phalanx, illustrating the effect of flexor tendon sheath attached to the plantar surface of the second phalanx and restraining the deep and superficial flexor tendons. For further explanation, see text.

to the joint; there remains, therefore, a positive moment that must be balanced by action of the plantar aponeurosis, possibly reinforced by the posterior tibial and long peroneus muscles.

During walking on a flat surface, the metatarsophalangeal joint sustains a dorsally directed force from the flexor tendons. If the heel is in contact with the ground, there is an additional reaction of the ground against the foot that increases the bending moment of the metatarsal bone. In quadrupedal posture, the load on the toes can be shifted by a slight angulation of the interphalangeal joints to the distal epiphyses of the metatarsals. Although the flexor tendons are thus arranged in the same configuration as in Fig. 28, electromyographical research in man shows that the flexors have a very low activity; apparently the tension in the plantar aponeurosis is sufficient to equilibrate the tarsometatarsal joints without the help of the flexors. The first metatarsal and corresponding phalanges deserve special attention, especially in those species where the great toe is distinctly abductable and opposable. The function of the plantar aponeurosis in this ray—equilibration of the tarsometatarsal joint—is augmented by the abductor and adductor muscles of the hallux; these also stabilize the metatarsophalangeal joint in a dorsoplantar direction. Loading in a mediolateral direction occurs when the animal uses the foot for a power grip. Obviously, the bending moments can be large in both dorsoplantar and mediolateral directions, and an adaptive curvature of the shaft in both directions is impossible. In this case, the required resistance against bending W (formula 16) is accomplished by modification of the cross section of the first metatarsal so that its diameter is different in successive transverse sections. In the mediolateral plane, the moment of resistance has its highest value at the most distal part of the bone; in a dorsoplantar direction, the cross section (and W with it) increases in a proximal direction.

The distribution of material types in the foot—bone on the dorsal side and muscular and connective tissue on the plantar side—points to the fact that, when loaded, the ray as well as the tarsal bones sustain bending. This results in compression stress on the dorsal, and tensile stress on the plantar side, and computations show that the neutral axis lies on the plantar side. The foot can be compared with a three-hinged arch with hinges at the metatarsophalangeal joint, the talocrural joint, and the point of support at the heel. The function of the plantar aponeurosis and the dorsal and plantar flexors can be appreciated by an analysis of a simplified model of the primate foot.

In Fig. 29, the metatarsal and tarsal divisions are illustrated as a beam with a bend at the talocrural joint. Let this construction be loaded by a weight (W) of 10 kg, the working line being concurrent with a vertical crural axis. The vertical reactions at A and B follow from the application of the law of moments and, in the scale used, A is 2 kg and B is 8 kg. The bending moments about ten consecutive points in the foot can be easily determined by computing the moments about A (point 0):

Point		Moment (kg cm)
0	$+ (2 \times 0)$	$= 0$
1	$+ (2 \times 1)$	$= 2$
2	$+ (2 \times 2)$	$= 4$
3	$+ (2 \times 3)$	$= 6$
4	$+ (2 \times 4)$	$= 8$
5	$+ (2 \times 5)$	$= 10$
6	$+ (2 \times 6)$	$= 12$
7	$+ (2 \times 7)$	$= 14$
8	$+ (2 \times 8) - (10 \times 0)$	$= 16$
9	$+ (2 \times 9) - (10 \times 1)$	$= 8$
10	$+ (2 \times 10) - (10 \times 2)$	$= 0$

In a second model (Fig. 30) the introduction of the intertarsal joint (C) between talus and calcaneus renders the construction unstable, as the points of support tend to spread under the load. This can be prevented by putting a tie (=plantar aponeurosis) between A and B. The force M_1 in the tie follows from the application of the law of moments about C and must be 8 kg. The graph of the bending moments shows

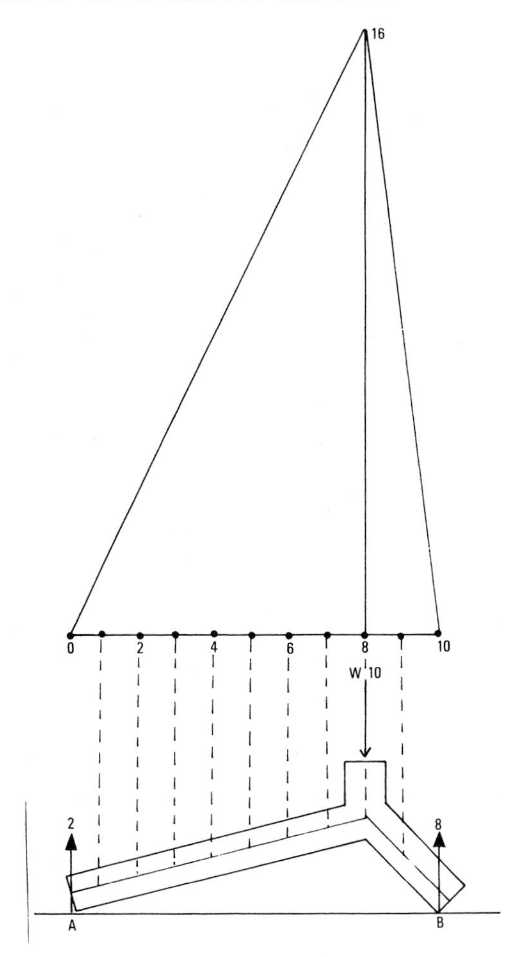

Fig. 29. The primate foot schematically rendered as a bent beam to illustrate the theoretical distribution of bending moments under a 10 kg load (W).

two positive peaks of 0.8 and one negative of -1.2, and thus the plantar aponeurosis considerably reduces the bending moments.

In a third model (Fig. 31), the line of action of the weight is shifted forward and accordingly the vertical reactions at A and B change: 4 kg at A and 6 at B. The weight causes a negative moment about C which is counteracted by a positive moment exerted by the force M_2 of the gastrocnemius. The resultant R of W and M_2 must pass through C and is 30 kg; M_2 is 20 kg. The tension in the plantar aponeurosis follows from the law of moments about C and is 16 kg. The graph shows an overall increase of these moments. In the next model (Fig. 32), the

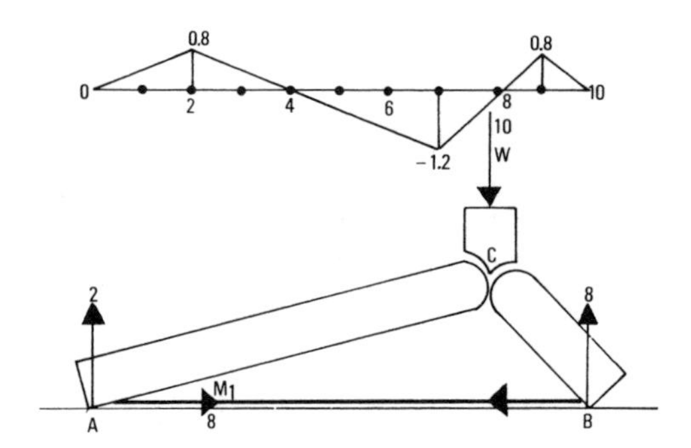

Fig. 30. The reduction of bending moments in the primate foot as a result of the action of the plantar aponeurosis $(A - B)$. For further explanation, see text.

posterior tibial muscle is added to the construction. Equilibrium at C requires a gastrocnemius force (M_2) of 13.6 kg and posterior tibial force (M_3) of 12.8; M_1 remains 12.8 kg. The graph shows that the addition of the posterior tibial further increases the bending moments. Putting the deep flexor of the toes in the next model (Fig. 33) modifies the bending moment diagram as follows: the plantar flexors (M_4) and plantar aponeurosis (AB) exert 8.6 kg each and the gastrocnemius (M_2) 10.5 kg in order to balance C. Obviously, the presence of the deep flexor diminishes the force required in the gastrocnemius (12.8 versus 10.5 kg), and therefore in primates with long toes and strong flexors the gastrocnemius need be only moderately developed; this is the biomechanical explanation of the "absence of the calf" which distinguishes the profile of the lower leg of monkeys from that of man. The bending moment diagram shows that contraction of the deep flexors reduces the bending moments in the proximal as well as in the distal part of the foot.

The analysis of these models demonstrates that two constructional principles diminish the bending moments and stresses in bones. Recall that the intensity of pure bending stress is given by

$$\sigma = Fue/I$$

and in the case of eccentric loading by

$$\sigma = F/P \pm Fue/I$$

Decrease of σ can be accomplished either by reducing Fu or by increasing I; in the latter case the decrease results from the fact that e figures as unit in the numerator but to the fourth power in the moment of inertia $(I = \frac{1}{4}\pi e^4)$ in the denominator. An increase in the diameter of the bone involves a linear increase of e but an increase to the fourth power of I

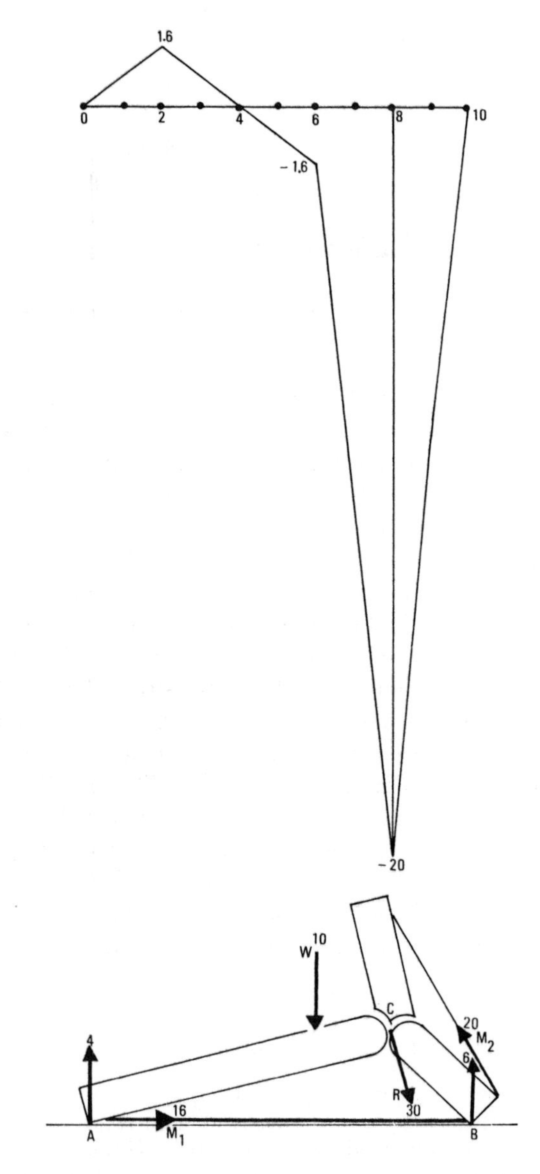

Fig. 31. The bending moments of the primate foot in which the line of action of the weight is shifted forward. M_2, force of the gastrocnemius muscle. For explanation, see text.

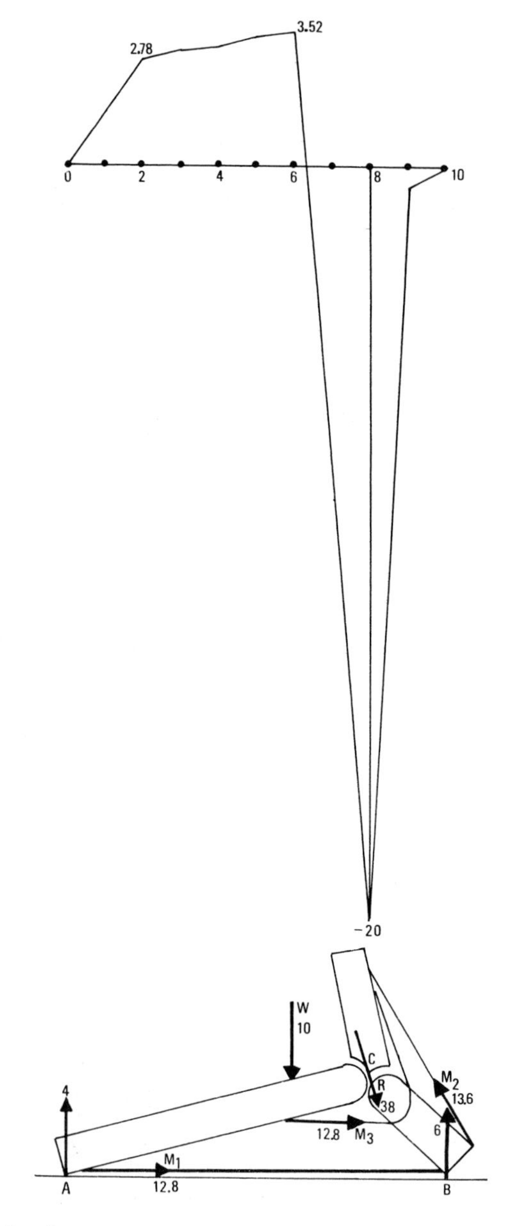

Fig. 32. The bending moments of the primate foot with the same actions as in Fig. 31, but with the addition of the posterior tibial muscle force (M_3). For explanation, see text.

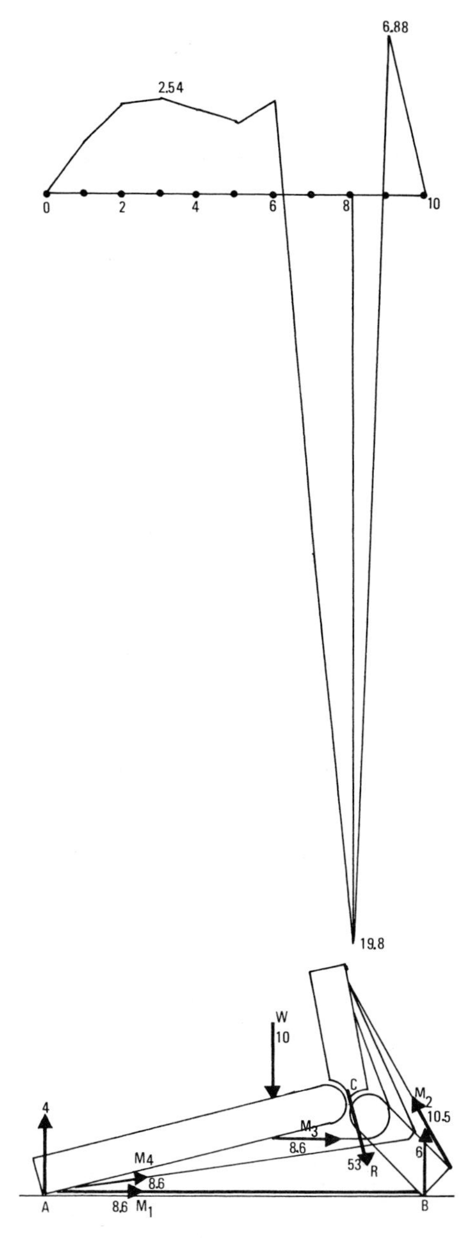

Fig. 33. The bending moments of the primate foot as in Fig. 32, but with the addition of the deep flexors of the toes (M_4). For explanation, see text.

so that the net value of Fue/I decreases. In the construction of the foot both solutions are employed: Fu is decreased by the balanced contraction of the plantar and dorsal flexors of toes and ankle; the moment of inertia is increased by enlargement of the cross-sectional area of the bones, such as in the case of the first metatarsal.

Although the construction of the shoulder and elbow joints in the forelimb is governed by the same mechanical laws as that of the hindlimb, the greater mobility of the forelimb and the versatility of the manipulations that can be carried out by the hands give a special character to the biomechanics of the forelimb. In quadrupedal progression, the shoulder joint sustains loading that varies intermittently with the swing and thrust phases of the gait. The joint is primarily stabilized by the resultant of the reaction of the humerus against the glenoid cavity $(N) = (-W)$ and the force exerted by the muscles of the musculotendinous cuff—supraspinatus, infraspinatus, teres minor, and subscapularis. When the arm is in the pendent position during upright positioning of the body, the sign of the resultant and the vector W changes to the opposite, and this situation is frequent in hanging–climbing species. In brachiating, the body is suspended by one or two arms and accordingly the arms, the shoulder girdles and parts of the thorax form an intermittently open and closed chain. In addition to the muscular "couples" that attach the shoulder blade to the trunk (Fig. 17), the weight of the body is also carried by the latissimus dorsi.

The elbow joint is morphologically rather uniform among primates; the shape of its articular surfaces essentially limits excursions to flexion and extension. In a bipedal primate, the joint is stressed by the resultant of the weight of the lower arm and hand and the force exerted by the extensors and flexors of the joint. In quadrupedal gait, the situation is more complicated since during the thrust phase, that part of the body weight carried by the forelimbs contributes to the resultant strain.

The various positions of the arms and trunk in brachiating species are of importance so far as the elbow joint is concerned. The resultant force that passes through the hypomochlion frequently alters its magnitude and direction and these determine the type and intensity of stress in the joint. In Fig. 34A, a brachiatory model is represented as a solid body without an elbow joint. When the weight is W and the surface area of a cross section of the arm measures F cm^2, the tensile stress in the section follows from $\sigma = W/F$. In a second example with an elbow joint (Fig. 34B), the position of the body with reference to its suspension and center of gravity is unaltered, i.e., in both models the body tends to swing forward until the center of gravity and the midpoint of the branch are on a

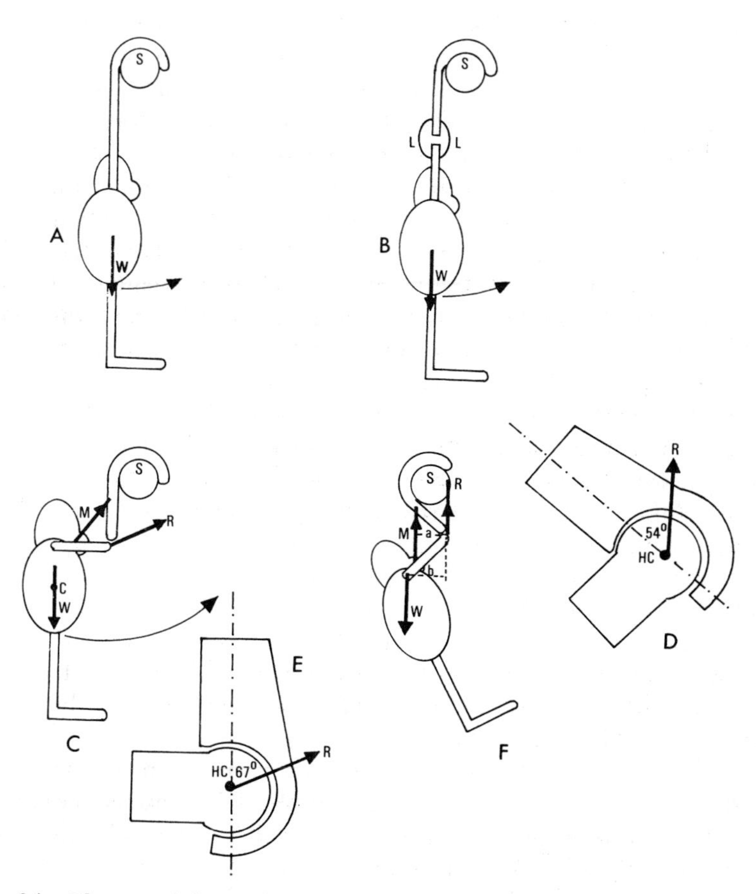

Fig. 34. Theoretical biomechanics of the elbow joint in a brachiating primate. *R,* the force vector resulting from *M,* flexor muscle contraction, and *W,* body weight. For further explanation, see text.

vertical line. If the distal and proximal segments of the arm were connected at the elbow only by collateral ligaments (LL), each ligament would sustain a pull of $\frac{1}{4}$ *W* and the joint would be surrounded by muscles which put the joint under net compression stress. The primary task of the muscles in this case is to relieve the collateral ligaments and the joint capsule. As the working line of the weight and the point of suspension are almost vertically beneath each other, the elbow joint is extended and no counterbalancing moment of the muscles is required. In the third model (Fig. 34C), the elbow is flexed 90°, and the center of gravity (*C*) is to the left side of the point of suspension (*S*). The body tends to swing forward until C and S are on the same vertical line. The joint can be kept in this position by the force M of forearm flexors. The resultant R of W and M passes through the hypomochlion of the joint. The next model (Fig. 34D) shows the elbow a little more flexed; as equilibrium demands that the moments about the joint are equal, $Ma = Wb$. Closer inspection of Fig. 34C and D shows that R has a different position in relation with the articular surfaces of the joint; this is clarified in Fig. 34E and F. In the former, the articular surface of the humeral condyle (HC) is more centrically loaded ($\alpha = 67°$); in the latter, the vector is shifted toward the periphery of the articular surface ($\alpha = 54°$). It can be mathematically shown (the precise technique lies beyond the scope of this contribution) that the more the resultant approaches the periphery of the articular surface the higher the intensity of compression stress (see Pauwels, 1965; Badoux, 1970). Microscopical analysis of the structure of the epiphyses of humerus and ulna reveals that the relative density of trabecular bone responds to this distribution of stress.

Hand

Further discussion of the forelimb will be confined to some aspects of the function of the hand in brachiation. Equilibration of the carpal and metacarpal joints is effected by similar principles as outlined for the ankle and foot. There are, however, some features of the hand of great apes that relate to biomechanical adaptations to brachiation. During the swing, the body is suspended by the hand and can be compared with a compound pendulum. When the body of the animal is moving in a segment of a circle with radius r and with velocity V, it has an acceleration toward the center of the circle equal to V^2/r. To produce this acceleration, a centrifugal force is required of $F = mV^2/r$, in which m is the mass of the animal. In the swinging brachiator, the radius of the swing r is the distance between the point of support at the branch and the center of gravity. In Fig. 35 where the mechanics at the point of support are depicted, the centrifugal force F is counteracted by $-F$, and this force can be resolved into a component $-F \sin \alpha$ and a frictional force $-F \cos \alpha$, the latter preventing the fingers from slipping from the branch. Apparently, the relative magnitude of the components of $-F$ depends on the magnitude of angle α and this, in turn, is related to the position of the hand with respect to the branch. This position can be influenced by the length of the accessory carpal bone (os pisiforme, p), and the

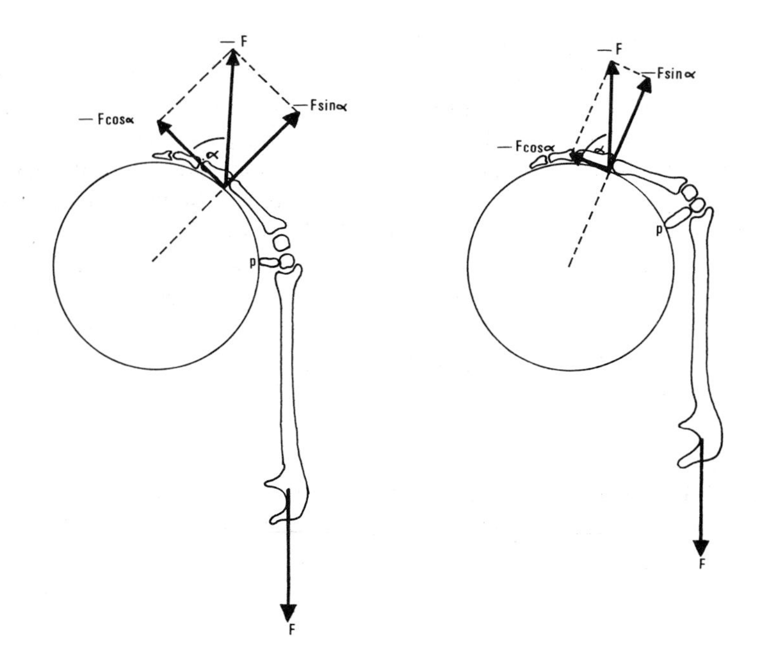

Fig. 35. The biomechanics of limb involving a relatively short os pisiforme (p) and fifth metacarpal (left) as in the orangutan and gibbon, and in forms such as the gorilla and chimpanzee with a relatively longer os pisiforme and fifth metacarpal (right). *F* represents the centrifugal force generated about the supporting branch (circle).

fifth metacarpal, both of which reveal an interesting range in dimensions among great apes.

TABLE I

	Length of		Index
	Os pisiforme (mm) (a)	Metacarpal V (mm) (b)	$\left[\frac{100 \times (a)}{(b)}\right]$
Gorilla	28	85	32.9
Chimpanzee	20	58	34.4
Orangutan	16	80	20.0
Symphalangus	8	40	20.0

The accessory carpal bones in comparison with the fifth metacarpal are relatively shorter in the gibbon and orangutan than in the gorilla and chimpanzee. The accessory carpal bone lodges the tendons of insertion of the ulnar carpal flexor and the minor digital abductor which play an important role in clenching the lateral part of the hand. In the gibbon and orangutan, the great length of the fifth metacarpal and corresponding finger facilitates the grip around thick branches so that no additional flexion of the wrist is necessary in order to maintain a safe support. In the other species, flexion of the carpus compensates for the shorter fingers in securing a safe grip of the hand; flexion is promoted by the os pisiforme which acts as a lever and increases the moment about the axis of carpal flexion.

References

Alt, F. (ed.) (1966). "Advances in Bioengineering and Instrumentation," Vol. 1. Plenum, New York.

Badoux, D. M. (1970). Some biomechanical aspects of the elbow joint in the horse during the normal gait. *Proc. Kon. Ned. Akad. Wetensch.* C73, 35–47.

Den Hartog, J. P. (1949). "Strength of Materials." Dover, New York.

Evans, F. G. (ed.) (1957). "Stress and Strain in Bones." Thomas, Springfield, Illinois.

Evans, F. G. (ed.) (1961). "Biomechanical Studies of the Musculo-Skeletal System." Thomas, Springfield, Illinois.

Evans, F. G. (ed.) (1966). "Studies on the Anatomy and Function of Bone and Joints." Springer Verlag, New York.

Frankel, V. H., and Burstein, A. H. (1970). "Orthopaedic Biomechanics." Lea and Febiger, Philadelphia, Pennsylvania.

Frost, H. M. (1967). "An Introduction to Biomechanics." Thomas, Springfield, Illinois.

Gray, J. (1944). Studies in the mechanics of the tetrapod skeleton. *J. Exp. Biol.* **20**, 88-117.

Gray, J. (1968). "Animal Locomotion." Weidenfeld and Nicholson, London.

Heberer, G. (ed.) (1965). "Menschliche Abstammungslehre, Fortschritte der Anthropogenie 1863–1964." Gustav Fischer Verlag, Stuttgart.

Howell, A. B. (1944). "Speed in Animals: Their Specialization for Running and Leaping." University of Chicago Press, Chicago, Illinois.

J. Biomechanics, Vol. I (1968). Continued. Pergamon Press, New York.

Kummer, B. (1959). "Bauprinzipien des Säugerskeletes." Georg Thieme Verlag, Stuttgart.

MacConaill, M. A., and Basmajian, J. V. (1969). "Muscles and Movements: A Basis for Human Kinesiology." Williams & Wilkins, Baltimore, Maryland.

Mohsenin, N. N. (1970). "Physical Properties of Plant and Animal Materials," Vol. 1. Gordon & Breach, New York.

Pauwels, F. (1965). "Gesammelte Abhandlungen zur funktionellen Anatomie des Bewegungsapparates." Springer Verlag, New York.

Preuschoft, H. (1969). Statische Untersuchungen am Fusz der Primaten. I. *Z. Anat. Entwicklungsgesch.* **129**, 285–345.

Slijper, E. J. (1946). Comparative Biologic-Anatomical Investigations on the Vertebral Column and Spinal Musculature of Mammals. *Kon. Ned. Akad. Wetensch.* **5** (Section 2, Part XLII), 1–128.

Tricker, R. A. R., and Tricker, B. J. K. (1967). "The Science of Movement." Mills and Boon, London.

Uhlmann, K. (1968). Hüft-und Oberschenkelmuskulatur, Systematisch und vergleichende Anatomie *Primatologia* **10**, 1–442.

Williams, M., and Lissner, H. R. (1967). "Biomechanics of Human Motion," Saunders, Philadelphia, Pennsylvania.

2

Pads and Claws in Arboreal Locomotion

MATT CARTMILL

Introduction

The supports on which an arboreal mammal sits, lies, or moves are the branches and trunks of trees, and these differ from the surface of the earth in four important ways; they are (1) discontinuous, (2) limited and variable in width, (3) mobile, and (4) oriented at all possible angles to the gravity vector. An arboreal mammal is correspondingly likely to fall and be injured in crossing from one support to the next, in balancing on narrow supports, or in moving on sloping or vertical supports. Different lineages of arboreal mammals have adapted to these dangers in different ways, each of which has certain implications for the animal's way of life. To appreciate the adaptive significance of primate locomotor morphology and behavior, we need to consider some of the nonprimate alternatives and their implications. The undeniable arboreal competence of most extant primates has sometimes led mammalogists to assume that animals conspicuously unlike primates in morphology and behavior must be imperfectly adapted to arboreal life. This assumption has prompted many dubious assertions: that claws are quite unsuitable for fast arboreal locomotion (Campbell, 1966, p. 66), that arboreal locomotion demands prehensile appendages (Grand, 1968), that tree squirrels are not arboreal "in the full sense of the term" (Lavocat, 1955), and so on. If we think of primates simply as mammals perfectly adapted to arboreal life, we cannot hope to identify the peculiarities of "their *specialized* arboreality which has little in common with that of squirrels or any other tree-living eutherian mammal" (Napier and Napier, 1967).

The Cheiridia of Arboreal Mammals

The grasping hind foot is probably the only locomotor specialization characteristic of the whole order Primates. A broadly divergent and prehensile first toe (Fig. 1) is found in all genera of extant primates except *Homo* and *Pongo* (Tuttle and Rogers, 1966), whereas the thumb is not significantly divergent or opposable (or pseudo-opposable) in at least fifteen extant genera. In all living primates, including the clawed forms (marmosets, *Callimico, Daubentonia*), the claw of the first toe is opened out into a transversely flattened nail overlying an enlarged apical pad; most primates have more or less similar nails on the other digits as well. With few exceptions, both hands and feet of primates are of the types described by Haines (1958) as "clasping" or "opposable"—that is, they display enlarged mm. contrahentes of the first and fifth digits, which arise from a median raphe and run transversely to their insertions on metapodials or proximal phalanges. Contracting, these muscles draw the margins of the hand or foot closer together. In most primates, the contrahens of digit I (adductor hallucis or pollicis) is greatly enlarged to augment the force exerted by the first digit against the four postaxial digits in grasping branches and other objects. The predominantly transverse fiber direction of the first contrahens may be accentuated by the appearance of one or more intercapitular heads of this muscle, arising from the deep transverse intermetapodial ligaments (Jouffroy and Lessertisseur, 1959). In the manus, this transverse orientation of the contrahentes is associated with the appearance of an axial hinge line in the carpal articulations (Altner, 1971).

Most of these generalizations apply equally to arboreal marsupials (with the exception of the secondarily arboreal *Dendrolagus*). A grasping hind foot is also found in certain climbing murids, some of which (e.g., *Hapalomys, Chiropodomys, Chiromyscus*) resemble primates and arboreal marsupials in lacking a claw on the pseudo-opposable first toe (Thomas, 1925; Ellerman, 1941; F. Petter, 1966). A general reduction of the claws of all the digits is seen in the arboreal diprotodonts *Tarsipes, Cercartetus,* and *Burramys* (Thomas, 1888; Jones, 1924; Warneke, 1967). Like *Tarsius,* these *Microcebus*-sized marsupials sport functional claws only on the toilet digits of the hind foot.*

The cheiridia of other arboreal mammals show little resemblance to those of primates. Arboreal Pholidota and Edentata can grasp branches

* *Dendrohyrax,* like more terrestrial hyracoids, has hooflike "nails" and retains a claw only on the pedal toilet digit; but its cheiridia are not prehensile in the primate sense, each having only three functional digits (Richard, 1964).

by volarflexing the great claws of the manus against an enlarged proximal pad, and the tree porcupines of the genus *Coendu* have developed a grasping pes by converting the tibial sesamoid and overlying pad into a surrogate hallux (Whipple, 1904; Böker, 1932; Jones, 1953); but these are clearly analogical and secondary convergences with the grasping cheiridia of primates. The contrahentes of tupaiids originate from raphes (Le Gros Clark, 1924a, 1926), but the primary raphe lies between contrahentes 2 and 4 (Haines, 1955), and digit I is not pseudo-opposable. Although the thumbs of tree squirrels, like those of many terrestrial rodents, are used in manipulation and bear a flattened nail, they are too small to function in locomotion and cannot be opposed to the postaxial digits around a support. All five toes of the tree squirrel foot (Fig. 1) spring from permanently adducted metacarpals and bear well-developed claws. Arboreal carnivorans have cheiridia which more or less closely resemble the feet of tree squirrels.

Posture and Locomotion on Horizontal Supports

Essentially terrestrial locomotor adaptations suffice for a small mammal moving on relatively large horizontal branches. The problem of equilibration becomes more critical on relatively slender branches. An animal with grasping extremities can avoid this problem by hanging underneath its support, and it has been suggested (Napier, 1967) that this represents part of the original adaptive significance of primate cheiridial prehensility.

However, animals with clawed extremities seem to get along well enough on slender supports, as long as these are roughly horizontal and fairly stable. *Sciurus carolinensis* (Burton and Burton, 1969), *Funambulus palmarum* (Phillips, 1935), *Nasua narica* (Chapman, 1929), *Paradoxurus* sp. (Ridley, 1895), and *Genetta* sp. (Leakey, 1969) are all able to run or walk atop a horizontally stretched cable; the urban squirrels that I have observed in Durham and Chicago habitually use telephone wires as elevated thoroughfares. A clawed mammal can hang beneath a horizontal cable or thin branch as easily as a primate can, as the extreme case of the sloths demonstrates. *Sciurus carolinensis* (Shorten, 1954), *Ratufa macroura* (Walker, 1964, p. 669), *Eutamias quadrivittatus* (Polyak, 1955), and other sciurine squirrels feed hanging from slender terminal branches by their clawed hind feet. The giant tropical squirrel *Ratufa*, a *Cebus*-sized animal reaching a weight of 2.5 kg in some species (Moore and Tate, 1965), is continually active among the small branches of the forest canopy, where it feeds, leaps acrobatically, and builds its nests (Wroughton, 1915; Phillips, 1935; Hill, 1949; Webb-Peploe, 1949; McClure, 1964; Prater,

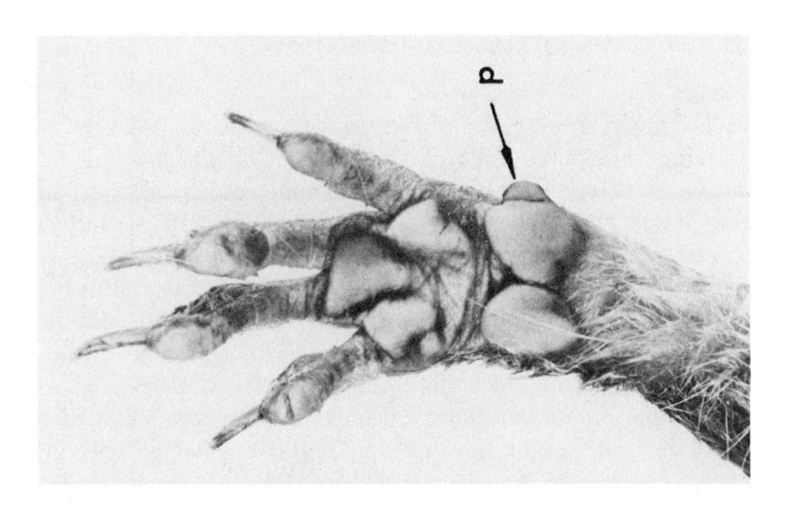
P

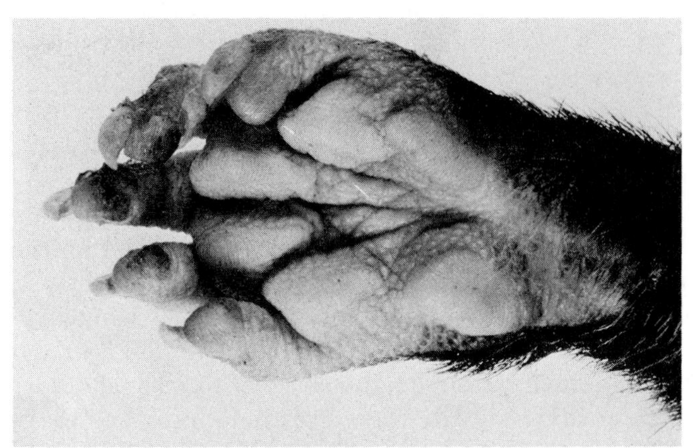

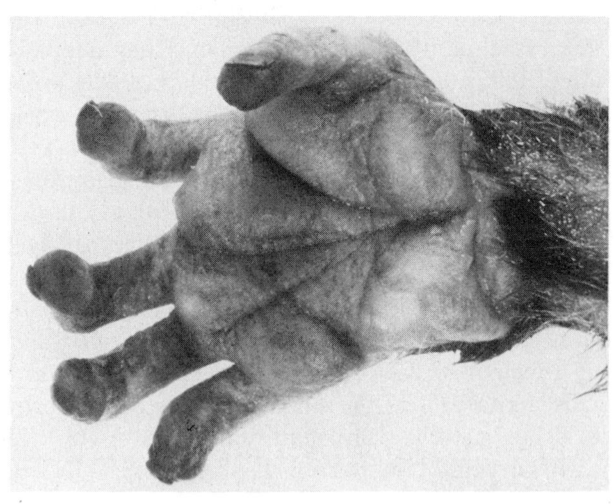

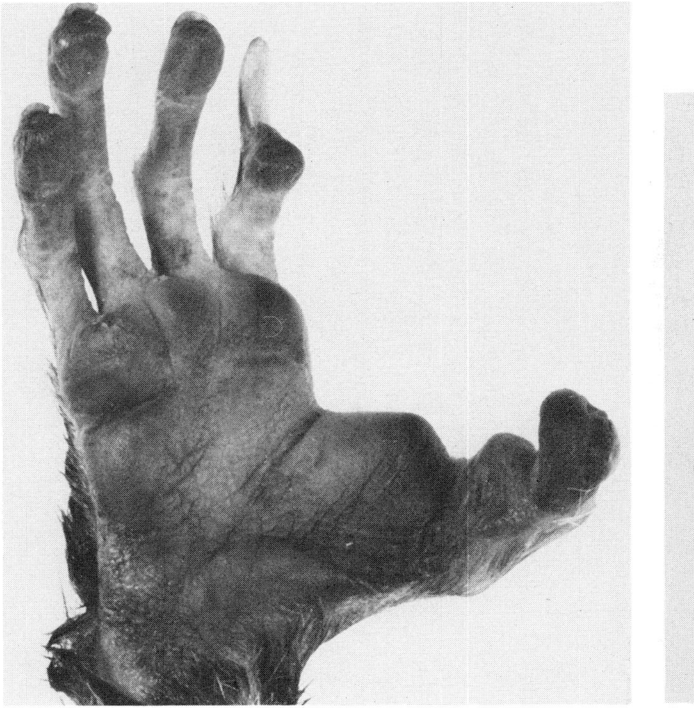

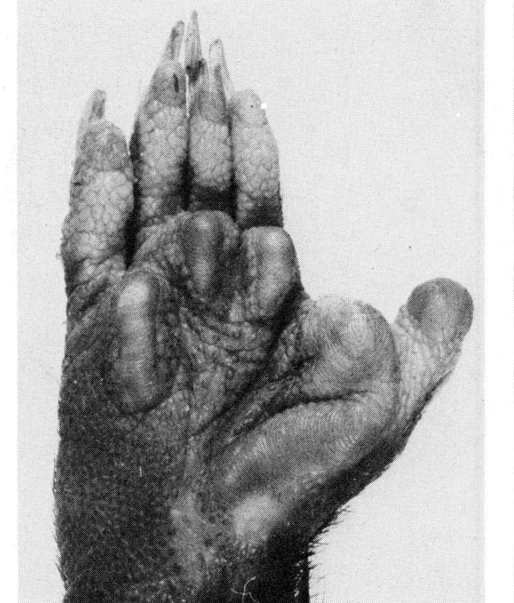

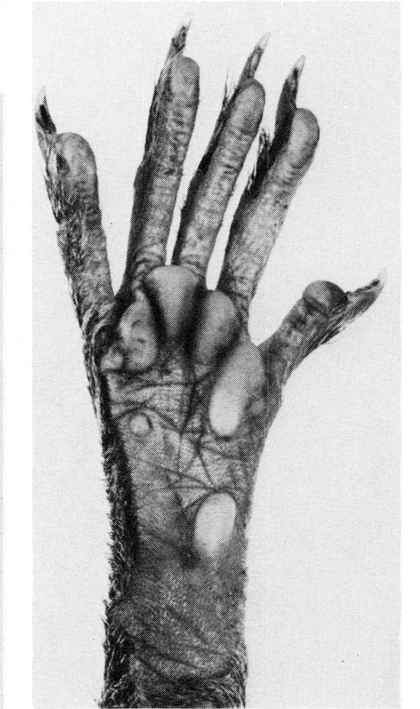

Fig. 1. Right hands (top row) and right feet (bottom row) of some arboreal mammals. Left, *Galago crassicaudatus,* showing typical primate morphology; the first digits (especially the hallux) are markedly divergent, and the partly coalesced volar pads are covered with papillary ridges. Center, *Didelphis marsupialis,* illustrating primitive marsupial condition; claws are retained on all digits except the hallux, the thumb and fifth finger are about equally divergent, and the volar pads are relatively discrete. Right, *Sciurus carolinensis,* showing typical sciurid features; all functional digits bear long, sharp claws, pads lack papillary ridges, and the thumb and first toe cannot be opposed to the postaxial digits. p, Pollex. (Photographs by W. L. Hylander.)

1965). Its hands and feet show no functional convergences with those of primates, apart from a tendency toward enlargement and fusion of the volar pads (Pocock, 1922). This tendency is most pronounced in the hand, where the fused thenar and first interdigital pads form a surrogate thumb used in manipulating food items (Hill, 1949).

Clawed arboreal mammals moving or posturing on narrow horizontal supports show certain behavioral adaptations which help to compensate for a lack of cheiridial prehensility. When feeding in the forest canopy, *Ratufa* does not sit erect on branches as smaller tree squirrels do; it balances transversely across the branch, its head and body hanging down below the support on one side and its tail on the other (Ridley, 1895; Hill, 1949; Harrison, 1951). This keeps the animal's center of gravity low and so reduces or eliminates its pitching moment around the axis of the branch. Arboreal viverrids carry the center of gravity closer to the support than terrestrial species do in walking on horizontal substrates (Taylor, 1970). It seems likely that the belly vibrissae of tree squirrels provide sensory input guiding similar locomotor behavior. On the other hand, lorises, with the most extreme cheiridial specializations for grasping found among mammals, can and do carry the center of gravity high when moving on horizontal supports (Bishop, 1964, Fig. 18). Although primate-like hands and feet facilitate this kind of locomotor behavior, they appear to have no special value for locomotion on horizontal supports per se.

Adaptations to Discontinuous Substrates

When crossing from one branch to the next, an arboreal mammal must bridge the gap from a support of proved reliability over to one that may break, bend precipitously, or prove unreachable. The danger of falling is reduced if the animal can keep a firm grip on the support behind while testing the one ahead. Therefore, the grasping ability of the hind foot is more important in locomotion than that of the hand, and the hallux is accordingly more divergent than the pollex in the majority of arboreal mammals having grasping extremities.

The security of the trailing end of the body can be enhanced by the development of a prehensile tail. Enders (1935) found that in *Didelphis*

> . . .the tail is loosely curled around the limb or vine along which the animal is passing, and it can be clamped around a support with telling effect if the limb is shaken. Should the animal lose its footing, the tail does much to prevent a fall. The "hands" of *Didelphis* make it very sure-footed on small vines, but less so on larger vines and limbs, and it is here that the tail is used most effectively.

Trichosurus, Pseudocheirus and other phalangeroids, *Marmosa,* and *Arctictis* use their prehensile tails similarly in moving cautiously from one support to the next (Ridley, 1895; Jones, 1921, 1924; Enders, 1935; Ogilvie, 1958). Although *Potos* runs acrobatically on large horizontal "highway" branches, it moves slowly and uses its prehensile tail whenever the substrate becomes discontinuous (Enders, 1935). All these prehensile-tailed animals practically never make leaps of any distance, and this also seems to be true of prehensile-tailed pangolins, anteaters, rodents, and viverrids. The only prehensile-tailed mammals (apart from some gliding diprotodonts) which often leap acrobatically in the forest canopy are ceboids.

The majority of arboreal carnivorans and rodents lack grasping specializations at the trailing end of the body; they ordinarily cross substrate discontinuities by leaping. Tree shrews also execute frequent short jumps (Vandenburgh, 1963), although they "do not take the big leaps from tree to tree that squirrels of a similar size do" in the same habitat (Ryley, 1914). Prosimians, with the conspicuous exception of the lorisines and reportedly also of *Cheirogaleus* spp. and *Daubentonia* (Petter, 1962a; Petter and Peyrieras, 1970), also execute acrobatic leaps from one branch to another. Any arboreal mammal that crosses between branches by leaping increases its danger of falling; it may fall short of the target branch, or leap over it, or have it break off on impact. Collins (1921) was the first to suggest that the optic and orbital convergence characteristic of primates was initially an adaptation to saltatory arboreal locomotion, enabling a leaping animal to use stereopsis in estimating the distance to the next branch. Subsequent primatologists have generally agreed that "orbital frontality . . . is an outgrowth of the sort of behavior seemingly requisite for arboreal survival" (Simons, 1962). However, arboreal squirrels, which have laterally directed eyes, habitually leap many body lengths from branch to branch in the canopy; leaps of over 6 meters are reported for *Ratufa* (Phillips, 1935; Hill, 1949).

It seems clear that grasping hind feet and prehensile tails permit their possessors to exercise more caution in crossing narrow gaps between supports; but, for an arboreal mammal that leaps across substrate discontinuities, clawed, squirrel-like cheiridia are not evidently inferior to prehensile cheiridia.

Nonhorizontal Supports and Frictional Forces

An animal standing on or hanging beneath a horizontal support is pressed against the support by its own weight: or, in other words, the gravity vector in this case is normal (perpendicular) to the contact surface and has no tangential component. On sloping supports, the force of gravity has both a normal and a tangential component, and on vertical supports the pull of gravity is wholly tangential to the contact surface. An arboreal mammal can remain on a sloping or vertical support only by virtue of the frictional forces between its body and the support's surface, which resist the tangential component of the gravitational force.

The forces of static friction (F) between two solid bodies are represented to a first approximation by the equation $F = \mu W$, where W is the normal force that keeps the two bodies in contact and μ is a constant, the coefficient of friction, determined empirically for each different combination of materials. The coefficient of friction of leather on clean wood is approximately 0.3–0.4, which is to say that the horizontal force needed to start a wooden block moving across a horizontal leather surface is equal to 30 or 40% of the block's weight.

Frictional force can also be expressed as a function of the real area of contact between two bodies. This area is not the same thing as the geometric or apparent area of contact; it represents the area of contact at the molecular level between submicroscopic asperities of the two surfaces. For most solids, the area of real contact is directly proportional to the normal force W. For theoretical purposes, we can treat the interface between an arboreal mammal and its support as a collection of point contacts.

When an animal clings to a vertical trunk, forces normal to the contact surface must be generated either by air pressure or by the animal's own muscular effort. Certain hyracoids are alleged to have cheiridia which function as suction cups and enable these animals to walk up and down flat vertical surfaces (Dobson, 1876), but this remains to be demonstrated (Richard, 1964). Similar, but equally unproved, abilities are sometimes claimed for tarsiers (e.g., by Schultz, 1969); in this connection it might be pointed out that corrugated and convex pads are not likely to make very satisfactory suction cups, especially on dry tree bark. Most arboreal mammals produce the necessary normal force by adducting two or more appendages deployed across some arc of the trunk or vertical branch to which they cling; these may be the digits of a single extremity on a support of small diameter, or opposing hands and feet on large trunks.

This situation is diagrammed in Fig. 2. It is evident that, except in the extreme case where the arc of deployment $\theta \geqslant 180°$, the force of adduction A will have both a normal component X and a tangential component Y. When the apposed appendages subtend a small central angle θ, proportionately more of the animal's total muscular effort will go into opposing the frictional force rather than augmenting it.

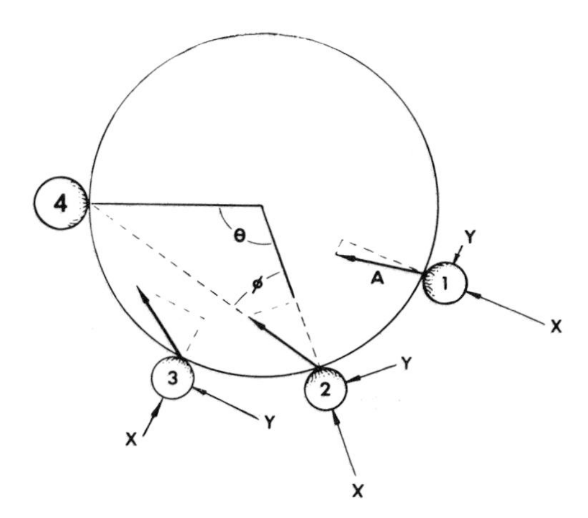

Fig. 2. Pad grip and relative support diameter: cross section through cylindrical support. As one pad (1) is moved progressively closer (2 and 3) to another (4) on the surface of the support, the normal component (X) of the constant force of adduction (A) grows smaller and the tangential component (Y) grows larger. Angle $\phi = \frac{1}{2}(180° - \theta)$. In the ideal case, the grip will slip when the central angle θ falls below the point at which tan ϕ equals the coefficient of friction of the pad on the support.

The minimal friction needed to keep the animal from slipping will just equal the tangential force. Therefore, $F_{min} = Y = \mu X$, and $\mu = Y/X = \tan (180° - \theta)/2$. Taking the coefficient of friction of leather on wood as an estimate of that of the animal's skin on the tree, we would conclude that any arboreal mammal relying on pad friction to cling to a vertical cylindrical trunk must subtend a central angle of at least 136° on that trunk. Even this is too generous an estimate given our assumptions, since we have not taken the animal's weight into account; the total tangential force to be resisted by friction at any contact point is the resultant of Y and some fraction of the animal's weight, and larger than either.

Factors Enhancing Pad Friction

Two factors, which are not taken into account in this estimate, act to increase the effective coefficient of friction of the volar pads on a support of large diameter. The first is interlocking of the pads' friction ridges with the surface of the support; the second is a more complex effect which follows from the elasticity of the pads' surface.

The volar pads of all primates and many nonprimate arboreal mammals are covered with dermatoglyphic friction ridges. These ridges project into minute concavities of substrate surfaces. As a result, tangential forces exerted in a direction perpendicular to the axes of the friction ridges will not effect displacement of the ridged pad without either (a) shearing the ridges off or (b) sliding each ridge up the face of the corresponding depression, a movement opposed by both frictional and normal forces. In either case, the effective coefficient of friction is increased to a limited extent by friction ridges. A similar imbrication on a larger scale is provided by the backward-pointing scales at the base of the tail in anomalurid rodents, who are reported to employ this patch of "friction skin" in climbing up tree trunks inchworm fashion (Sanderson, 1937).

Whipple (1904) suggested that "the direction of ridges is at right angles with the force that tends to produce slipping, or to the resultant of such forces when these forces vary in direction." In most prosimians and arboreal marsupials, the striation of the apical pads is parallel to the digital axis (Jones, 1924; Dankmeijer, 1938; Biegert, 1959, 1961). Kidd (1907) urged that this fact controverts Whipple's thesis, since axially oriented ridges cannot oppose the backward thrust of the digits in locomotion. However, in many arboreal mammals with axial striation of the apical pads, the digits of the grasping foot (and to a lesser extent those of the hand) are placed almost perpendicular to the axis of the branch along which the animal moves (Fig. 3). In arboreal mammals lacking widely divergent first digits, the digits are put down roughly parallel to the branch axis, and apical striation if present is predominantly transverse, e.g., in *Potos* and tupaiids (Kidd, 1907; Cummins and Midlo, 1943; Haines, 1955; Biegert, 1961). These facts support Whipple's thesis. For an animal with widely divergent first digits, apical striation parallel to the digital axis also resists the animal's tendency to slide downward on vertical supports. However, such striation cannot resist that component of the tangential force which is generated by the digital flexor muscles. It therefore does not affect our estimate of the minimal central angle required for a secure pad grip.

Amonton's laws of friction state that the frictional forces between two solids are (1) independent of the area of apparent contact and (2) proportional to the normal force. The first of these laws is thought to hold for all solid surfaces except those which lack submicroscopic asperities over large areas (e.g., fracture planes in certain crystals). This implies that enlargement of volar pads cannot augment pad friction, despite assertions to the contrary in the primatological literature (Whipple, 1904; Pocock, 1918; Le Gros Clark, 1959). Although automotive engineers report that the skid resistance of rubber tires improves slightly when tread width is increased, the theoretical reason for this is not entirely clear. Part of the improvement may be due to increase in losses of energy dissipated in

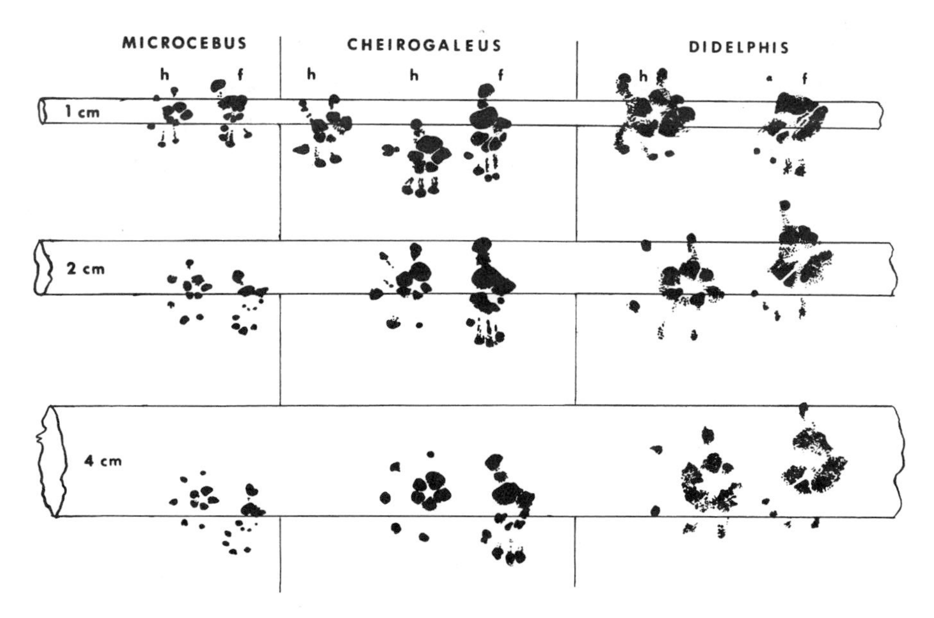

Fig. 3. Prints of left cheiridia (h, hand; f, foot) of *Microcebus murinus, Cheirogaleus medius,* and subadult *Didelphis marsupialis* on horizontal supports. Prints were developed by dusting the animals' tracks on paper-wrapped dowels with xerographic toner. They are shown here as traced from the unwrapped paper, correctly oriented on outlines of the supports (horizontal bands) as seen from above. Support diameters show scale. Two handprints are traced from *Cheirogaleus* tracks on 1-cm support, showing variation in placement of digit II.

elastic deformation of the rubber; such losses due to internal hysteresis account for up to 90% of the rolling or sliding friction between a solid object and a rubber surface (Bowden and Tabor, 1967).

Amonton's second law does not apply for highly elastic substances like rubber which undergo no plastic deformation under moderate compression. The area of real contact between a convex surface of some such elastic solid and a plane surface of some other substance is not proportional to the force normal to their interface, but to approximately the two-thirds power of the normal force (Bowden and Tabor, 1954). As a result, the coefficient of friction increases as the normal force diminishes. The convex surfaces of mammalian volar pads are covered with epidermal stratum corneum, composed chiefly of cystine-poor "soft" keratins which have few disulfide cross-linkages and are accordingly highly elastic (Ward and Lundgren, 1954; Fraser, 1969). We can therefore infer increase of the coefficient of friction between pad and substrate with reduction of the normal force. This must permit stable clinging to tree trunks with smaller values of θ than would otherwise be possible; as X diminishes, the increasing frictional coefficient of the elastic pad on the support will partly compensate for the correlated increase in Y.

To obtain an estimate of the influence of these factors on the coefficient of pad friction, twenty-two human subjects were asked to grasp and hold up a smooth 69-gm cardboard cylinder 11 cm in diameter, using thumb and one finger placed as close together as possible on the surface of the cylinder. The mean minimal central angle at which the grip could be maintained proved to be 124° (S.D. = 12.8°). All unknown factors (surface interlocking, surface elasticity, pressure-sensitive adhesion, air pressure, etc.) combine to increase the coefficient of human pad friction on a smooth wood-fiber surface to an average of 0.53, an increase of only about 24% over our initial estimate.

Claw Grip and Support Diameter

Even if the pads of other primates should prove to have frictional characteristics superior to our own, it is clear that if the central angle that a clawless arboreal mammal subtends on a vertical support falls below some critical value, the animal must fall as well. Clawed arboreal mammals are under no such limitation. A squirrel, for instance, can cling to and run along a flat vertical surface (Fig. 4); this is equivalent to clinging to a tree trunk of infinite diameter, on which the animal subtends a central angle of zero degrees.

When a claw penetrates the surface of a support, the volar aspect of the imbedded part of the claw forms some angle with the substrate surface (Fig. 5A). Adduction force is not exerted against the support's surface, but against the new interface beneath the volar surface of the claw. The result is that the effective diameter of the support is reduced, the extent of the effective reduction depending on the magnitude of the angle of penetration (Fig. 5B). So long as this angle equals or exceeds 90°, the animal's grip is secure, no matter what its central angle or frictional coefficient on the support may be. In effect, clawed arboreal mammals can adjust support diameter at their convenience.

It follows that small mammals with grasping extremities and reduced claws should have more difficulty moving on large vertical supports than do clawed mammals of similar size. This is supported by behavioral observations. *Loris tardigradus* can climb thick trunks if the bark is sufficiently irregular to offer handholds (Petter and Hladik, 1970); but these animals are incompetent on smooth-barked trees as small as 10 cm in diameter (Subramoniam, 1957), and trees in humid tropical forest typ-

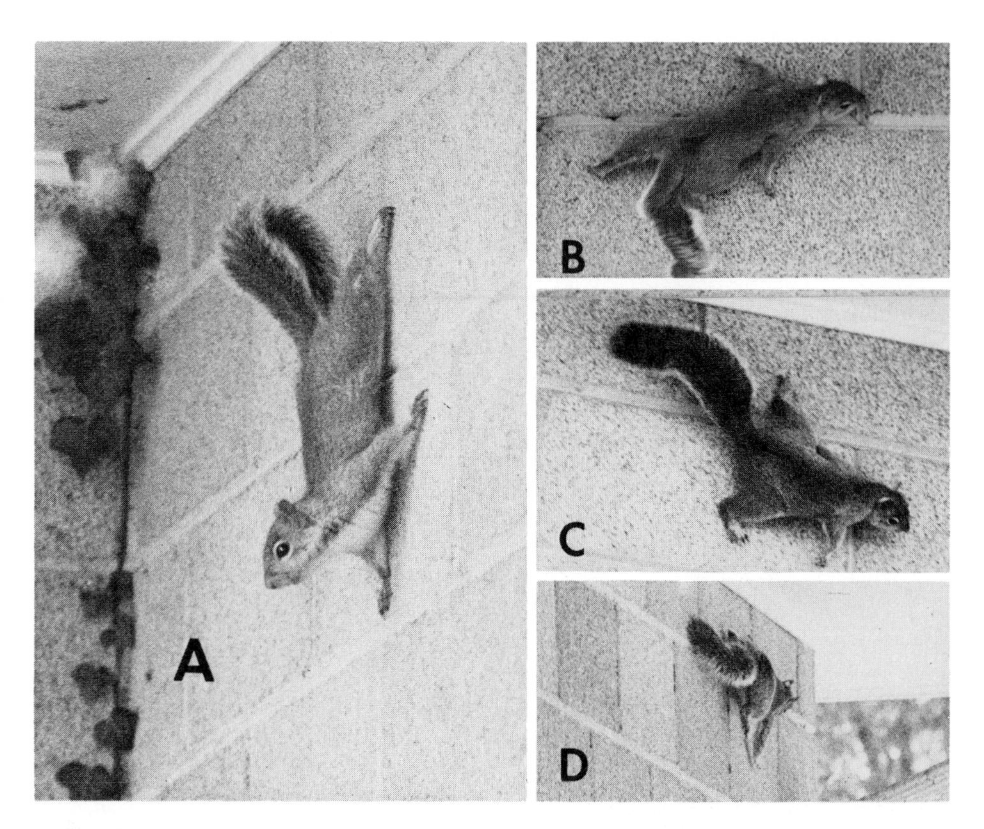

Fig. 4. *Sciurus carolinensis* on a vertical cinder-block wall. (A) Clinging head downward to the wall; the supinated left pes points upward above the animal. (B–D) Running horizontally across the wall, fleeing the photographer. The animal adopts a diagonal-couplets gait (most clearly seen in B); the asymmetrical gaits that squirrels ordinarily use, in which both forelimbs are protracted together, would result in unmanageable yawing of the unsupported end of the body.

ically bear very smooth, thin bark offering few fissures or crevices (Richards, 1966). Captive *Nycticebus coucang* at the Duke University Primate Facility prove incompetent on smooth vertical supports with a 14-cm diameter (personal observations). Small-branch substrates are also preferred by *Microcebus* (Petter, 1962a) and by most of the African lorisids (Charles-Dominique, 1971).

Moreover, clawed primates appear more competent than their clawless relatives of similar size in locomotion on thick vertical supports. *Daubentonia's* claws "allow it to climb up great trunks or, when it attacks the wood with its teeth, to cling firmly in positions no lemur could adopt"

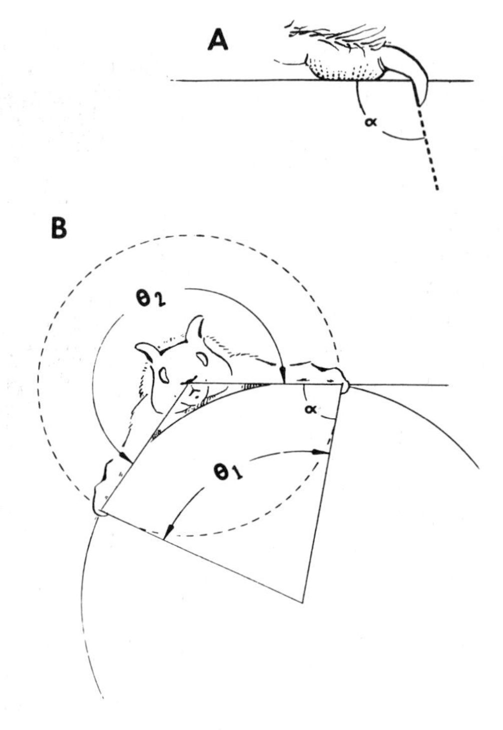

Fig. 5. Claw grip in arboreal mammals. (A) Clawed digit gripping flat surface, showing angle of penetration (α). The size of the embedded part of the claw is exaggerated here. (B) Clawed mammal clinging to cylindrical support. Solid circle, actual support surface tangent to animal's pads; dashed circle, effective support surface tangent to volar aspect of embedded claws; θ_1, actual central angle; θ_2, effective central angle.

(Petter and Peyrieras, 1970).* The specialized pointed nails of the needle-clawed galago, *Galago elegantulus,* allow it to eat sap and resin from large smooth trunks and branches, which other galagos do not climb on (Charles-Dominique, 1971). Similar marked specializations for clinging to large vertical supports appear in *Phaner* and in *Microcebus coquereli,* which also feed on sap and resin (Petter *et al.,* 1971). The claws of the marmosets *Callithrix* and *Saguinus* facilitate several squirrel-like locomotor habits, including running spirally up and down large tree trunks (Bates, 1863; Krieg, 1930; Thorington, 1968; Hershkovitz, 1969).

* "Ses griffes puissantes . . . lui permettent en outre de grimper aux gros troncs ou de s'agripper fortement dans des positions impossibles à adopter pour les Lémurs lorsqu'il attaque le bois avec ses dents."

Locomotion on Vertical Supports

While there is some reason to believe that, for small mammals, claws are adaptively superior to flattened nails in climbing on large vertical trunks or branches, there seems no reason why this should be true for larger arboreal mammals. Other things being equal, larger mammals must subtend larger central angles on the trunks which they climb. Why, then, are chimpanzees or men not as nimble as squirrels in running up and down tree trunks? Physiological, rather than anatomical, factors make this impossible. Since metabolic rates vary inversely with body weight, the increase in oxygen consumption demanded by vertical locomotion represents a much larger fraction of the resting metabolism in large animals than in small ones (Taylor *et al.*, 1972); the efficiency of vertical climbing relative to that of horizontal locomotion goes down as body weight increases. For this reason, animals that often move cursorially on vertical trunks must be small; being small, they must subtend small central angles on those trunks; and therefore, their digits bear sharp claws which permit them to subtend effectively larger central angles without any inefficient increase in body weight.

A four-footed animal clinging to a vertical support must maintain its position by continual muscular effort. Because the animal's center of gravity (CG) is not coplanar with its points of support, its weight produces rotatory moments around axes traversing various pairs of support points. In the case of an animal climbing with an asymmetrical gait, we can distinguish two principal moments due to the animal's weight (Fig. 6): a moment around an axis (*A*) passing through the contact surfaces of the two superior limbs, and a moment around an axis (*B*) passing through the contact surfaces of the two inferior limbs. The unresisted moment around *A* will produce an inward rotation, bringing the animal's CG closer to the trunk; the moment around *B* tends to rotate the CG outward. A comparable situation is that of a man standing on a vertical ladder. If his feet slip, he will swing forward into the ladder; if his hands slip, he will topple away from the ladder and fall. The moment around *B*, then, must be kept to a minimum and resisted as strongly as possible by the climbing animal. The moment around *A* is less important.

Assuming a constant weight, the gravitational moment around *B* can be kept to a minimum only by minimizing the moment arm, either (1) by shifting the CG ventrally or (2) by approximating the lower feet so that *B* is drawn closer to the line of gravity. We might therefore expect that an animal climbing up a trunk would embrace the trunk with widely

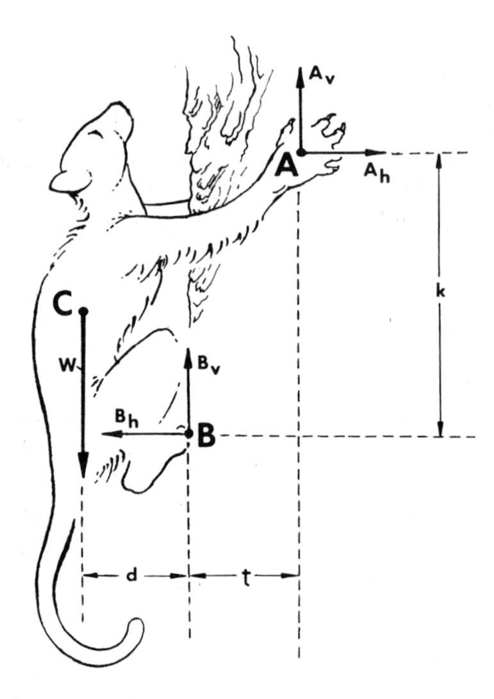

Fig. 6. Mammal on vertical support. *C*, Center of gravity; *W*, weight of animal. *A* and *B* are lateral projections of axes defined in text. Horizontal (A_h, B_h) and vertical (A_v, B_v) components of forces through *A* and *B* are indicated. At equilibrium, $Wd + A_v t = A_h k$, and $W(d + t) = B_v t + B_h k$.

spread upper limbs (thus maximizing the central angle that they subtend and bringing the CG as close as possible to the tree) and keep the lower feet close together below the CG (thus minimizing the gravitational moment around *B*). *Presbytis entellus* (Ripley, 1967), *Nasua narica* (Kaufmann, 1962), *Callosciurus prevosti* (Banks, 1931), and many other primates, carnivorans, and rodents climb trees in this way.

The adducted hind limbs provide much or most of the propulsive thrust in climbing. If the hind feet subtend a small central angle, adduction can generate little force normal to the contact surface. Frictional forces between the hind feet and the support must therefore be slight, and the feet will tend to slip downward. This presents particular difficulties for a large arboreal mammal which cannot rely on claw grip, e.g., for a large anthropoid. The normal force at the lower contact points (B_h: Fig. 6) can be augmented by bringing *A* and *B* closer together and leaning away from the tree; since $W(d + t) = B_v t + B_h k$ (Fig. 6), an increase in *d* and decrease in k implies an increase in B_h. The result is an equal increase in A_h, however, which cannot be tolerated unless the upper grip is exceptionally secure—implying that the upper limbs (if clawless) must subtend a large central angle. Therefore, habitually tree-climbing mammals of very large size might be expected to have either claws (e.g., *Helarctos*, Pfeffer, 1969) or relatively long and powerful forelimbs (e.g., *Pan*). This provides at least a partial explanation for the fact that the relative length of the forelimb increases as body size goes up among the catarrhines (Biegert and Maurer, 1972). A chimpanzee climbing a tree augments pedal friction by leaning away from the support; this behavior can be mimicked by a man wearing long hooks strapped to his wrists (Kortlandt, 1968).

The postures and behavior that serve in ascending will serve also in descending if the animal backs down the tree tail first. However, most arboreal mammals descend vertical supports head first, improving their chances of spotting dangers in their path. The biomechanical problems that this presents to a typical quadruped are evident to anyone who has watched a domestic cat trying to climb down a tree. In head-first descent, the propulsive muscles of the hind limb—triceps surae, hamstrings, and the deeper glutei—cannot act as they do in head-first ascent to help support the animal's weight. To keep the plantar surfaces applied to the support, the feet must be kept well craniad, between the animal's belly and the support; but this moves the pelvis away from the support, shifting the CG outward and increasing the gravitational moment arm. Moreover, a clawed arboreal mammal descending head first finds that its claws are all pointing the wrong way; they cannot be dug in at the proper angle to resist the craniad tug of gravity.

Head-first descent therefore depends on an ability to supinate or laterally rotate the extended hind limb so that the plantar surface of the volarflexed foot can be applied to the surface with the digits pointing caudad (Fig. 4A). This ability is especially crucial for an animal depending on claw grip. When the hind limb is thus supinated, the CG is kept close to the support and the digital flexor muscles can act to help resist the gravitational pull and slow the descent. This posture of descent is adopted by *Sciurus carolinensis* (Shorten, 1954), *Tamiasciurus hudsonicus* (Hatt, 1929), *Callosciurus finlaysoni* (Walker, 1964), *Ratufa macroura* (ibid., p. 669), *Funambulus palmarum* (Pfeffer, 1968, p. 194), *Petaurus breviceps* (Morcombe, 1968), *Bassariscus astutus* (Grinnell *et al.*, 1937; Trapp, 1972), *Nasua narica* (Kaufmann, 1962), *Nandinia binotata* (Taylor, 1970), *Cryptoprocta ferox* (Albignac, 1970), *Felis wiedi*

(Leyhausen, 1963), and *Tupaia glis* (Jenkins, this volume). It is also seen in *Galago crassicaudatus* (Bishop, 1964, Fig. 25a) and in other prosimians capable of hanging by the hind feet to feed.

Cats, mongooses, and other carnivorans with permanently pronated hind limbs sometimes enter trees in pursuit of prey. Incapable of controlled head-first descent, they either attempt to back down, as cats and many herpestines often do (Dücker, 1971), or else proceed, like *Herpestes sanguineus*, by "rushing up and usually falling down" (Taylor, 1970). Any clawed mammal with permanently pronated hind limbs is probably, like most species of *Felis*, capable of only limited arboreal activity in the lower strata of the forest.

The approximately 170° of supination necessary for controlled head-first descent are achieved differently in different groups of arboreal mammals. In arboreal marsupials, the fibula articulates with the femur above and the talus below via cartilaginous articular disks, which permit the fibula to rotate slightly around both its longitudinal and transverse axes; in supination, the distal fibula is thought to shift backward or posteriorly relative to the distal tibia, thus rotating the foot laterally as a whole (Barnett and Napier, 1953; Barnett, 1954). In galagos and tarsiers, the elongated navicular and the cuneiforms rotate as a unit together with the cuboid around an axis passing through the talonavicular and calcaneocuboid joints—an arrangement mimicking that seen in the hominoid forearm (Hall-Craggs, 1966). In *Felis wiedi*, the only species of cat known to be capable of hind limb supination, the rotatory movement takes place between knee and ankle; the tibia and fibula cross each other when the animal descends trees head first (Leyhausen, 1963). In fresh cadavers of *Tupaia glis* and *Sciurus carolinensis*, I find that approximately 90° of the requisite supination takes place at the subtalar joints, the foot rotating as a unit around the common axis of the talocalcaneonavicular articulations. This type of rotation, which is also found among arboreal procyonids (Trapp, 1972), is ordinarily augmented by lateral rotation at the hip joint; however, Jenkins (this volume) finds that almost 180° of rotation may occur at the subtalar joints of living tree shrews descending vertical supports.

Adaptive Significance of Grasping Extremities

From the foregoing, it is reasonable to conclude that clawed squirrel-like hands and feet are better adapted than primatelike grasping hands and feet to clinging to or climbing on thick trunks and branches, and equally well adapted to running along horizontal supports, hanging beneath slender supports, leaping from tree to tree, and many other arboreal locomotor activities considered simply as locomotor activities. We have no longer any warrant for concluding that "compared with claws . . . terminal digital pads . . . provide a much more efficient grasping mechanism for animals which find it necessary to indulge in arboreal acrobatics" (Le Gros Clark, 1959, p. 174). We are, therefore, entitled to reject that model of primate evolution which envisions the strepsirhine prosimians as having evolved out of treeshrew-like clawed mammals under the selective pressures imposed by arboreal life per se.

Why, then, do not all primates have squirrel-like hands and feet? Or, if we admit that the grasping hallux is an advantage in certain situations, why do not prosimians, cebids, and catarrhines have hands and feet like a marmoset's, combining the best features of both adaptive extremes? Three alternative explanations are available, each of which is tied up with a different general theory of the origin and differentiation of the order Primates. We will consider them in their historical order.

Primate Traits as Primitive Retentions

Elaborating and refining Dollo's (1899) argument for a didelphidlike arboreal way of life in the ancestral marsupials, Matthew (1904) proposed that the earliest eutherians were also primarily adapted to life in the trees. This conclusion was drawn mainly from the absence of cursorial or saltatory specializations in the known postcranial remains of Paleocene mammals, although Matthew also claimed that a few vestigial primatelike traits, notably a "more or less opposable" hallux and pollex, could be seen in the earliest creodonts and condylarths. In his last study of the subject, Matthew (1937) concluded that the early placentals were generalized tree-dwelling forms with mobile limbs, divergent but not fully opposable first digits, and sharp claws, ". . . differing from arboreal Primates chiefly in that these have carried certain arboreal specializations still further." Whether or not this is an accurate description of the ancestral eutherians, it implies a model of primate differentiation essentially like Le Gros Clark's; it assumes that the "sharp small claws" (Matthew, 1937, p. 326) of the primates' insectivoran ancestors were lost because flattened nails and enlarged apical pads are better suited to the demands of arboreal locomotion. This, as we have seen, is not a defensible assumption.

A more radical model of early primate evolution, stemming in part from Matthew's work and in part from the German morphological tradition of Gegenbaur and Klaatsch, was worked out by Jones (1916) and Böker and his co-workers (Böker, 1927, 1932; Böker and Pfaff, 1931; Panzer, 1932). They concluded that the primate cheiridial traits were primitive for the class Mammalia, and that all nonprimate mammals had become specialized by going through a phase of more or less irreversible adaptation to terrestrial locomotion. During this supposed terrestrial phase in each lineage, claws were developed from the primitive flat nails, and the primitive full opposability of the first digits was lost. By this view, the primate traits require no particular explanation; they are mere primitive retentions.

Matthew's arguments for the opposability of the thumb in the ancestral placentals were convincingly answered by the rebuttals of Gidley (1919) and Haines (1958), reviewed by Jenkins elsewhere in this volume. More recently, Lewis (1964) has advanced comparative evidence for the opposability of the primitive hallux. Among extant mammals, the musculature of the foot differs from the reptilian pattern in two peculiar respects. (1) In reptiles, the peroneal muscles, the crural equivalents of m. extensor carpi ulnaris, insert on the base of metatarsal V. In mammals, part of this musculature forms a m. peroneus longus, whose tendon of insertion extends medially across the sole to the base of metatarsal I. (2) A m. flexor accessorius or quadratus plantae, of unknown antecedents, appears in the mammalian foot; primitively, it runs from the lateral surface of the calcaneus to insert into the tendon of the fibular digital flexor. Lewis proposes that both these peculiarities of mammals were originally correlated with the development of a grasping hallux. The peroneus longus' insertion permits it to aid in powerful adduction of the widely divergent hallux; and the flexor accessorius compensates for the divergence of the hallux by correctly aligning the pull of the long flexor tendon to the hallux. Since the peroneus longus and flexor accessorius are found in monotremes as well as therians, Lewis' interpretation implies that the common ancestors of the class Mammalia, by no later than the Late Triassic (Hopson and Crompton, 1969), had primate- or didelphidlike grasping extremities.

Little fossil evidence bears directly on this theory. No foot bones of Triassic or Jurassic mammals have as yet been described. The first metatarsal of the Cretaceous multituberculate *Eucosmodon* was small and not widely divergent from the other four (Granger and Simpson, 1929). Footprints of the supposed mammal "*Ameghinichnus*" from the Jurassic of Patagonia (Casamiquela, 1961) were left by a digitigrade animal with short, slightly divergent but nonopposable first and fifth digits. The notion that the ancestral mammals had didelphidlike opposable halluces remains tenable, but the evidence is not conclusive.

Other comparative anatomical evidence, however, militates against the idea that the basal mammals were adapted to living in trees. The eye of a mammal, like that of a snake, is a simplified, stripped-down version of the reptilian eye. Early in mammalian history, color vision, the nictitating membrane, and the reptilian (and avian) mechanism of accommodation were lost (Stibbe, 1928; Walls, 1942; Polyak, 1955). Although some mammals have analogous adaptations, these have been redeveloped independently out of rather different material. Loss of color vision strongly implies that the ancestral mammals were nocturnal; loss of the power of accommodation indicates that precise focusing of the retinal image was unimportant in their way of life; and the disappearance of the reptilian nictitans apparatus suggests that they were micropic. The inferred degeneration of the visual apparatus in the basal mammalian stock is correlated with a great elaboration of the apparatus of olfaction, a trend which was already manifest in the latest cynodonts (Simpson, 1927, 1928; Hopson, 1969). In short, the comparative anatomy of the sensory organs leads us to conclude that primitive mammals were nocturnal beasts, with small eyes and a snout full of olfactory receptors, that made their living catching insects on the forest floor much as similar living insectivorans do. Degenerative eyes and elaborate olfactory apparatus seem unlikely to be primary adaptations to running through the treetops.

It seems likely in any event that the digits of the early mammals bore sharp claws, not primatelike nails. Panzer (1932) and Böker (1932) contended that since claws are structurally more complex than nails, they are therefore more specialized; and that since evolution is irreversible, the less-specialized nail could not have evolved from the claw. The argument is patently shoddy, and, led by Le Gros Clark (1936), most contemporary workers have returned to Boas' (1884) original view of the claw as primitive. This view is supported by the available fossil evidence, i.e., by the claw-shaped distal phalanges of cynodont therapsids.

The conclusion seems warranted that the early therians had somewhat tupaiidlike hands and feet, with clawed digits, a mobile but nonopposable thumb, and a somewhat divergent first toe. No evidence suggests that primitive placental mammals had as divergent a first toe as that of *Cheirogaleus* (Fig. 3), for example. If early placentals were active in the trees, they were evidently able to employ a claw grip, but not a specialized pad grip employing digital opposition; and the hypothesis that the primate adaptations are primitive retentions may accordingly be rejected.

The "Vertical Clinging and Leaping" Theory

A second body of theory, persuasively elaborated in several recent publications by J. R. Napier and his associates, postulates that the ancestral Cretaceous primates were small, clawed, semiarboreal insectivores with "the build, and probably the gait, of squirrels or other rodents" (Napier, 1970, p. 102). During the Paleocene, one or more lineages of these tupaiidlike ancestors adopted a "vertical clinging and leaping" locomotor habit like that of *Tarsius* or *Propithecus*, perhaps as an adaptation to escaping from arboreal predators (Napier and Walker, 1967). This adaptation was evidently highly successful, since for ". . . all the fossil postcranial bones reputably assigned to Eocene prosimians . . . all the skeletal characters point to these animals being Vertical Clingers and Leapers" (Napier and Walker, 1967). One possible indication of this is the appearance of the grasping foot in several Eocene lineages; the non-opposable thumb and divergent, "pseudo-opposable" first toe of tarsiers and similarly constituted Eocene prosimians "are adaptations for the vertical clinging posture where the weight of the body is supported principally by the feet" (Napier, 1967).

During the Eocene, several lines of primates began to specialize in eating fruit and leaves. These grow at the tips of fine branches. Small mammals have no difficulty in balancing on these slender supports, but the larger Eocene primates were forced to develop more specialized prehensile extremities to move about safely in terminal branches (Napier, 1967). However, "the limitations imposed by the vertical clinging and leaping habit upon feeding behaviour are . . . quite marked; the larger forms, especially, are at a disadvantage when feeding in a small branch milieu" (Napier and Walker, 1967). Although vertical clingers and leapers can exploit the resources of the terminal branches by hanging by their prehensile hind limbs, "to possess the faculty of suspension by all four limbs is infinitely more effective" (Napier, 1970, p. 103). Accordingly, the thumb became divergent to enhance the grasping power of the hands. As the grasping forelimbs became more important in suspensory postures, they increased in relative size, and this trend resulted in the abandonment of the vertical clinging and leaping habit in favor of quadrupedal locomotion in a great many primate lineages (Napier, 1967).

There are several problems with this thesis, one of which emerges from the preceding discussion of claw grip. If vertical clingers and leapers are at a disadvantage in a small branch milieu, they are presumably primarily adapted to clinging to large vertical supports. This is not a reason for developing a grasping hallux; on the contrary, it would be an excellent reason for retaining the clawed, squirrel-like hands and feet of the hypothetical ancestor. A reasonable inference is that the vertical clinging and leaping adaptations seen in such clawless Eocene prosimians as *Notharctus* have been acquired secondarily, and that the claws were lost during an earlier phase in which trunks and large vertical branches were *not* the preferred locomotor substrate.

The combination of a grasping first toe with a nonopposable thumb is not correlated with vertical clinging and leaping; as shown above, it is found among didelphids (Fig. 3) and other arboreal mammals that clearly have no indriidlike or tarsierlike ancestors. It makes more sense to interpret the grasping foot as an adaptation to prolonged or cautious movement among slender branches and vines, in didelphids and early prosimians alike. Similarly, although the divergence of the primate thumb correlates roughly with body size or relative support diameter, it shows no correlation with locomotor categories. The vertically clinging and leaping indriids have rather gibbonlike hands, quite unlike the more primitive hands of the quadrupedal cheirogaleines (cf. Figs. 3 and 7). In indriids, as in gibbons, the thumb is set off from the other digits by a deep first interdigital cleft; this is probably an adaptation for climbing large vertical supports, as it appears to be in gibbons (Van Horn, 1972). The specialized hands of *Propithecus* are also well adapted to the fine terminal branches where these large vertical clingers hang and forage for leaves (Jolly, 1966). *Lepilemur* has similar foraging habits (Charles-Dominique and Hladik, 1971). Among Madagascar prosimians, the vertical clinging and leaping habit correlates with a complex of visceral, dental, and behavioral specializations for feeding on leaves in the periphery of trees; we have no warrant for saying that *Lepilemur* and the indriids are at a disadvantage when feeding in a small branch milieu, since that is, in fact, where they feed. The evidence suggests that the vertical clinging and leaping habit is a secondary specialization among Lemuriformes, and that the ancestral Madagascar lemurs were quadrupeds resembling *Cheirogaleus* or *Microcebus*. Martin's (1972) metrical study of prosimian calcanei supports this conclusion.

If we reject the hypothesis that the remote ancestors of most or all living primates were vertical clingers and leapers, how can we accommodate the conclusion that all Eocene primates display specializations for vertical clinging and leaping? This conclusion is more sweeping than the fossil evidence warrants. The known parts of the omomyid hind limb, for instance, resemble their homologs among the extant cheirogaleines in such respects as the overall elongation of the tarsus (Mivart, 1873; Simpson, 1940) and some aspects of the morphology of the proximal femur (Cartmill, 1972). Since it would be a mistake to infer vertical clinging and leaping habits from the large medial epicondylar index found in *Saimiri sciureus* (Napier and Davis, 1959), parallel inferences about fossil adapids (Day and Walker, 1969) are not sound. Some of the other arguments

Fig. 7. *Propithecus verrauxi* climbing on slender supports. The forelimbs of this vertically clinging and leaping prosimian are well adapted to hanging and feeding from slender terminal branches. (Cf. Bishop, 1964, Fig. 23a.) (Photograph courtesy of J. Buettner-Janusch.)

advanced in favor of the Napier-Walker hypothesis are subject to similar objections (Cartmill, 1970). *Notharctus* and *Smilodectes* appear to have been rather indriidlike, and some necrolemurines show similarities to *Tarsius;* going further than this seems unwarranted at present.

Predatory Adaptations in Primitive Primates

In tropical forests, insects are concentrated in three strata: (1) the leaf litter and humus of the forest floor, (2) the upper canopy, and (3) the dense shrub layer of clearings and the forest margins. Forest floor insects are preyed upon by long-snouted predators like *Solenodon, Microgale, Tenrec, Erinaceus,* and *Nasua,* which rely little on vision and greatly on smell, hearing, and vibrissal contact to locate the insects they encounter while nosing through the leaf litter (Kaufmann, 1962; Eisenberg and Gould, 1966, 1970; Poduschka, 1969). Forest-floor predators may climb to nose out insects concealed in the dense crowns of palms and similar plants (Kaufmann, 1962; Heim de Balsac and Vuattoux, 1969), but their search procedures are poorly adapted to stalking insects amid the slender branches of the canopy and shrub layers. The insects which abound in these strata are attacked by small predators like *Microcebus, Loris, Nycticebus, Galago,* and *Tarsius,* which locate and track prey with their eyes and seize it with their hands (Le Gros Clark, 1924b; Harrisson, 1962; Bishop, 1962, 1964).* Similar predatory patterns are seen among chameleons (where the tongue substitutes for the hand) and among some of the smaller arboreal marsupials like *Cercartetus* and *Caluromys.* All these animals have broadly overlapping visual fields, a predatory adaptation convergent with the similar adaptations of cats and owls (Cartmill, 1972).

These small visually guided predators also have grasping hind feet. The utility of this trait to their way of life is evident. Although squirrels can move and forage among the terminal branches of the canopy and shrub layer, they spend little time actually feeding there; having cut loose a food item from a slender branch, a squirrel will ordinarily either retreat with it to a larger branch or trunk to feed, or drop it to the gound to be gathered later (Ridley, 1895; Hatt, 1929; Hutton, 1949; Shorten, 1954; Polyak, 1955). Squirrels running and leaping along terminal branches from one tree to the next do not exhibit the slow caution that grasping feet or a prehensile tail make possible; they bound swiftly from one unstable perch to another, without making futile attempts to dig in their claws while they balance atop slender twigs.† By hanging bipedally below terminal branches, squirrels can pause and forage in the periphery of a tree, but in this milieu their sojourns are relatively brief and their locomotion is acrobatic and fairly noisy. Lorises, by contrast, move with silent deliberation

* The African lorisines rely on olfaction in detecting prey to a greater extent than the Asian *Loris,* but this seems to be a secondary specialization; the diet of the African forms is composed largely of foul-smelling insects for which there is no competition from the sympatric galagines (Charles-Dominique, 1971).

† Per contra, a clawless primate like *Cercopithecus mona campbelli* can simply gallop down sloping trunks too thick and steep for it to grasp securely (Bourlière *et al.,* 1970).

among terminal branches for hours, stalking and seizing the insects that they spy (Petter and Hladik, 1970). Their grasping feet enable them to clamber up vertical twigs too narrow to embrace, and to resist any rolling moments induced by gravity when they creep toward an insect along a horizontal twig or lift their hands from the support to catch their prey. Similarly morphology in other predatory prosimians, didelphids, and chameleons facilitates similar visually directed predation. Since grasping extremities first appear in the primate fossil record among Eocene families whose smaller representatives also manifest large, convergent orbits, it is reasonable to conclude that these novelties were functionally related, and that the last common ancestor of the extant primates was a small predatory animal superficially resembling *Microcebus* or a convergent marsupial like *Marmosa* or *Cercartetus*.

Like its alternatives, this third interpretation of the original adaptive value of the primate locomotor specializations encounters several difficulties. One possible objection is that there is no reason to think that *Chiromyscus, Hapalomys*, and other small rodents with grasping hind feet are preying on insects among terminal branches. When something is learned about the habits of these rodents, the theory may have to be modified or discarded.* A more pressing objection is that the simplest interpretation of the primate fossil record suggests that ". . . the insectivore-primate transition was probably initiated at the end of the Cretaceous or earlier by behavioral and physiological adaptations (to) . . . a predominantly frugivorous-herbaceous diet" (Szalay, 1969, p. 323). The principal reason for thinking this is that the cheek teeth of the earliest primates resemble those of condylarths and primitive rodents in such features as reduction of the stylar shelf, expansion of the talonid basin, and general reduction in cusp height by comparison with the presumed ancestral condition (Wood, 1962; Szalay, 1968). Yet even if this interpretation is correct, it remains to be demonstrated that animal protein ceased to provide a substantial proportion of the diet of the basal primates; however the order is defined, its earliest members include animals too tiny to have been exclusively frugivorous.

A suggestive parallel can be found among didelphid marsupials. The

* There is evidence that the grasping feet and dental peculiarities of *Hapalomys* and *Chiropodomys* may be adaptations to feeding on and nesting in bamboo (Musser, 1972). The stems of bamboo are slender, vertical, smooth, and siliceous, and clawed digits would be of little use to an animal attempting to climb them. It is unlikely, however, that the grasping extremities of primates originated as part of an adaptive shift to a diet of bamboo.

subfamily Microbiotheriinae is distinguished from the Didelphinae by features of the molar dentition, including expansion of the talonid basin, corresponding enlargement of the protocone, reduction of the stylar shelf, decrease of trigonid height, and a general deemphasis of transverse shear (Simpson, 1935a; Reig, 1955). These dental traits are seen in the extant woolly opossum, *Caluromys philander,* which some would assign to the Microbiotheriinae (Reig, 1955; Biggers *et al.,* 1965). Although the present inadequate field data suggest that this canopy-dwelling opossum may eat proportionately more fruit than undoubted didelphines, captive animals eagerly accept insects, and, unlike other didelphids, use their hands to capture and kill prey (Hall and Dalquest, 1963). By comparison with other opossums, *Caluromys* shows greater frontation and approximation (but not convergence) of the orbits, more extensive periorbital ossification, a smaller snout, and more pronounced flexion of the face on the braincase (Cartmill, 1970). In these traits of morphology and behavior, *Caluromys* is convergent with extant predatory prosimians. It also displays relatively large canines and a lowered mandibular condyle—traits usually not associated with herbivory. A definitive functional interpretation of its dentition, or those of early primates, will depend on improvements in our understanding of the relations between dental morphology and the mechanical properties of various animal and vegetable foodstuffs.

Reduction of Claws: Phylogeny and Functional Significance

In a mammal whose first toe has become divergent enough to oppose the other four toes in grasping slender branches, the claw of the first toe can add little to the security of the animal's claw grip on thick supports. Since the four postaxial digits can exert no more force against the first digit in a stable grip than the first digit can exert against them, efficient prehension demands that the first digit be supplied with a relatively large slip of the common digital flexor tendon; but this reduces the strength of the claw grip correspondingly. Two evolutionary pathways are then likely. (1) If the animal's way of life emphasizes prolonged cautious locomotion among slender branches, the first digit will become proportionately more powerful, and claw grip will be proportionately enfeebled. This appears to be true for such marsupials as *Marmosa, Didelphis,* and *Petaurus,* which despite their well-developed claws are reported to have difficulty climbing on thick trunks and to seek by preference slender branches and lianas (Fleay, 1933; Enders, 1935). (2) If the pressures of natural selection favor redevelopment of an efficient claw grip, the divergent first digit will become proportionately reduced.* This appears to have happened in the evolution of the marmoset hallux, which is small, seemingly regressive, and furnished with rather feeble intrinsic musculature (Beattie, 1927; Böker, 1932; Midlo, 1934; Hill, 1957, 1959; Biegert, 1961), although it is still functional in locomotion on small branches (Rothe, 1972). The thumbs of marmosets (like the halluces of squirrels) have remained relatively large, suggesting that these clawed digits never became widely divergent in the first place.

The lessened importance of the claw of the divergent first toe renders it susceptible to loss; it is invariably the first claw to suffer reduction or elimination in lineages developing grasping specializations of the hind feet. Believers in probable mutation effects may wish to invoke one here. A positive advantage for claw loss is suggested by Hershkovitz (1970), who proposes that claws "interfere with full phalangeal flexure" in a grasping foot. I take this to mean roughly that a sharp-clawed mammal opposing digit I to digits II–V around very slender supports will be continually stabbing itself in the digits or volar pads whenever its perforating flexors powerfully contract.

In most marsupials, the claw on the first toe has been lost altogether; among caenolestids (Osgood, 1921) and primates, it has been converted into a flattened nail or nail-like claw. The utility of nails is poorly understood. In a study of the chimpanzee hind limb, Preuschoft (1970) proposed that the nail, by preventing distortion of the distal part of the apical pad, increases the area of the pad over which a load can be applied; this is said to be desirable "because of the limited strength of the skin." While this analysis may well be applicable in the case of a giant primate like the chimpanzee, it is difficult to see how it might apply to the well-developed nails of, e.g., *Perodicticus*. Moreover, nail length is not in general correlated with length of the apical pads; the relatively large apical pads of tarsiers, for instance, are supported by relatively miniscule nails (Le Gros Clark, 1936; Hill, 1955) of no apparent mechanical significance.

Whatever its original value may have been, the hallucal nail, like the divergent and grasping hallux, seems to have been present in the last common ancestor of the extant primates. It is not yet known whether any of the Plesiadapoidea (*sensu* Szalay, 1972a), generally but not universally accepted as the earliest representatives of the order Primates,

* A parallel trend among birds is seen in the woodpeckers (Picidae); in genera that do most of their foraging on large vertical trunks, the primitively opposed hallux is either rotated back into parallel with the second digit or reduced to a functionless vestige (Bock and Miller, 1959).

possessed a divergent or clawless hallux; but claws were evidently present on at least some of the other digits of *Plesiadapis* (Simpson, 1935b; Russell, 1964). Widely differing inferences have been drawn from this fact. Szalay (1972b), in an attack on the Napier-Walker "vertical clinging and leaping" hypothesis, argues roughly as follows:

1. Claws catch on bark and impede "sudden and bold jumps" of the kind tarsiers, galagos, and indriids make.
2. Claws will therefore be selected against in arboreal species with a saltatory locomotor habit.
3. But primitive primates (e.g., *Plesiadapis*) had claws.
4. Therefore, primitive primates were not vertical clingers and leapers.

This argument is sound enough if premises (1–3) are accepted; but the conclusion (4) is not damaging to Napier's argument, which, as shown here, accepts this conclusion as a premise. More importantly, Szalay's first premise is false, as Thorington's (1968) observations of vertical clinging and leaping in free-ranging *Saguinus midas* demonstrate. Mechanisms which automatically withdraw the claws from the substrate when the digital flexors relax have been described for a variety of clawed arboreal mammals (Schaffer, 1905).

Martin (1972) has likewise suggested that, if *Plesiadapis* is representative of early primates, the Napier-Walker hypothesis is not tenable. Rejecting the Napier-Walker hypothesis in its extreme form, Martin proposes that the last common ancestor of all undoubted primates, living and fossil, must at least have exhibited "hindlimb-dominated arboreal locomotion"—that is, it must have possessed grasping hind feet with an enlarged and divergent hallux, and have tended to rely on the hind limb more than the fore for arboreal propulsion and support. Since there is no evidence that this applies to *Plesiadapis* and its relatives, Martin inverts Szalay's argument and proposes that "the difficulty is, of course, overcome if the Plesiadapidae are recognized as quite unrelated to Primates." He goes on to offer the ingenious suggestion that the claws of *Daubentonia* have been derived from keeled nails like those of *Phaner, Galago elegantulus,* and other sap-eating strepsirhines, and to revive the idea that flattened nails may be primitive for primates and perhaps for mammals in general.

Two facts contravene this interpretation. The first is that the nails of *Cebus albifrons* display in vestigial form the terminal matrix and deep stratum characteristic of true claws (Thorndike, 1968), indicating that their gross morphology represents a secondary modification of a marmoset-like ancestral condition. The second is that the same features persist in the grooming claws of *Tarsius* (Le Gros Clark, 1936), where they must surely be vestigial retentions lacking any present functional significance. We may conclude that primitive sharp claws were present on all digits (with the possible exception of the first toe) of the last common ancestor of the cebids, callithricids, and tarsiids and that they had therefore not been lost in the basal primates. To assume that the basal primates lacked claws requires that the lineage leading to *Cebus albifrons* had lost its claws by the Eocene, reacquired them during the Oligocene, and begun to lose them again during the later Tertiary. This is possible, but unparsimonious. It is still less parsimonious to assume that claws have been acquired *de novo* in many different mammalian orders; it is extremely difficult, on this interpretation, to account for the presence of claws in didelphids and other arboreal marsupials, which display hindlimb-dominated arboreal locomotion in a pronounced form. The comparative anatomical evidence indicates that the hands and feet of the last common ancestor of the extant primates must have resembled those of an opossum; claws have been lost independently in four or five parallel lineages of primates.

Among arboreal marsupials, claw loss has proceeded farthest in diminutive forms like *Tarsipes, Cercartetus,* and *Burramys.* This does not conform to Hershkovitz' expectation, based on comparative studies of the Ceboidea, that claws should be reduced progressively as body size increases. This principle may apply here if we rephrase it in terms of relative support diameter. The habitat of *Burramys* is unknown, but the principal habitat of *Tarsipes* is the sandplain scrub of southwestern Australia, and *Cercartetus* is widely distributed in heath and sclerophyll forest communities with a well-developed stratum of shrubby undergrowth (Ride, 1970).* Small bush-dwelling animals in these shrub-dominated floral communities have little need of claws, since their environment presents them with few supports of very large diameter; the claws of these pigmy possums have accordingly become vestigial, possibly for the reasons advanced by Hershkovitz. Claw loss evidently can result either from a trend toward increased size in animals inhabiting the higher strata of tropical forest, or from restriction to the lower strata of a relatively treeless heath or scrub floral community. Functional claws appear to have been lost in the various ceboid lineages for the first reason. The absence of claws in the Madagascar lemurs other than *Daubentonia* may reflect their derivation from a small bush-dwelling species like

* Note added in proof: Dixon (1971) reports that *Burramys* inhabits alpine shrub and heath communities at the edge of the tree line, in which the largest trees are stunted snow gums (*Eucalyptus coriacea*) about 3 meters high.

Microcebus murinus, whose distribution and ecology in the southwest of Madagascar resemble those of *Cercartetus* (Petter, 1962b). In any event, the terminal matrix and deep stratum appear to have been present in the ancestral species which colonized Madagascar, since they are retained in the aye-aye.

Prehensile Specializations of the Hand

Prehensile specializations of the hand are secondary to those of the hind foot. As Figure 3 shows, the relatively unspecialized hands of *Microcebus, Cheirogaleus,* and *Didelphis* show little functional opposition of the thumb on horizontal supports, particularly those of larger diameter. However, even in these animals there is a preferred orientation of the hand, which is usually set down so that the axis of the support passes through or just to either side of the second digit; this is also true of galagos (Bishop, 1964), and can be taken as the primitive condition for the order Primates. Further specialization of the hand for prehension may follow one of three pathways: the hand may be restructured so that the axis of a grasped branch must fall through digit II, or between I and II, or between II and III. The first alternative, in which the second digit contributes nothing to the grip, has been adopted by the lorisine genera (Bishop, 1964). The second alternative, in which the thumb opposes the other four digits, is that seen in the larger Madagascar lemurs and many catarrhines. The third condition, generally called schizodactyly or zygodactyly, appears in a relatively unspecialized form among cebids (Erikson, 1957); among callithricids, *Callithrix jacchus* becomes schizodactyl on narrow supports (Rothe, 1972), but *Saguinus oedipus* reportedly does not (Sonek, 1969). Schizodactyly is characteristic of most of the arboreal diprotodont marsupials, and is especially pronounced in the koala, whose first and second fingers form a pair of markedly divergent "thumbs" opposing digits III–V. Although Haines (1958) classified the hand of *Didelphis* in his unspecialized "clasping" category, the living opossum shows a marked schizodactyly on slender horizontal supports (Fig. 3). While there seems to be no reason to accept Erikson's (1957) hypothesis that schizodactyly is a primitive retention from the ancestral mammals, it may well be a heritage trait of the marsupial infraclass.

Summary

Clawed hands and feet like those of squirrels are as well-adapted to most sorts of arboreal activity as are the clawless, prehensile hands and feet characteristic of primates. On tree trunks and other nonhorizontal supports of relatively large diameter, claw grip is demonstrably superior to pad grip. There is no justification for believing that the primate condition is a primitive retention from the Triassic ancestors of the Mammalia. The primate specializations therefore demand explanation. Primatelike specializations are found among small marsupials that stalk and manually seize insect prey among the slender branches of bushes and low trees in tropical forests and woodlands. A similar way of life is characteristic of many extant prosimians. Prehensile specializations of the hind foot, which facilitate cautious and well-controlled movements on slender branches during the stalk and pounce, are part of the adaptations of these animals; similar specializations have been developed independently by chameleons, which have similar habits. For this and other reasons, it seems likely that the last common ancestor of the extant primates was a small predatory animal, resembling *Marmosa* or *Cercartetus,* which had claws on all digits (with the possible exception of the first toe) and which foraged for insects and fruit in the understory of the forest. Reduction of claws and development of grasping specializations of the hand have occurred independently, in different ways and for different reasons, in various primate lineages.

Acknowledgments

I am grateful to John Buettner-Janusch, William L. Hylander, Arthur L. Klein, William Longley, C. Owen Lovejoy, Robert J. Russell, and Robert W. Sussman for their advice and assistance. My research is supported by General Research Support Grant RR-5405-09 from the General Research Support Branch, Division of Research Facilities and Resources, National Institutes of Health. Some of the animals used in this study were made available to the Duke University Primate Facility through the kind efforts of Dr. R. Paulian and M. Roederer of the Institut de Recherche Scientifique à Madagascar (now Office de la Recherche Scientifique et Technique d'Outre-Mer), M. Ramanantsoavina (Eaux et Forêts), and M. Ramalanjoana, Vice-President of the Government of the Malagasy Republic. The Duke University Primate Facility is supported by Grants RR-00388 and GM-13222 from the United States Public Health Service, by Grant GB-4000 from the National Science Foundation, and by the Duke University Primate Facility Fund.

References

Albignac, R. (1970). Notes éthologiques sur quelques carnivores malgaches: le *Cryptoprocta ferox* (Bennett). *Terre Vie* **24**, 395–402.

Altner, G. (1971). Histologische und vergleichend-anatomische Untersuchungen zur Ontogenie und Phylogenie des Handskeletts. *Folia Primatol.* 14 (Suppl.), vi, 106 pp.

Banks, E. (1931). A popular account of the mammals of Borneo. *J. Malay. Br. Roy. Asiat. Soc.* **9**(2), 1–139.

Barnett, C. H. (1954). The structure and function of fibrocartilages within vertebrate joints. *J. Anat.* **88**, 363–368.

Barnett, C. H., and Napier, J. R. (1953). The form and mobility of the fibula in metatherian mammals. *J. Anat.* **87**, 207–213.

Bates, H. W. (1863). "The Naturalist on the River Amazons." London, John Murray.

Beattie, J. (1927). Anatomy of the common marmoset (*Hapale jacchus* Kuhl). *Proc. Zool. Soc. London,* pp. 593–718.

Biegert, J. (1959). Die Ballen, Leisten, Furchen und Nägel von Hand und Fuss der Halbaffen. *Z. Morphol. Anthropol.* **49**, 316-409.

Biegert, J. (1961). Volarhaut der Hände und Füsse. *Primatologia* **2** Part 1, No. 3, 1–326.

Biegert, J., and Maurer, R. (1972). Rumpfskelettlänge, Allometrien und Körperproportionen bei catarrhinen Primaten. *Folia Primatol.* **17**, 142–156.

Biggers, J. D., Fritz, H. I., Hare, W., and McFeely, R. A. (1965). Chromosomes of American marsupials. *Science* **148**, 1602–1603.

Bishop, A. (1962). Control of the hand in lower primates. *Ann. N.Y. Acad. Sci.* **102**, 316–337.

Bishop, A. (1964). Use of the hand in lower primates. *In* "Evolutionary and Genetic Biology of Primates" (J. Buettner-Janusch, ed.), Vol. 2, pp. 133–225. Academic Press, New York.

Boas, J. E. V. (1884). Morphologie der Nägel der Säuger. *Morphol. Jahrb.* **9**, 389–400.

Bock, W. J., and Miller, W. DeW. (1959). The scansorial foot of the woodpeckers, with comments on the evolution of perching and climbing feet in birds. *Amer. Mus. Nov.* **1931**, 1–45.

Böker, H. (1927). Die Entstehung der Wirbeltiertypen und der Ursprung der Extremitäten. *Z. Morphol. Anthropol.* **26**, 1–58.

Böker, H. (1932). Beobachtungen und Untersuchungen an Säugetieren während einer biologisch-anatomischen Forschungsreise nach Brasilien im Jahre 1928. *Morphol. Jahrb.*, **70**, 1–66.

Böker, H., and Pfaff, R. (1931). Die biologische Anatomie der Fortbewegung auf dem Boden und ihre phylogenetische Abhängigkeit vom primären Baumklettern bei den Säugetieren. *Morphol. Jahrb.*, **68**, 496–540.

Bourlière, F., Hunkeler, C., and Bertrand, M. (1970). Ecology and behavior of Lowe's Guenon (*Cercopithecus campbelli lowei*) in the Ivory Coast. *In* "Old World Monkeys" (J. R. Napier and P. H. Napier, eds.), pp. 297–350. Academic Press, New York.

Bowden, F. P., and Tabor, D. (1954). "The Friction and Lubrication of Solids." Oxford University Press, London.

Bowden, F. P., and Tabor, D. (1967). "Friction and Lubrication." Methuen, London.

Burton, J., and Burton, M. (1969). Tight-rope artist. *Animals* **11**, 454–455.

Campbell, B. (1966). "Human Evolution: An Introduction to Man's Adaptations." Aldine, Chicago.

Cartmill, M. (1970). The orbits of arboreal mammals: a reassessment of the arboreal theory of primate evolution. Ph.D. dissertation, University of Chicago, Chicago, Illinois.

Cartmill, M. (1972). Arboreal adaptations and the origin of the order Primates.

In "Functional and Evolutionary Biology of Primates" (R. H. Tuttle, ed.), pp. 97–122. Aldine-Atherton, Chicago.

Casamiquela, R. M. (1961). Sobre la presencia de un mamífero en el primer elenco (icnológico) de vertebrados del Jurásico de la Patagonia. *Physis* **22**, 225–233.

Chapman, F. M. (1929). "My Tropical Air-Castle," Appleton, New York.

Charles-Dominique, P. (1971). Éco-éthologie des Prosimiens du Gabon. *Biol. Gabon.* **7**, 121–228.

Charles-Dominique, P. and Hladik, C. M. (1971). Le *Lepilemur* du sud de Madagascar: écologie, alimentation et vie sociale. *Terre Vie* **1**, 3–66.

Collins, E. T. (1921). Changes in the visual organs correlated with the adoption of arboreal life and with the assumption of the erect posture. *Trans. Ophthalmol. Soc. U.K.* **41**, 10–90.

Cummins, H., and Midlo, C. (1943). "Finger Prints, Palms and Soles: An Introduction to Dermatoglyphics," 1961 reprint. Dover, New York.

Dankmeijer, J. (1938). Zur biologischen Anatomie der Hautleisten bei den Beuteltieren. *Morphol. Jahrb.* **82**, 293–312.

Day, M. H., and Walker, A. C. (1969). New prosimian remains from early Tertiary deposits of southern England. *Folia Primatol.* **10**, 139–145.

Dixon, J. M. (1971). *Vict. Nat.* **88**, 133–138.

Dobson, G. E. (1876). On peculiar structures in the feet of certain species of mammals which enable them to walk on smooth perpendicular surfaces. *Proc. Zool. Soc. London*, pp. 526–535.

Dollo, L. (1899). Les ancêtres des Marsupiaux étaient-ils arboricoles? *Trav. Stat. Zool. Wimereux* **7**, 188–203.

Dücker, G. (1971). Gefangenschaftsbeobachtungen an Pardelrollern *Nandinia binotata* (Reinwardt). *Z. Tierpsych.*, **28**, 77–89.

Eisenberg, J. F., and Gould, E. (1966). The behavior of *Solenodon paradoxus* in captivity, with comments on the behavior of other Insectivora. *Zoologica* **51**, 49–58.

Eisenberg, J. F., and Gould, E. (1970). The tenrecs: A study in mammalian behavior and evolution. *Smithson. Contrib. Zool.* **27**, 1–137.

Ellerman, J. R. (1941). "The Families and Genera of Living Rodents," Vol. II. Brit. Mus. Natur. Hist., London.

Enders, R. K. (1935). Mammalian life histories from Barro Colorado Island, Panama. *Bull. Mus. Comp. Zool. Harvard Univ.* **78**, 383–502.

Erikson, G. E. (1957). The hands of the New World primates with comparative functional observations on the hands of other primates. *Amer. J. Phys. Anthropol.* **15**[N.S], 446.

Fleay, D. (1933). A beautiful phalanger. *Vict. Nat.* **50**, 35–40.

Fraser, R. D. B. (1969). Keratins. *Sci. Amer.* **221** (2), 86–96.

Gidley, J. W. (1919). Significance of divergence of the first digit in the primitive mammalian foot. *J. Wash. Acad. Sci.* **9**, 273–280.

Grand, T. I. (1968). The functional anatomy of the lower limb of the howler monkey (*Alouatta caraya*). *Amer. J. Phys. Anthropol.* **28**, 163–182.

Granger, W., and Simpson, G. G. (1929). A revision of the Tertiary Multituberculata. *Bull. Amer. Mus. Natur. Hist.* **56**, 601–676.

Grinnell, J., Dixon, J. S., and Linsdale, J. M. (1937). "Fur-Bearing Mammals of California: Their Natural History, Systematic Status, and Relation to Man," Vol. I. Univ. Calif. Press, Berkeley.

Haines, R. W. (1955). The anatomy of the hand of certain insectivores. *Proc. Zool. Soc. London* **125**, 761–777.

Haines, R. W. (1958). Arboreal or terrestrial ancestry of placental mammals. *Quart. Rev. Biol.* **33**, 1–23.

Hall, E. R., and Dalquest, W. W. (1963). The mammals of Veracruz. *Univ. Kans. Publ. Mus. Natur. Hist.* **14**, 165–362.

Hall-Craggs, E. C. B. (1966). Rotational movements in the foot of *Galago senegalensis*. *Anat. Rec.* **154**, 287–294.

Harrison, J. L. (1951). Squirrels for bird-watchers. *Malay. Nature J.* **5**, 134–154.

Harrisson, B. (1962). Getting to know *Tarsius*. *Malay. Nature J.* **16**, 197–204.

Hatt, R. T. (1929). The red squirrel: its life history and habits, with special reference to the Adirondacks of New York and the Harvard Forest. *Bull. N.Y. State Coll. Forest.* (*Syracuse*), **2**(1-B), 1–146.

Heim de Balsac, H., and Vuattoux, R. (1969). *Crocidura douceti* H. de B. et le comportement arboricole des Soricidae. *Mammalia* **33**, 98–101.

Hershkovitz, P. (1969). The Recent mammals of the Neotropical region: a zoogeographic and ecological review. *Quart. Rev. Biol.* **44**, 1–70.

Hershkovitz, P. (1970). Notes on Tertiary platyrrhine monkeys and description of a new genus from the late Miocene of Colombia. *Folia Primatol.* **12**, 1–37.

Hill, W. C. O. (1949). The giant squirrels. *Zoo Life* (*London*) **4**, 98–100.

Hill, W. C. O. (1955). "Primates. Comparative Anatomy and Taxonomy," Vol. II: Haplorhini: Tarsioidea. Univ. Edinburgh Press, Edinburgh.

Hill, W. C. O. (1957). "Primates. Comparative Anatomy and Taxonomy," Vol. III: Pithecoidea, Platyrrhini—Hapalidae. Univ. Edinburgh Press, Edinburgh.

Hill, W. C. O. (1959). The anatomy of *Callimico goeldii* (Thomas), a primitive American primate. *Trans. Amer. Phil. Soc.* **49**(5), 1–116.

Hopson, J. A. (1969). The origin and adaptive radiation of mammal-like reptiles and nontherian mammals. *Ann. N. Y. Acad. Sci.* **167**, 199–216.

Hopson, J. A. and Crompton, A. W. (1969). Origin of mammals. *Evol. Biol.* **3**, 15–72.

Hutton, A. F. (1949). Notes on the snakes and mammals of the High Wavy Mountains, Madura District, South India. *J. Bombay Natur. Hist. Soc.* **48**, 681–694.

Jolly, A. (1966). "Lemur Behavior," Univ. Chicago Press, Chicago, Illinois.

Jones, F. W. (1916). "Arboreal Man." E. Arnold, London. 1964 reprint, Hafner, New York.

Jones, F. W. (1921). On the habits of *Trichosurus vulpecula*. *J. Mammal.* **2**, 187–193.

Jones, F. W. (1924). The Mammals of South Australia, Part II, containing the bandicoots and the herbivorous marsupials. British Science Guild, Adelaide.

Jones, F. W. (1953). Some readaptations of the mammalian pes in response to arboreal habits. *Proc. Zool. Soc. London* **123**, 33–41.

Jouffroy, F. K., and Lessertisseur, J. (1959). Réflexions sur les muscles contracteurs des doigts et des orteils (contrahentes digitorum) chez les primates. *Ann. Sci. Natur. Zool.* **12**(1), 211–235.

Kaufmann, J. H. (1962). Ecology and social behavior of the coati, *Nasua narica*, on Barro Colorado Island, Panama. *Univ. Calif. Publ. Zool.* **60**, 95–222.

Kidd, W. (1907). "The Sense of Touch in Mammals and Birds, with Special Reference to the Papillary Ridges." Adam and Charles Black, London.

Kortlandt, A. (1968). Handgebrauch bei freilebenden Schimpansen. *In* "Handgebrauch und Verständigung bei Affen und Frühmenschen" (B. Rensch, ed.), pp. 59–102, Hans Huber, Bern.

Krieg, H. (1930). Biologische Reisestudien im Südamerika. Die Affen des Gran Chaco und seiner Grenzgebiete. *Z. Morphol. Oekol.* **18**, 760–785.

Lavocat, R. (1955). Sur un squelette de *Pseudosciurus* provenant du gisement d'Armissan (Aude). *Ann. Paleontol.* **41**, 77–89.

Leakey, L. S. B. (1969). "Animals of East Africa." National Geographic Society, Washington, D. C.

Le Gros Clark, W. E. (1924a). The myology of the three-shrew (*Tupaia minor*). *Proc. Zool. Soc. London,* pp. 461–497.

Le Gros Clark, W. E. (1924b). Notes on the living tarsier (*Tarsius spectrum*). *Proc. Zool. Soc. London,* pp. 217–223.

Le Gros Clark, W. E. (1926). On the anatomy of the pen-tailed tree-shrew (*Ptilocercus lowii*). *Proc. Zool. Soc. London,* pp. 1179–1309.

Le Gros Clark, W. E. (1936). The problem of the claw in primates. *Proc. Zool. Soc. London,* pp. 1–24.

Le Gros Clark, W. E. (1959). "The Antecedents of Man: An Introduction to the Evolution of the Primates." Univ. Edinburgh Press, Edinburgh.

Lewis, O. J. (1964). The evolution of the long flexor muscles of the leg and foot. *Intern. Rev. Gen. Exp. Zool.* **1**, 165–185.

Leyhausen, P. (1963). Über südamerikanische Pardelkatzen. *Z. Tierpsych.* **20**, 627–640.

McClure, H. E. (1964). Some observation of primates in climax dipterocarp forest near Kuala Lumpur, Malaya. *Primates* **5**, 39–58.

Martin, R. D. (1972). Adaptive radiation and behaviour of the Malagasy lemurs. *Phil. Trans. Roy. Soc. London* **B264**, 295–352.

Matthew, W. D. (1904). The arboreal ancestry of the Mammalia. *Amer. Natur.* **38**, 811–818.

Matthew, W. D. (1937). Paleocene faunas of the San Juan Basin, New Mexico. *Trans. Amer. Phil. Soc.* **30**, 1–510.

Midlo, C. (1934). Form of the hand and foot in primates. *Amer. J. Phys. Anthropol.* **19**, 337–389.

Mivart, St. George (1873). On *Lepilemur* and *Cheirogaleus* and on the zoological rank of the Lemuroidea. *Proc. Zool. Soc. London,* pp. 484–510.

Moore, J. C., and Tate, G. H. H. (1965). A study of the diurnal squirrels, Sciurinae, of the Indian and Indochinese subregions. *Fieldiana* (*Zool.*) **48**, 1–351.

Morcombe, M. K. (1968). Mammals. *In* "Animals of the World: Australia" (G. P. Whitley, C. F. Brodie, M. K. Morcombe, and J. R. Kinghorn, eds.), pp. 60–97. Paul Hamlyn, London.

Musser, G. G. (1972). The species of *Hapalomys* (Rodentia, Muridae). *Amer. Mus. Nov.* **2503**, 1–27.

Napier, J. R. (1967). Evolutionary aspects of primate locomotion. *Amer. J. Phys. Anthropol.* **27**, 333–342.

Napier, J. R. (1970). "The Roots of Mankind." Smithsonian Inst. Press, Washington.

Napier, J. R., and Davis, P. R. (1959). The fore-limb skeleton and associated remains of *Proconsul africanus. Fossil Mammals of Africa No.* **16**. British Mus. Natur. Hist., London.

Napier, J. R., and Napier, P. H. (1967). "A Handbook of Living Primates." Academic Press, London.

Napier, J. R., and Walker, A. C. (1967). Vertical clinging and leaping—a newly recognized category of locomotor behavior of primates. *Folia Primatol.* **6**, 204–219.

Ogilvie, C. S. (1958). The binturong or bear-cat. *Malay. Nature J.* **13**, 1–3.

Osgood, W. H. (1921). A monographic study of the American marsupial, *Caenolestes*. *Field Mus. Publ. (Zool.)* **14**, 1–162.

Panzer, W. (1932). Beiträge zur biologischen Anatomie des Baumkletterns der Säugetiere. I. Das Nagel-Kralle-Problem. *Z. Anat. Entwicklungsgesch.* **98**, 147–198.

Petter, F. (1966). *Dendroprionomys rousseloti gen. nov., sp. nov.*, rongeur nouveau du Congo (Cricetidae, Dendromurinae). *Mammalia* **30**, 129–137.

Petter, J.-J. (1962a). Ecological and behavioral studies of Madagascar lemurs in the field. *Ann. N. Y. Acad. Sci.* **102**, 267–281.

Petter, J.-J. (1962b). Recherches sur l'écologie et l'éthologie des Lémuriens malgaches. *Mem. Mus. Nat. Hist. Natur. (Zool.)* **27**, 1–146.

Petter, J.-J., and Hladik, C. M. (1970). Observations sur le domaine vital et la densité de population de *Loris tardigradus* dans les forêts de Ceylan. *Mammalia* **34**, 394–409.

Petter, J.-J., and Peyrieras, A. (1970). Nouvelle contribution à l'étude d'un Lémurien malgache, le Aye-Aye (*Daubentonia madagascariensis* E. Geoffroy). *Mammalia* **34**, 167–193.

Petter, J.-J., Schilling, A., and Pariente, G. (1971). Observations éco-éthologiques sur deux Lémuriens malgaches nocturnes: *Phaner furcifer* et *Microcebus coquereli*. *Terre Vie* **25**, 287–327.

Pfeffer, P. (1968). "Asia: A Natural History." Random House, New York.

Pfeffer, P. (1969). Considérations sur l'écologie des forêts claires du Cambodge Oriental. *Terre Vie*, pp. 3–24.

Phillips, W. W. A. (1935). "Manual of the Mammals of Ceylon." Colombo Museum, Ceylon.

Pocock, R. I. (1918). On the external characters of the lemurs and of *Tarsius*. *Proc. Zool. Soc. London*, pp. 19–53.

Pocock, R. I. (1922). On the external characters of the beaver (Castoridae) and of some squirrels (Sciuridae). *Proc. Zool. Soc. London*, pp. 1171–1212.

Poduschka, W. (1969). Ergänzungen zum Wissen über *Erinaceus e. roumanicus* und kritische Überlegungen zur bisherigen Literatur über europäische Igel. *Z. Tierpsych.*, **26**, 761–804.

Polyak, S. (1955). "The Vertebrate Visual System." Chicago Univ. Press, Chicago, Illinois.

Prater, S. H. (1965). "The Book of Indian Animals." Bombay Natural History Society and Prince of Wales Museum of Western India, Bombay.

Preuschoft, H. (1970). Functional anatomy of the lower extremity. *In* "The Chimpanzee" (G. H. Bourne, ed.), Vol. 3, pp. 221–294. Univ. Park Press, Baltimore, Maryland.

Reig, O. A. (1955). Noticia preliminar sobre la presencia de microbiotherinos vivientes en la fauna sudamericana. *Invest. Zool. Chil.* **2**, 121–130.

Richard, P. B. (1964). Notes sur la biologie du daman des arbres *Dendrohyrax dorsalis*. *Biol. Gabon.* **1**, 73–84.

Richards, P. W. (1966). "The Tropical Rain Forest." Cambridge Univ. Press, Cambridge.

Ride, W. D. L. (1970). "A Guide to the Native Mammals of Australia," Oxford Univ. Press, Melbourne.
Ridley, H. N. (1895). The mammals of the Malay Peninsula. *Natur. Sci.* **6**, 23–29, 89–96.
Ripley, S. (1967). The leaping of langurs: a problem in the study of locomotor adaptation. *Amer. J. Phys. Anthropol.* **26**, 149–170.
Rothe, H. (1972). Beobachtungen zum Bewegungsverhalten des Weissbüscheläffchens *Callithrix jacchus* Erxleben, 1777, mit besonderer Berücksichtigung der Handfunktion. *Z. Morphol. Anthropol.* **64**, 90–101.
Russell, D. E. (1964). Les mammifères Paléocènes d'Europe. *Mem. Mus. Nat. Hist. Natur. (Paris)* **C13**, 1–321.
Ryley, K. V. (1914). Bombay Natural History Society's Mammal Survey of India. Report No. 14. *J. Bombay Natur. Hist. Soc.* **22**, 710–725.
Sanderson, I. T. (1937). "Animal Treasure." Macmillan, New York and London.
Schaffer, J. (1905). Anatomisch-histologische Untersuchung über den Bau der Zehen bei Fledermäusen und einigen kletternden Säugetieren. *Z. Wiss. Zool.* **83**, 231–284.
Schultz, A. H. (1969). "The Life of Primates." Weidenfeld & Nicolson, London.
Shorten, M. (1954). "Squirrels." Collins, London.
Simons, E. L. (1962). Fossil evidence relating to the early evolution of primate behavior. *Ann. N. Y. Acad. Sci.* **102**, 282–294.
Simpson, G. G. (1927). Mesozoic Mammalia. IX. The brain of Jurassic mammals. *Amer. J. Sci.* **14**, 259–268.
Simpson, G. G. (1928). "A Catalogue of the Mesozoic Mammalia in the Geological Department of the British Museum." Brit. Mus. Natur. Hist., London.
Simpson, G. G. (1935a). Note on the classification of recent and fossil opossums. *J. Mammal.* **16**, 134–137.
Simpson, G. G. (1935b). The Tiffany fauna, Upper Paleocene. II. Structure and relationships of *Plesiadapis*. *Amer. Mus. Nov.* **816**, 1–30.
Simpson, G. G. (1940). Studies on the earliest primates. *Bull. Amer. Mus. Natur. Hist.* **77**, 185–212.
Sonek, A., Jr. (1969). Functional anatomy of the weight-bearing and prehensile hand of a quadrupedal primate (*Saguinus oedipus*). Ph.D. Thesis, University of Oregon, Eugene, Oregon.
Stibbe, E. P. (1928). A comparative study of the nictitating membrane of birds and mammals. *J. Anat.* **62**, 159–176.
Subramoniam, S. (1957). Some observations on the habits of the slender loris, *Loris tardigradus* L. *J. Bombay Natur. Hist. Soc.* **54**, 387–398.
Szalay, F. S. (1968). The beginnings of primates. *Evolution* **22**, 19–36.
Szalay, F. S. (1969). Mixodectidae, Microsyopidae, and the insectivore-primate transition. *Bull. Amer. Mus. Natur. Hist.* **140**, 193–330.
Szalay, F. S. (1972a). Cranial morphology of the early Tertiary *Phenacolemur* and its bearing on primate phylogeny. *Amer. J. Phys. Anthropol* **36**, 59–76.
Szalay, F. S. (1972b). Paleobiology of the earliest primates. *In* "The Functional and Evolutionary Biology of Primates" (R. H. Tuttle, ed.), pp. 3–35, Aldine-Atherton, Chicago, Illinois.
Taylor, C. R., Caldwell, S. L., and Rowntree, V. J. (1972). Running up and down hills: some consequences of size. *Science* **178**, 1096–1097.

Taylor, M. E. (1970). Locomotion in some East African viverrids. *J. Mammal.* **51**, 42–51.

Thomas, O. (1888). "Catalogue of the Marsupialia and Monotremata in the Collection of the British Museum (Natural History)." British Museum Natur. Hist., London.

Thomas, O. (1925). The mammals obtained by Mr. Herbert Stevens on the Sladen-Godman Expedition to Tonkin. *Proc. Zool. Soc. London,* pp. 495–506.

Thorington, R. W. (1968). Observations of the tamarin *Saguinus midas. Folia Primatol.* **9**, 95–98.

Thorndike, E. E. (1968). A microscopic study of the marmoset claw and nail. *Amer. J. Phys. Anthropol.* **28**, 247–262.

Trapp, G. R. (1972). Some anatomical and behavioral adaptations of ringtails, *Bassariscus astutus. J. Mammal.* **53**, 549–557.

Tuttle, R. H., and Rogers, C. M. (1966). Genetic and selective factors in reduction of the hallux in *Pongo pygmaeus. Amer. J. Phys. Anthropol.* **24**, 191–198.

Vandenbergh, J. G. (1963). Feeding, activity and social behavior of the tree shrew, *Tupaia glis,* in a large outdoor enclosure. *Folia Primatol.* **1**, 199–207.

Van Horn, R. N. (1972). Structural adaptations to climbing in the gibbon hand. *Amer. Anthropol.* **74**, 326–334.

Walker, E. P. (1964). "Mammals of the World." Johns Hopkins Press, Baltimore, Maryland.

Walls, G. L. (1942). "The Vertebrate Eye and its Adaptive Radiation," 1963 reprint. Hafner, New York.

Ward, W. H., and Lundgren, H. P. (1954). The formation, composition, and properties of the keratins. *Advan. Protein Chem.* **9**, 243–297.

Warneke, R. M. (1967). Discovery of a living *Burramys. Aust. Mammal Soc. Bull.* **2**, 94–95.

Webb-Peploe, C. S. (1949). Field notes on mammals of South Tinnevelly, South India. *J. Bombay Natur. Hist. Soc.* **46**, 629–644.

Whipple, I. L. (1904). The ventral surface of the mammalian chiridium with special reference to the conditions found in man. *Z. Morphol. Anthropol.* **7**, 261-368.

Wood, A. E. (1962). The early Tertiary rodents of the family Paramyidae. *Trans. Amer. Phil. Soc.* **52** [N.S.], 1–261.

Wroughton, R. C. (1915). Bombay Natural History Society's Mammal Survey of India. Report No. 19. *J. Bombay. Natur. Hist. Soc.* **24**, 96–110.

3

Tree Shrew Locomotion and the Origins of Primate Arborealism*

FARISH A. JENKINS, JR.

Introduction

The family Tupaiidae (tree shrews) ranges from India and southern China throughout most of southeastern Asia. Tree shrews are found on Hainan, Sumatra, Java, Borneo, and Mindanao (Phillipines) in addition to other smaller island groups, but they are not found southeast of Wallace's Line. Their typical habitat is montane and tropical rain forest, although few species are as strictly arboreal as their common name implies.

Squirrel-sized or smaller, tree shrews are represented by five genera (*Anathana, Dendrogale, Ptilocercus, Tupaia,* and *Urogale*) and about fourteen species. *Ptilocercus* is sufficiently distinct to warrant a separate subfamily (Ptilocercinae), the other genera being grouped in the Tupaiinae (Lyon, 1913). Although there is some disagreement on species taxonomy in *Tupaia* (e.g., contrast Martin, 1968a with Napier and Napier, 1967), the major taxonomic question concerns the phylogenetic relationship of tupaiids as a group. Early anatomical studies, particularly those of Carlsson (1922) and Le Gros Clark (e.g., 1924, 1925, 1926, 1960), presented evidence that tupaiids were more closely related to prosimians

* This study was supported by grants from the National Science Foundation (GB-13662 and GB-30724) and National Institutes of Health (5-S05-RR-07046-06). The author is grateful to M. W. Sorenson and J. G. Fleagle for reviewing the manuscript, to L. Meszoly for preparation of Figs. 2 and 7, and to A. H. Coleman for photographic assistance.

than to insectivores. Recent reexamination of this evidence, in addition to histological and biochemical studies, has not confirmed this relationship. Thus the current taxonomic status of tree shrews is uncertain (for a review of literature on this question, see Luckett, 1969, and Sorenson, 1970). Some authors rank tupaiids in a separate order Tupaioidea (Martin, 1966), while others retain them among the Primates (Napier and Napier, 1967) or Insectivora (Hill, 1953).

The uncertainty surrounding tupaiid phylogeny is a consequence of an inadequate fossil record. At present early Tertiary mammals are represented principally by teeth and fragmentary jaws. Dental similarity between tupaiids, marsupials, and insectivores makes it unlikely that an early Tertiary fossil tupaiid will be positively identified without the recovery of more of a skull than the dentition. A number of fossil genera have been proposed as tupaioid but these taxonomic assignments are doubtful (McKenna, 1966). Van Valen (1965) indentifies *Adapisoriculus minimus* and *A.? germanicus* from the Paleocene as tupaiids, although other authors had variously identified the genus as insectivore or marsupial. Szalay (1968), however, considers that these species "may or may not be tupaiids" and proposes that another Paleocene (Lutetian) genus, *Messelina,* is at least as similar to tupaiids as *Adapisoriculus* in lower dentition and more similar in upper dentition. Thus there is no consensus as to which fossil forms, if any, represent the tree shrew lineage.

Despite taxonomic ambiguities and the apparent remoteness of a tupaiid–primate relationship, there are several reasons why tree shrews merit inclusion among studies of primate locomotion. First, understanding anatomical and behavioral aspects of arboreal versus terrestrial activity in different tree shrew species may clarify similar phenomena among true primates. Second, tree shrews represent one possible "model" of a primitive primate or placental mammal. Gregory (1910, 1913) was among the first to allude to tupaiids in this sense; the increase in knowledge of early mammals, primates, and tree shrews since that time has not diminished the potential usefulness of this concept. Critics of the "model" approach to paleontology usually cite differences or specializations which purportedly invalidate comparison between the fossil species and the living "model." In this case, however, use of tree shrews as an experimental "model" does not imply equivalency with early Tertiary insectivores or primates, for differences of varying degree are well known; rather, there are sufficient similarities to support the expectation that a better understanding of form–function problems in tree shrews may contribute to understanding similar problems among early insectivores and primates. Simpson (1965) summarized the point:

. . . The tupaioids arose, and still stand, somewhere between the earliest placental (nominally insectivore) stem and that of the Primates. Their reference to one group or the other is in part arbitrary or semantic. Use of them to represent the earliest primate or latest preprimate stage of evolution is as valid and useful, and subject to as much caution, as is any use of living animals to represent earlier phylogenetic stages.

As McKenna (1966) pointed out, one trend of current opinion is to regard "tupaiids as leptictid-like insectivores (*sensu lato*) with special similarity among primates to Malagasy lemurs, *Adapis* and *Notharctus*. Among living nonprimates the tupaiids are apparently the closest primate relatives, and these conclusions in no way lessen the value of tupaiids to primatology."

Tree Shrew Locomotion: Behavioral Observations

From the very first zoological descriptions (Diard, 1820; Raffles, 1821), tupaiids have been associated with a squirrel-like habitus in name ("tree shrew"; the generic root "tupai" is the Malay word for squirrel) as well as description. At one time the tupaiid–squirrel comparison was satisfactory as a first approximation, but the association is vague and even misleading in view of the diversity of habitats among both squirrel and tree shrew species. However, there is some evidence that a mimicry relationship exists between certain species of *Tupaia* and squirrels of the genera *Sciurus* and *Funambulus* (Banks, 1931).

Discussions of tupaiid locomotion usually attempt to define their habitat, and in this context have tended to focus on the question of an arboreal versus a terrestrial existence. As is well known in primatology, this distinction is seldom absolute; the question usually involves the relative percentage of time spent on the ground, in trees, or in some intermediate habitat (e.g., undergrowth). Little such data have been gathered for free-ranging tree shrew species, possibly because of their small size and shyness, and notwithstanding their apparent tendency to territorial restriction. Some reports of the preponderance of terrestrial over arboreal activity in tree shrews may be biased because of the animal's greater visibility at ground level; conversely, other reports may be biased in favor of an arboreal existence because of an arboreal escape reaction when confronted with an observer. Since most accounts of tree shrew activity are brief commentaries based on incidental rather than programmed observations, they should not be regarded invariably as definitive. However, it is certain that considerable locomotor differences exist between tree shrew genera, and even between species of *Tupaia*.

Observations on Tree Shrews in the Wild

Relatively few species are considered to be principally arboreal in the sense of seldom descending to the ground and being especially adept in arboreal locomotion. Kloss (1903, 1911) reported that *Tupaia glis longicauda* from East Perhentian Island (off the east coast of the Malay Peninsula) and *T. nicobarica* from the Nicobar Islands are truly arboreal. Bartels (1937) implied that *T. javanica* is arboreal but did not document his observation; Müller (quoted in Horsfield, 1851) claimed that this species nests some distance above the ground but feeds both on the ground and in trees. The only other certainly arboreal tupaiine is *T. minor* (Le Gros Clark, 1927). Davis (1962) reported that this species occupies small trees and vines of the lower middle story between 5 and 50 feet above the ground, and similarly Lim (1969) referred to *T. minor* as a canopy and undercanopy animal which ". . . is often seen running upside down along thin branches of tall trees." Although Lim (1969) claimed that this species seldom descends to the ground, Banks (1931) reported that they are often seen running on the ground and on fallen tree trunks. *Tupaia gracilis,* somewhat larger but otherwise similar to *T. minor,* is apparently also arboreal on the basis of a few field observations (Davis, 1962). The only other certainly arboreal tupaiid is *Ptilocercus lowii* (the feather- or pen-tailed tree shrew); its nocturnal habits are unique among tree shrews (Le Gros Clark, 1926; Davis, 1962; Muul and Lim, 1971; Lim, 1967, 1969).

Two tupaiines are predominantly terrestrial. The Phillipine tree shrew, *Urogale everetti,* apparently nests in the ground or in cliffs and is active on the forest floor; nevertheless it is reputed to be an excellent climber (Wharton, 1950). *Tupaia (Lyonogale) tana,* the so-called terrestrial tree shrew, has similar habits (Banks, 1931; Davis, 1962). The extent to which tanas will climb is uncertain. Davis (1962) observed that they ". . . are rarely seen even in the lowest branches of the trees. When pressed they do not resort to climbing, but escape on the ground." However, observations on captive specimens (see below) are equivocal in regard to their climbing ability and propensity.

On the basis of relatively limited field observations, most other tree shrew species appear to be in some measure both arboreal and terrestrial. Most accounts favor the view that the majority of activity is on the forest floor or at least in low undergrowth. Banks (1931) observed that ". . . though they can and do run about in the trees most species spend the greater part of their time on the ground running over and under fallen tree trunks." Of *Tupaia (glis) ferruginea,* Robinson and Kloss (1909) write that ". . . it is quite exceptional to see one anywhere than on the ground, among the roots of trees and on low bushes." Similar descriptions were made by Banks (1931), Bartels (1937), Davis (1962), Lim (1969), and Ridley (1895), the latter noting that *T. glis*, when alarmed, will partly ascend a tree but will usually descend if further pressed to escape. Pfeffer (1969) claims that the arboreal-terrestrial behavior of the species varies somewhat according to the forest type: "Le Tupaïa, qui en forêt dense ou en zone broussailleuse (forêts secondaires jeunes) est aussi bien terrestre qu'arboricole (sans monter toutefois dans les strates supérieures), est essentiellement terrestre et inféodé aux arbres morts et aux termitières en forêt claire."

Information is almost totally lacking for *Dendrogale murina* and *D. melanura* (the smooth-tailed tree shrews). The type specimen of *D. melanura* was procured in stunted montane jungle, and this species appears to be restricted to higher elevations. *Anathana wroughtoni*, the Madras tree shrew, is known to climb when pursued, and Verma (1965) suggested that certain anatomical features relate to an arboreal existence; definitive studies on these species' habits, however, remain to be made.

Observations on Captive Tree Shrews

Most studies on captive tupaiids confirm field observations as to terrestrial versus arboreal activity patterns. One exception is noteworthy, however. The terrestrial tree shrew (*Tupaia tana*) has been reported in the wild to be almost exclusively terrestrial (Banks, 1931; Davis, 1962). Sorenson and Conaway (1964) found that their captive tanas were clumsy climbers relative to other species, and that they often fell when descending. Yet in view of the fact that Sorenson and Conaway's tanas seldom ran and appeared to be rather lethargic—an unusual pattern for tree shrews generally and tanas in particular (Banks, 1931; Davis, 1962)—the observed pattern may have been abnormal. However, Sorenson and Conaway (1966) noted that tanas will engage in long chases that often result in a reversal of hierarchy. The captive tanas of Banks (1931) ". . . climbed well, were thoroughly at home on horizontal branches and slept at night in the top of their cage." Similarly, Schlott (1940) observed that "Die Tanas sind echte Baumtiere und in ihren Bewegungen unglaublich gewandt und jäh. Sie halten sich fast durchweg im Gezweige ihres Kletterbaumes auf . . . Auf den Erdboden kommen sie eigentlich nur, um zu trinken oder Nahrung zu holen." Steinbacher (1940) also reported that captive tanas are skillful climbers. Although the nature of the discrepancy in the foregoing accounts remains to be resolved, it appears nevertheless that *T. tana* has more ability in arboreal activity than is usually exhibited in the wild. The behavior of *Urogale everetti* is probably comparable. This species nests in the ground and is active on the forest floor but is also a good climber (Wharton, 1950). Captive specimens, although spending most of their time on the cage floor, climb frequently and skillfully (Polyak, 1957), yet to what extent free-ranging animals exhibit this behavior is uncertain.

In a comparative behavioral study of four species of *Tupaia* in captivity, Sorenson and Conaway (1964) confirmed field reports that *T. minor* is the most arboreal and *T. tana* the least. *T. gracilis* tends also to be arboreal, but less so than *T. minor. T. (glis) longipes* appears to be behaviorally intermediate; although it readily leaps up to 8 feet with excellent depth perception and sleeps in raised nest boxes, it often feeds and rests at ground level. Sorenson (1970), in a thorough behavioral study of eight tree shrew species, ranked each species according to its apparent affinity for arboreal activity. *Tupaia minor* and *T. gracilis* are the most arboreal (1), and *T. tana* is the most terrestrial (5). In comparison to *T. glis* and *T. montana,* which are intermediate (3), *T. longipes* is behaviorally more arboreal (2), and *T. chinensis* and *T. palawanensis* more terrestrial (4). It is interesting, however, that all species are capable of walking inverted along the underside of branches, although the arboreal forms exhibit this behavior most often. Sorenson observed other locomotor differences between species. At lower speeds, *T. gracilis* tends to hop whereas *T. longipes* uses a walking gait; *T. chinensis* moves more slowly and is less nervous than *T. longipes.* When descending vertically, *T. minor* and *T. gracilis* (as well as *T. glis*—personal observation) turn the feet around to a posteriorly directed position; in this posture, the claws are used as hooks to hang by (Fig. 6). *Tupaia chinensis* and *T. tana* do not reverse their feet in this situation as much as other species do, a trait that Sorenson believes may reflect a more terrestrial adaptation.

Other observers of captive *T. glis* also conclude that this species is both arboreal and terrestrial, but the precise balance of this dual behavior is not clear. Hofer's (1957) detailed study concluded that "Nach allen unseren Beobachtungen möchte ich *Tupaia glis* nur eine sehr beschränkte arboricole Lebensweise zubilligen." However, Kaufman's (1965) *T. glis,* although spending most of their time on the ground, were frequently active in 15 feet tall trees in a large enclosure. Particularly interesting are Vandenbergh's (1963) data on five animals (Table I) which indicate that there may be considerable lability in arboreal–terrestrial behavioral patterns. Escape reactions apparently differ between males and females, the males ascending for display and vocalization, the females fleeing at ground

TABLE I
FREQUENCY OF ARBOREAL AND TERRESTRIAL ACTIVITY OF *Tupaia glis*[a]

Sex	Time spent: On ground (minutes)	Time spent: In trees (minutes)	Time in trees (%)	Mean changes per hour	Duration of observation (hour)
♂	315.5	464.5[b]	59.6	27.8[c]	13
♂[d]	78.8	161.2[b]	76.2	6.3	4
♀	509.3[b]	270.7	34.8	16.2	13
♀[d]	401.2[b]	138.8	26.6	12.6	10
♀	768.1[b]	11.9	1.5	0.9	13

[a] From J. G. Vandenberg, 1963, *Folia Primatol.* 1, 199–207. Courtesy S. Karger and the author.

[b] Significantly more time at <0.001 level (X^2 test).

[c] Significantly more changes per hour than any other tupaia at <0.05 level (t test).

[d] Male and female died during the study.

level to secluded nesting sites (Sprankel, 1961; Vandenbergh, 1963). Resting on branches and preference for elevated nest boxes is not uncommon for captive animals.

In summary, the evidence for locomotor behavior in both captive and wild tree shrews indicates a moderate diversity of habitat preference. Clearly some tree shrew species are more arboreal than others; other species are more or less terrestrial. Perhaps the most significant fact is that all tree shrew species can climb, and at least occasionally, if not frequently, do so. Conversely, even the most arboreal species may be found on ground level. In tupaiids, the range of adaptive types is probably too subtle to be understood in terms of the gross categories of arborealism and terrestrialism, and as Sorenson (1970) has pointed out, there is need for further study of the behavior and ecology of wild tupaiids.

Locomotor Patterns of *Tupaia glis*

Rapid, scurrying movements punctuated by short pauses, often described as "nervous behavior," have been noted in almost every description of tree shrew activity. Sprankel (1961), Vandenbergh (1963), and Hofer (1957) described this behavior in detail, especially with regard to the agility and acrobatics of captive specimens.

The six *Tupaia glis* used in the present study were maintained in cages approximately 45 × 75 × 75 cm. In general, their locomotor behavior conformed to that reported in previous studies; therefore, further detailed description based on visual impressions will be omitted. However, several characteristics of their locomotor behavior are worth emphasizing. Individuals of *Tupaia glis* are active climbers, moving about on branches and the cage mesh as frequently as they do on the cage floor. Commonly, they establish a repetitious locomotor pattern which involves some acrobatic maneuver such as springing off a vertical wall or turning about on a narrow branch. *Tupaia glis* is particularly adept at rapid locomotion in an environment which necessitates abrupt changes in direction or elevation, and is capable of upward leaps of approximately one meter.

Footfall Patterns and Speed

According to Hildebrand (1967), symmetrical gaits (footfalls of each pair of feet spaced evenly in time) are common in walking primates but are apparently rare in tree shrews. Hildebrand's conclusions, based on limited observations of terrestrial locomotion in *Tupaia* (*glis*) *belangeri* and *Urogale everetti*, may now be supplemented with data from over 600 feet of film of terrestrial and arboreal locomotion in six captive specimens of *T. glis*. Footfall patterns are not specific for arboreal or terrestrial surfaces and vary principally with speed. Movement during exploratory activity is interrupted by momentary pauses, and the pattern of footfalls is highly asymmetrical (Fig. 1A). The asymmetry is accentuated by the

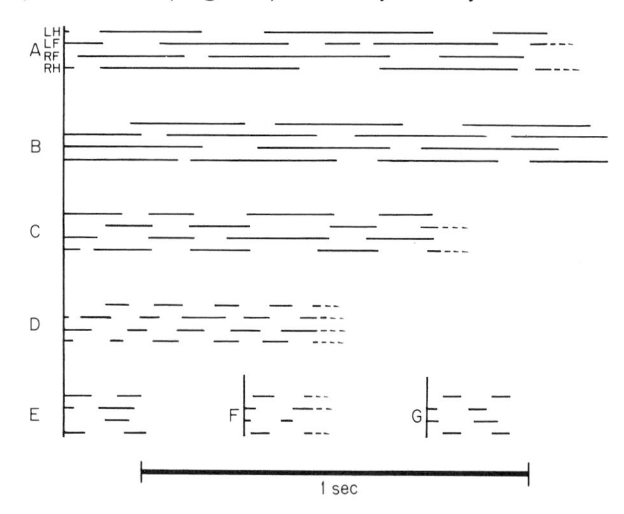

Fig. 1. Typical gait patterns of *Tupaia glis* at various speeds. Lines LH, LF, RF, and RH represent foot substrate contact of the left hind, left front, right front, and right hindfoot, respectively. Gaps in the lines represent periods when a foot is off the substrate. A 1-second interval is shown for scale. (A) Halting walking on a flat surface. (B) Slow walk on a horizontal branch 14 mm in diameter. (C) Fast walk on a flat surface. (D) Bounding run on a horizontal branch 17 mm in diameter. (E–G) Gait variations at top speed, exhibited on both terrestrial and arboreal surfaces that are more or less horizontal.

tendency for any given foot to make two half-steps in sequence (see LF, Fig. 1A). In a nonexploratory situation, such as a linear pathway along a narrow branch, the pattern may approach symmetry (Fig. 1B) and approximate Hildebrand's category of a fast, lateral-sequence, diagonal-couplets walk. With an increase in speed, the foot-substrate contact time decreases, but the footfall pattern remains essentially the same (Fig. 1C) up to speeds of approximately 150–175 cm/second. With greater speed, *T. glis* begins a bounding hop with the hindfeet functioning more or less synchronously (Fig. 1D). At speeds greater than approximately 225 cm/second, a phase occurs in which no foot is in contact with the substrate (Fig. 1E–G); the contact pattern of the forefeet may vary from more or less synchronous (Fig. 1E) to disynchronous (Fig. 1F), but the hindfeet are almost always nearly synchronous. Two male *T. glis*, chased over a 7.5 meter linear course to provoke maximum speeds, averaged approximately 13.6 km/hour over thirteen trials; the slowest speed was 10.4 km/hour, the fastest 18.3 km/hour. This rate is comparable to that reported for the red squirrel (*Tamiasciurus hudsonicus*, 14.5 km/hour) but slower than the maximum speed of the grey squirrel (*Sciurus carolinesis*), 27.2 km/hour (Layne and Benton, 1954).

Postural Patterns in Locomotion

The following description of skeletal movement is based principally on cineradiography at speeds from 48 to 150 frames/second, but also upon black and white cine, single exposure X-ray and photographical films. For descriptive purposes, the locomotory stride of a limb is divided into four phases. The initial propulsive movement is phase I; the limb, ceasing forward movement and contacting the substrate, begins a propulsive thrust. At phase III the limb is maximally extended at the completion of propulsive thrust. Phase II represents the sequence intermediate between phase I and III in the propulsive thrust. During phase IV the limb is off the substrate and moves forward to begin another propulsive thrust.

Limb Movements

During exploratory activity (footfall pattern as in Fig. 1B and C), the forelimb in phase I is commonly positioned as in Fig. 2A and B (right side). The scapula is inclined dorsoposteriad, with the scapular spine at about 35° to the vertical (Fig. 2A, right side). The distal end of the humerus is abducted 30° in dorsoventral view (Fig. 2B) and depressed 45° in lateral view. The forearm intersects the horizontal at 30°; the manus initially contacts the substrate anterolateral to the glenoid. Variation from this common posture principally involves the degree of humeral depression and concomitantly the orientation of the forearm. The humerus may depress as much as 55° or may be essentially horizontal during phase I, with the forearm approaching the horizontal or vertical planes, respectively. During phase II the distal humerus is adducted and elevated, so that by phase III the long axis is within about 15° of the parasagittal and the distal end is above the level of the glenoid (Fig. 2A and B, left side). The position of the manus in phase III is variable; seen in dorsoventral view, it may lie directly behind the elbow joint (Fig. 2B), posteromedial or posterolateral to the joint, or directly beneath the joint.

When resting, *Tupaia glis* adducts the humerus closer to parasagittal than is normal for locomotor postures (Fig. 3), but in other respects this stance resembles a phase II posture.

During phase I the mechanical axis of the femur (a line from head through condyles) typically is abducted from the parasagittal by 20 to 30° (Fig. 2B, left side) and is approximately horizontal (Fig. 2A). The crus, more or less parallel with the femur, is directed posteromedially and is inclined at 30° to the horizontal. The pes, in dorsoventral view, is usually slightly anterior to the hip joint (Fig. 2B, left side). Variations of this basic postural pattern principally involve the femoral orientation relative to horizontal; in some cases, the distal femur may be 10° or more above the proximal, in others, 25° below. The resulting variation in length of the stride is accompanied by variable foot positions which may be either medial or lateral to the femoral axis. During phase II the femur abducts and depresses relative to the hip joint. At phase III the distal

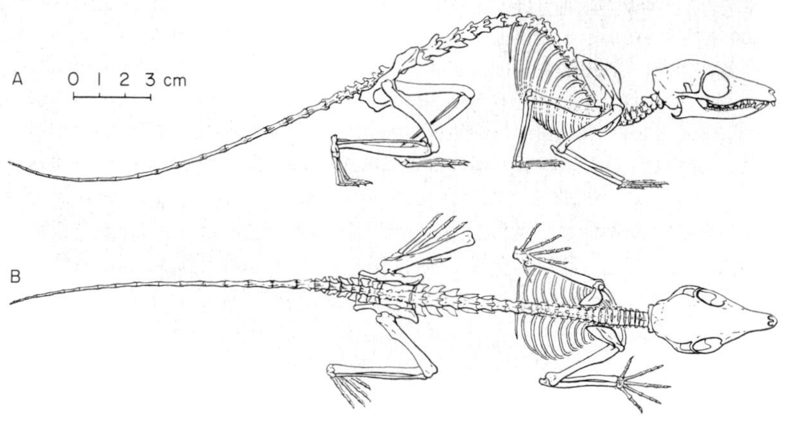

Fig. 2. A typical skeletal posture of *Tupaia glis* in lateral (A) and dorsal (B) views. This figure is based on radiographical films of living specimens.

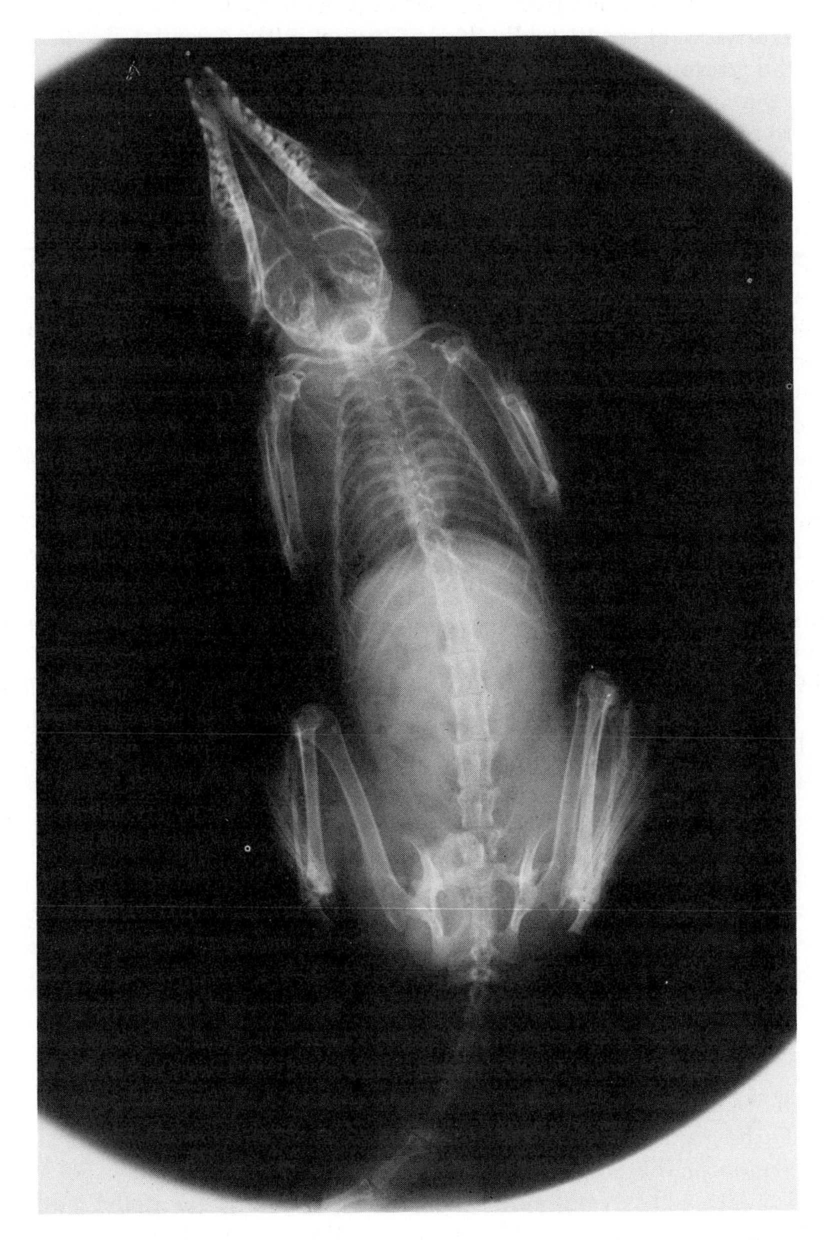

Fig. 3. Radiograph (dorsoventral projection) of *Tupaia glis* in a standing rest posture on a flat surface.

femur may be 50° or more below the proximal end and, seen in dorsoventral view (Fig. 2B, right side), is commonly oriented from 40 to 50° to the parasagittal plane. The crus is almost invariably parasagittal and horizontal in orientation. The proximodistal axis of the pes shifts to a more abducted orientation during phase III (Fig. 2B), and the principal locus of extension (dorsiflexion) is at the metatarsophalangeal joints (Fig. 2A, right side).

During the bounding run used for maximum speed, limb excursions of *Tupaia glis* are somewhat modified. In the forelimb, the adduction-abduction pattern of the humerus in the parasagittal plane remains essentially the same as at slower speeds, but the distal humerus frequently moves through 90° (Fig. 4), resulting in greater protraction of the manus and hence longer bounds. Similarly, the femur has a greater anteroposterior excursion, often through more than 100° (Fig. 4A–C). Seen in dorsoventral view, the pattern of femoral abduction from phase I to the middle of phase II approximates the phase I—II—III pattern characteristic of slower speeds; in particular, the mechanical axis of the femur lies at 20 to 25° to the parasagittal in phase I, and by phase II has abducted an additional 20–25°. By phase III, however, the femur again is in an adducted position, lying at about 20° to the parasagittal. The biomechanical significance of this pattern is not yet known.

Gambarian and Oganesian (1970) experimentally measured the relative forces sustained by fore and hindlimbs in diverse small mammals. In galloping mammals the total muscle weight of fore and hindlimbs is approximately the same, and both pairs of limbs participate equally in propulsion. In mammals using a "primitive rebounding jump," hindlimb weight is appreciably greater than that of the forelimb; the hindlimbs provide most of the propulsive thrust with the forelimbs acting principally as shock absorbers. The locomotor patterns of *Tupaia glis* appear to conform to Gambarian and Oganesian's category of the "primitive rebounding jump," and on this basis a functional differentiation of the forelimbs from hindlimbs may be postulated.

Manus

The manus of *Tupaia glis* is best classified as convergent (see Napier, 1961), for as the digits flex the claws tend to converge on the palm (for a discussion of the bony anatomy of the *Tupaia* "Spreizhand," see Altner, 1971). Although several authors have suggested that the pollex (digit I) is in some measure opposable, no substantial documentation of this possibility has been made and pollical movement appears to be simple convergence. In a nonweight-bearing posture (either in locomotor phase

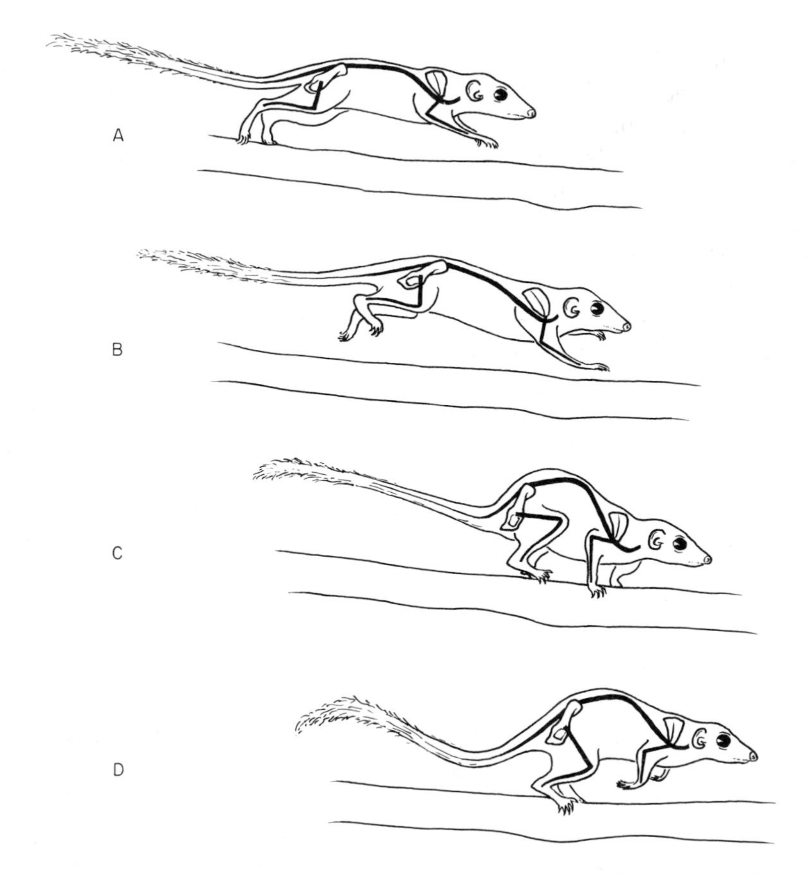

Fig. 4. (A–D) Sequential phases of the bounding run used by *Tupaia glis* at top speed. The sites of greatest curvature along the vertebral column (heavy dark line) are the lumbothoracic and cervicothoracic regions. Flexion and extension are pronounced only at the middorsal region (T11–T12, T12–T13, T13–L1 intervertebral articulations). Note also the orientations of the scapula and pelvis. As *T. glis* runs along a branch, each manus tends to be supinated (C) and each pes inverted (D), with the result that the branch is clasped between the feet. Based on radiographical and cine records.

IV or during stationary relaxation), the manus may be straight or slightly flexed at the carpus; the digits are usually not flexed. The longitudinal axis of digit III more or less coincides with that of the forearm. Digits II and IV symmetrically diverge from III at angles of less than 10°. Digit I, the pollex, is slightly more divergent (20°) than digit V (15°). In the weight-bearing manus on a flat surface, digital divergence may span an arc up to 150°. Digit III is usually directed exactly anteriad, and represents the central axis of the manus; however, slightly abducted and adducted manual positions are not uncommon, in which cases the central axis lies between digits II and III or III and IV, and digit V usually lies at 50 to 60°. The pollex is the most divergent and also the most variable in position, lying at least 60° and often as much as 85° from the axis of digit III. Thus, at maximum divergence, the axes of digits I and V are widely spread (as much as 150°); this finding modifies the report by Bishop (1964, Figs. 3 and 15d; see Fig. 15c, however) based on minimal divergence.

A characteristic digital posture, particularly in arboreal situations, is sharp flexion at the proximal interphalangeal joint; the distal (ungual) phalanx and the metacarpal remain relatively parallel to the substrate, and in this posture the holding effect of the claw is maintained. In an arboreal grip, the pollex frequently is abducted at the metacarpophalangeal joint (Fig. 5, top). This abduction, which in extreme cases is nearly 90°, appears to be passive and not the result of muscular control. In certain positions the body weight and the firm implantation of the claw apparently combine to produce forces acting perpendicularly to the digital axis. A similarly abducted posture is occasionally seen in other digits, notably IV and V and the hallux, but the degree of abduction is relatively small.

Tupaia glis uses the same manual posture when walking or running on relatively large branches (diameter greater than 2 cm) as on a flat surface. The palm is placed flatly on the substrate and at the end of phase III the palm and then the digits are lifted. However, the curvature of smaller branches (less than 2 cm diameter) forces the manus into varying degrees of supination which is accompanied by spreading and slight flexing of the digits. When running at maximum speed on small branches, *Tupaia glis* supinates each manus as much as 90°. As the contact of both hands is nearly synchronous, the effect is to grip the branch between the right and left manus (Fig. 4C), rather than to run on top of it.

Bishop (1964), studying prosimian hand orientation on slender, branch-like substrates, found the patterns used by *Tupaia glis* to be the most variable of all species studied (0.6 cm doweling was used for the tree shrews, larger doweling as appropriate for other species). Although by this criterion *T. glis* lacks a clearly defined manual grip pattern, Bishop cited the possible function of the large, protruding hypothenar pad for digital opposition. In Bishop's observations, the manus was almost always positioned across the branch in such a way that the branch lay between some of the digits and the hypothenar pad. The radiographical and anatomical data gathered in the present study give further support to

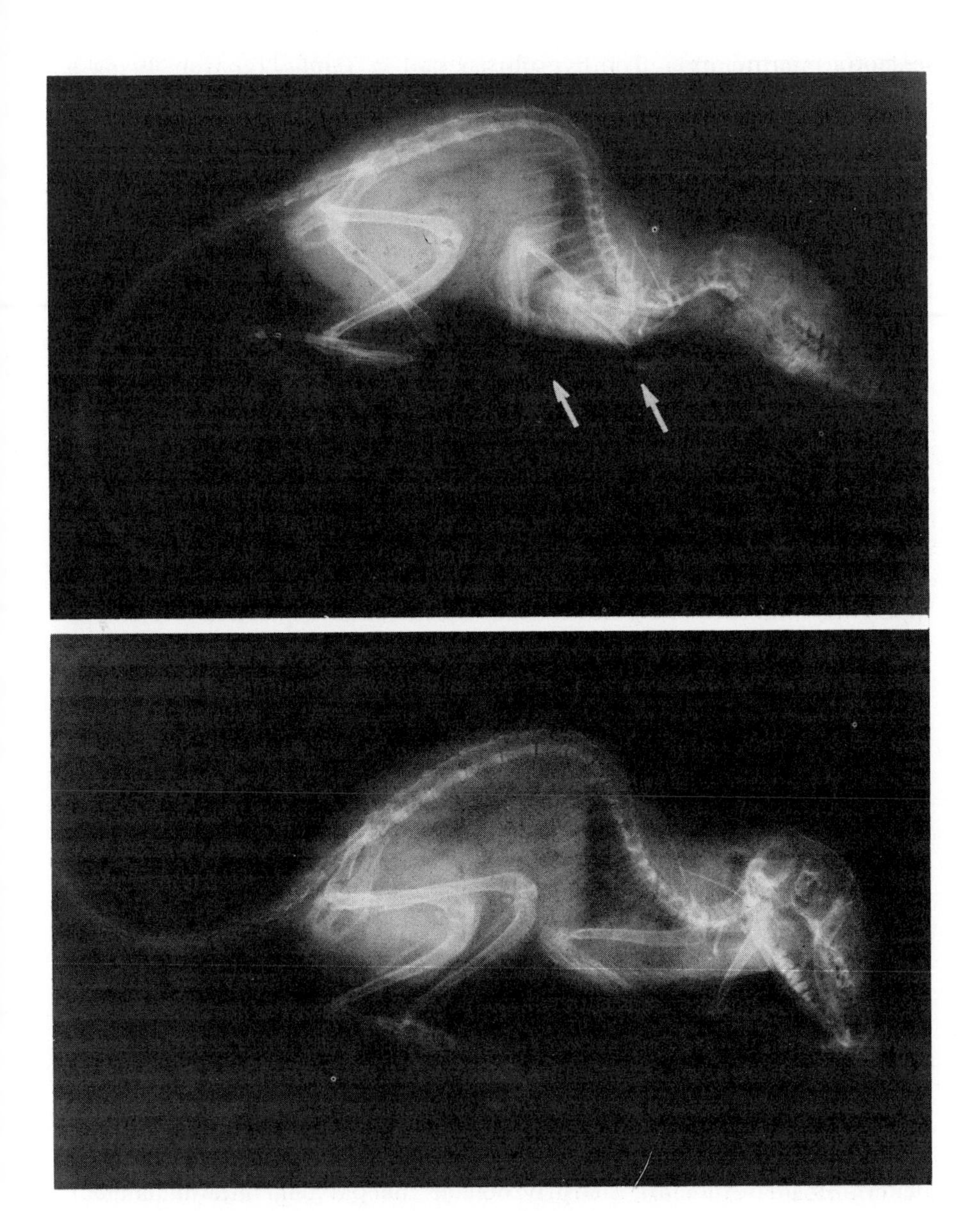

Fig. 5. Radiographs (lateral projection) of *Tupaia glis* on a horizontal branch 14 mm in diameter; in both cases the tree shrew had momentarily paused while walking along the branch. Note the sharp lumbothoracic and cervicothoracic curvatures. Arrows point to the large pisiform and hyperabducted pollex as described in text.

Bishop's interpretation. The hypothenar pad in *T. glis,* far from being a passive cushion, is buttressed at its proximal end by the pisiform bone and is traversed by several muscles. Articulating proximally at the carpus with the ulna and scapholunate and attached distally to the carpal ligament, the rodlike pisiform projects well into the hypothenar pad (Fig. 5). The long axis of the pisiform is inclined somewhat proximally away from the wrist, and this position probably is maintained largely by activity of the flexor carpi ulnaris, for the pisiform is embedded in its tendon as a sesamoid. Undoubtedly the bone is capable of movement which would alter the conformation and pliancy of the hypothenar pad. In addition to the flexor carpi ulnaris, other muscles acting on the pisiform are the abductor digiti minimi (with lateral and distal force components), palmaris brevis (the few slips from the carpal ligament would engender a medial force), and possibly the flexor digiti minimi brevis (although not directly attached, this muscle has an origin from the carpal ligament and thereby might contribute to medially directed force). The anatomy appears to be essentially the same in *T. tana* and *Ptilocercus lowii* (Haines, 1955). In *T. glis,* at least, the tendon of the flexor carpi ulnaris envelops only the base (articular end) of the pisiform, leaving the remainder to protrude palmarward. For a sesamoid such morphology would be inexplicable except on the basis that the nontendinous portion is acting as a bony strut. Both the anatomy and functional employment of the hypothenar pad support the conclusion that it serves to oppose digital flexion, with the result that the manus is capable of adapting to, if not actively gripping, a cylindrical or uneven substrate.

Pes

The median, proximodistal axis of the pes of *Tupaia glis* passes through digit III to the heel. The foot is narrow and elongate, with linear plantar pads along the medial and lateral sides. Digital divergence is less than in the manus. When relaxed, digits II and IV diverge only 5° from III (and the proximodistal axis). Digit V diverges about 10° and the hallux about 20°. During movement on flat surfaces or large branches, divergence may increase up to 10° for digits II and IV, and 25° for V. The hallux, which has a certain degree of independent movement, is discussed in detail below. As in the manus, sharp flexion at the proximal interphalangeal joints is associated with use of the claws in securing a grip.

Locomotory movements of the pes involve considerable plantar flexion in an abducted posture. At phase I, the foot is placed almost flatly on the ground, only the heel remaining slightly elevated; the amount of pedal abduction, measured at the intersection of the proximodistal axis of the foot with a parasagittal plane, varies from 10 to 25°. The digits are not usually flexed. By phase III the plantar aspect of the foot beneath the tarsus and metatarsus is positioned more or less vertically. Approximately 90° of extension occurs at the metacarpophalangeal joints and the digits are flexed at the proximal interphalangeal joints. Most striking, however, is the fact that the proximodistal axis of the foot is commonly at 45 to 75° to a parasagittal plane, and thus the foot is in a more abducted posture in phase III than in phase I. Abduction is accompanied by slight inversion. These movements occur as the result of the simultaneous elevation and forward movement of the ankle which passes to the medial side of the digits instead of directly above them. Thus the foot "twists" in place, that is, assumes a more abducted posture relative to the direction of movement.

The abducted, inverted posture of the pes is characteristic of arboreal locomotion, particularly on vertical surfaces and on branches less than 2 cm in diameter. On branches, plantar contact is limited to the metatarsus and digits; the heel remains well elevated, even in phase I. *Tupaia glis* places its feet on the upper sides of the branch with approximately the same degree of abduction as on other surfaces. However, the curvature of the branch requires that the foot be inverted with the plantar surface facing medioventrally instead of ventrally (as on flat surfaces). Not only does inversion ensure maximal plantar contact, but it also promotes stability when the tree shrew is moving at top speed. In this instance, the hindfeet tend to move synchronously, and by means of inversion are capable of grasping the sides of the branch which is thus held between two hindfeet (Fig. 4D). This posture, analogous to the manual supination used by the forelimb at top speeds, is usually accompanied by independent movements of the hallux to be discussed below.

Tupaia glis employs a hyperabducted foot posture when descending vertical or steep surfaces such as a tree trunk. The hindlimbs are usually sprawled out behind with the femora well abducted and the tibiae more or less parallel to the substrate. Normally, with the limbs in such a position the volar surface of the feet would face dorsad; however, through hyperabduction the volar surface is brought to bear against the substrate, and the claws are employed as hooks to control or prevent descent. Similar behavior is known in other arboreal mammals (see Cartmill, this volume). Radiography of living *T. glis* demonstrates hyperabduction occurring principally by rotation of both the calcaneus and the remaining pes about the head of the astragalus (Fig. 6). The subtalar (astragalocalcaneal) joint system is the major articulation involved in this movement.

The most striking feature of pedal function in *Tupaia glis* is the inde-

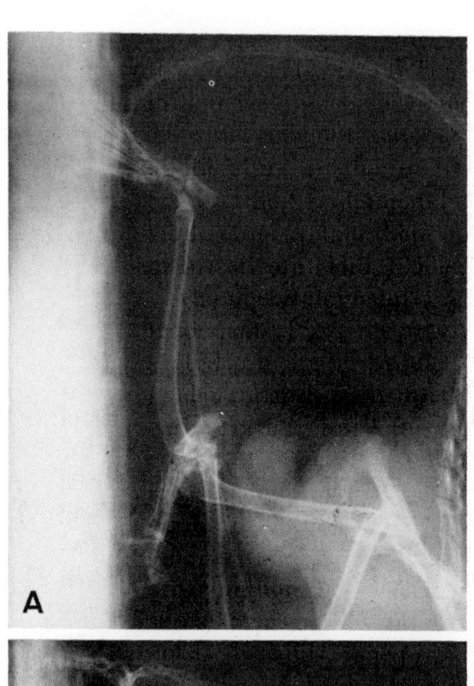
A

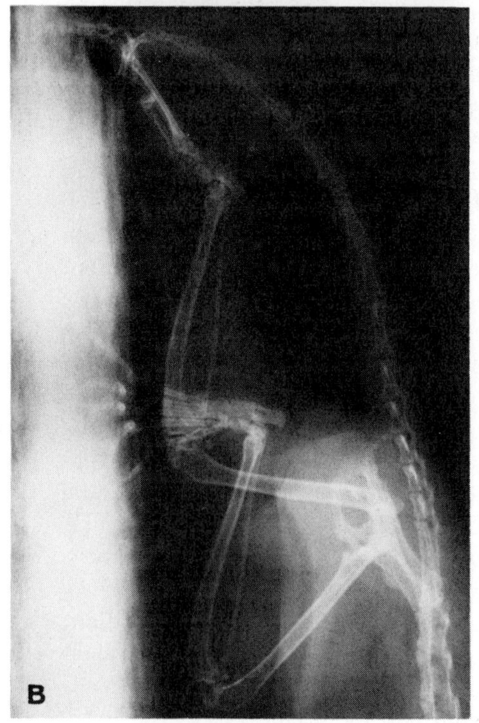
B

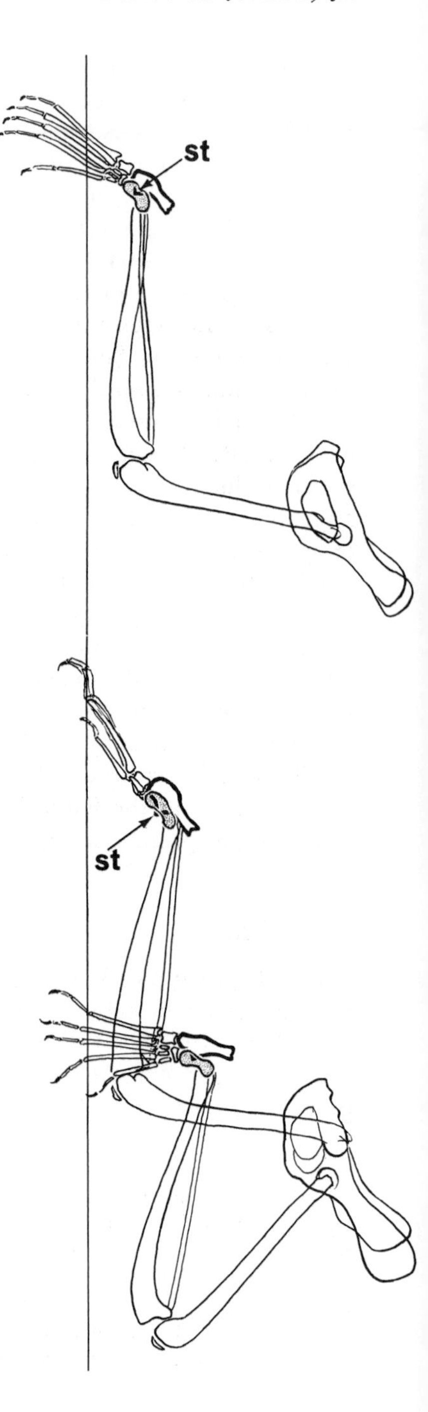
st
st

pendent action of the hallux on uneven or arboreal substrates. During late phase IV, the hallux is maximally extended and abducted (Fig. 7C). Thus the hallux is positioned to lie across the top of the branch (Fig. 7D and E). In this posture, the nail and phalanges prevent slippage off the side of the branch, and possibly the branch is actively gripped between hallux and the remaining digits. The hallux maintains this posture throughout phases II and III, and can be clearly seen in its abducted position even when the foot is maximally inverted at the end of phase III (Fig. 7H). Although the tarsometatarsal joint is saddle-shaped and probably permits some flexion–extension and abduction–adduction, cinematical and radiographical evidence shows that these movements are limited. Metatarsal I is bound to the palmar fascia by a thick aponeurotic band which limits abduction but which also may help to stabilize the hallux when abducted and bearing weight. The principal site of movement appears to be the metatarsophalangeal joint. The distal articular surface of metacarpal I is an asymmetrical condyle; the lateral aspect, being much more parallel to the shaft than the medial, permits phalangeal abduction. Crossing this joint are the following muscles which control the independent movement of the hallux: a large abductor hallucis, one tendon each from the extensor digitorum communis and extensor hallucis longus (the latter crossing the dorsolateral aspect and thus having an abducting component), the large tendon of the flexor hallucis longus, and the flexor hallucis brevis. The relative size and degree of morphological differentiation of the hallical muscles from those of the remaining digits confirms the present interpretation of the functional importance of the hallux.

Axial Skeleton

Tupaia glis typically has 7 cervical, 13 thoracic, 6 lumbar, 3 sacral and 24 (± 1 to 3) vertebrae. Although the vertebral column is capable of assuming a nearly straight configuration under special circumstances (such as when climbing on vertical surfaces), the usual posture involves pronounced dorsoventral curvatures. The cervical series normally de-

---

Fig. 6. Radiographs (lateral projection) of *Tupaia glis* making a vertical descent along a bark surface. Illustrated are the different degrees of pedal abduction used by the tree shrew to secure a grip. In the line diagram the calcaneus is indicated by a heavy line, and the astragalus by stippling. (A) Hyperabduction; the calcaneus and remaining pes begin to rotate about the head of the astragalus. (B) Extreme hyperabduction; note the change in calcaneal outline from (A) to (B), and especially the shift in the triangular sustentaculum tali (st).

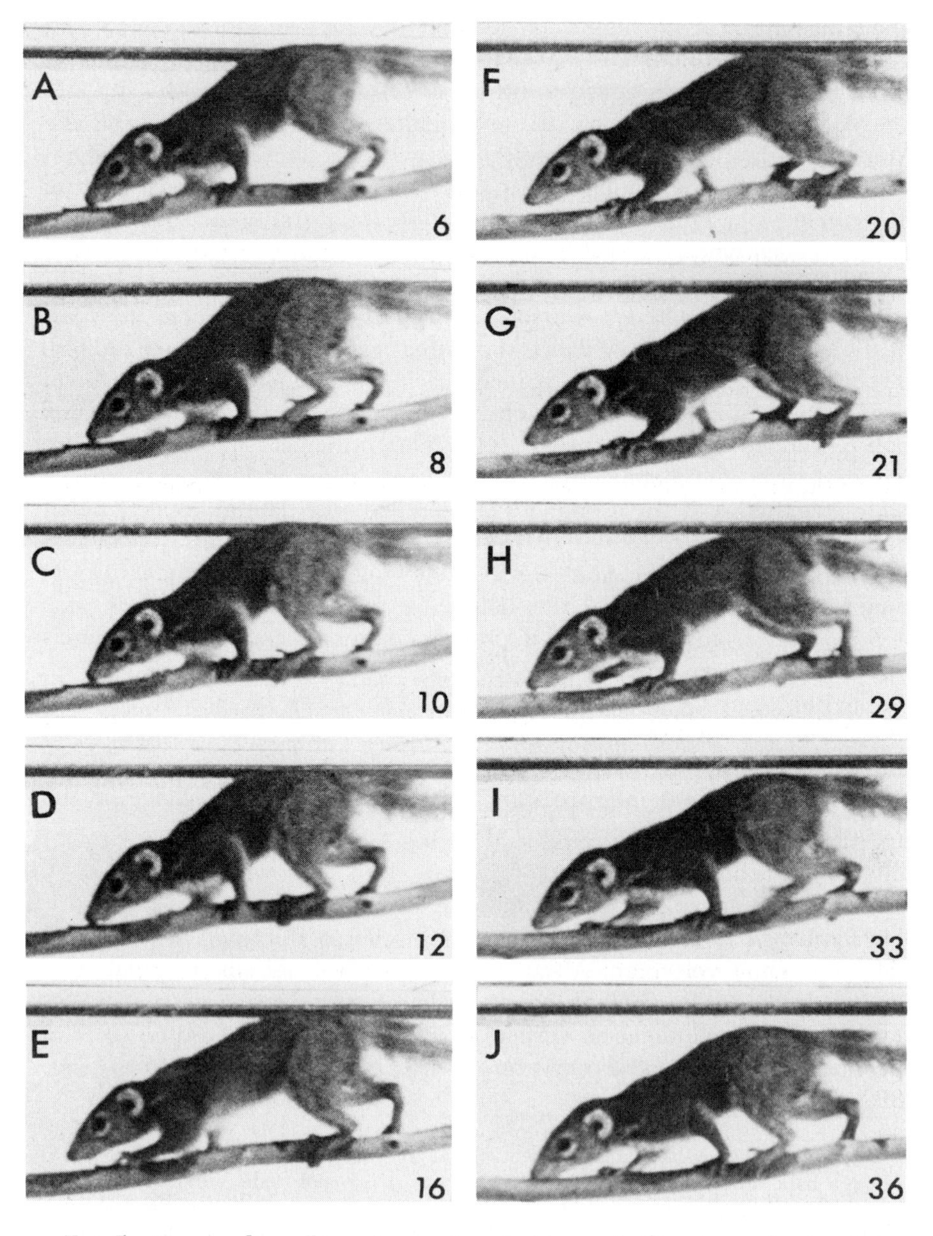

Fig. 7. Tupaia glis walking along a branch 14 mm in diameter and using independent movements of the hallux to secure the pedal grip. For further details, see text. This record is taken from a 16-mm movie film exposed at 64 frames/second; individual frame numbers are indicated.

scends from the skull at angles of 35 to 55° to the horizontal (Figs. 2 and 5). The reversed thoracic inclination of 30 to 40° creates a sharp flexure in the cervicothoracic region (Figs. 2 and 5). The lumbar series usually inclines posteroventrad at angles of 15 to 25° from horizontal, an orientation shared with the sacrum and proximal caudals.

The tail functions in a variety of postures but is not prehensile. At relatively slow gaits, the tail is held off the ground with the proximal half at approximately the same horizontal level as the sacrum. The distal half is highly mobile and is carried either in an upwardly curved arc, or more or less straight with ventral inclination (Fig. 8), or with distal tip turned downward, the remainder of the tail then being straight. Tail-flicking is frequent, particularly during excitement or at the onset of locomotion; in this movement the distal half is brought forward over the base of the tail in a single rapid twitch. When perched on a branch, *T. glis* frequently holds the tail in a downwardly curved arc. During rapid locomotion, the tail apparently trails flaccidly and undulates dorsoventrally in response to body movement (Fig. 4).

Pronounced flexion and extension of the trunk accompanies locomotion. The region of maximum mobility lies at the T11–T12, T12–T13, and T13–L1 intervertebral articulations. The remainder of the thoracic spine has a small amount of mobility in comparison, and the lumbar series does not appear to engage in any significant flexion. The striking feature of *Tupaia glis* in a slow, exploratory walk is the apparent elongation and contraction of the trunk; the head and shoulders appear to move ahead and stop, followed by the hindquarters in similar pattern. During "contraction," the back in the middorsal region is noticeably arched (Fig. 8B and D). Flexion at T11–T12, T12–T13, and T13–L1 is principally responsible for this pattern, and is related to limb movement. As the left hindlimb begins to advance during early phase IV, flexion in the middorsal region is only slight (Fig. 8A). By late phase IV (or early phase I) on the left side, the right hindlimb is completing its propulsive thrust (phase III); simultaneous flexion at the middorsal spine permits the pelvis to advance without corresponding movement at the forelimbs (Fig. 8B). Spinal extension returns as one forelimb moves through phase IV and the contralateral limb enters phase II (Fig. 8C). During the bounding run used for maximum speed, hind- and forelimb pairs function more or less synchronously; spinal flexion, although more pronounced, still appears to be principally restricted to the intervertebral articulations between T11 and L1. The lumbar series remains rigid and does not contribute to even the most extreme flexion observed. Maximum flexion approximately coincides with the initial contact of the hindfeet with the substrate (Fig.

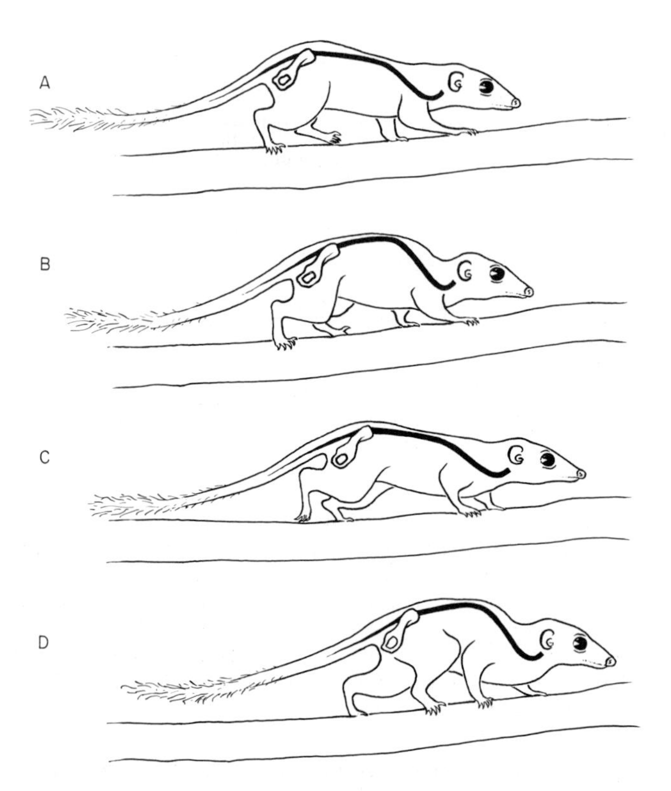

Fig. 8. (A–D) Sequential phases of the walk in *Tupaia glis* illustrating the coordination of flexion-extension of the vertebral column (heavy dark line) and limb movements. Based on radiographical and photographical records.

8C), and maximum extension with phase III of the hindfeet (Fig. 8A).

The anatomical adaptations whereby truncal flexion is concentrated largely at the T11–T12–T13–L1 invertebral articulations are not obvious. The diaphragmatic and the anticlinal vertebra (T10), which commonly has been regarded as the site of maximum spinal mobility, in fact, is only peripherally associated. There are, however, four features which relate to the dorsoventral mobility of this region.

1. In lateral view, the vertebral bodies of T12, T13, and L1 are trapezoidal, the dorsal (neural canal) aspect of the body being longer than the ventral length (Figs. 5 bottom and 9). Elsewhere in the vertebral column the bodies in sagittal section are either rectangular or resemble a parallelogram; apparently flexion is in part restricted by the compression of the intervertebral disc that accompanies approximation of the ventral edges of adjacent vertebral bodies. However, if the ventral edges of

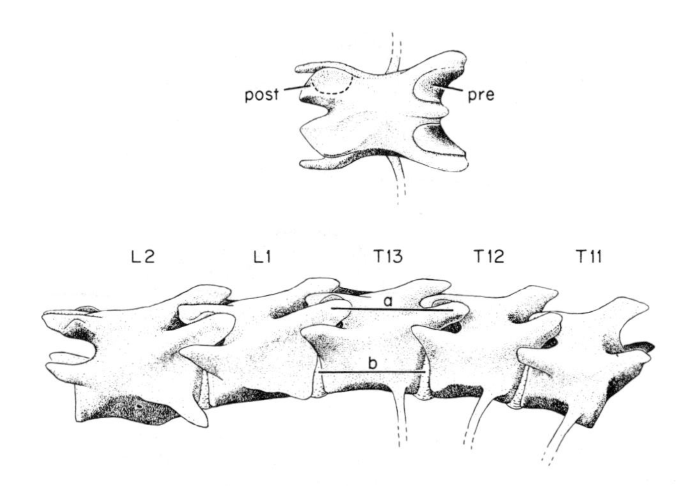

Fig. 9. Vertebral specializations of *Tupaia glis* that contribute to localization of spinal flexion and extension at the L1–T13, T13–T12, and T12–T11 joints. Below, vertebrae L2 through T11 are shown in lateral view. Note the wedge-shaped bodies at T13 and T12; the greater distance between articular surfaces (a) than that between adjacent nuclei pulposi (b). Top, the thirteenth thoracic vertebra is shown in dorsal view. Note the anterior edge of the prearticular surface (pre) is curved sharply ventrally, and the anteroposterior length of this surface is longer than that of the postarticular (post) surface.

adjacent bodies are recessed and the intervertebral disc is correspondingly wedge-shaped, flexion is facilitated by an increase in compressible substance (disc) and a decrease in noncompressible substance (ventral edges of adjacent vertebral bodies). This type of mechanism, well known in the human spine, appears to be operating at the T11–T12, T12–T13, and T13–L1 invertebral joints of *Tupaia glis.*

2. Another feature related to spinal flexion at T11 through L1 is the fact that the distances between the centers of pre- and postarticular surfaces of each vertebra are greater than the body length (measured between the centers of adjacent nuclei pulposi). Based on measurements made from single X-ray films of stationary individuals, this difference appears to be on the order of 10 to 20% at T12 and T13, but is usually less and sometimes nil at T11 and L1. Measurements from articulated as well as disarticulated vertebral columns of embalmed individuals yielded similar ratios. In one specimen this difference was approximately 25% at T13. The significance of this feature is that in the neutral position, with the centers of the articular surfaces on pre- and postarticular processes apposing, the vertebral column is in slight flexion.

3. At most intervertebral articulations in *T. glis* the anteroposterior

length of a zygapophyseal prearticular surface is comparable to that of a postarticular surface of the preceding vertebra and the displacement between them (and the intervertebral movement) is relatively small. Dissection and ligamentous preparations, however, reveal that the prearticular cartilaginous surfaces are normally about 25% longer than the postarticular surfaces at the L2–L1, L1–T13, and T13–T12 joints. The T12–T11 articular surfaces vary between specimens in this regard. The functional result of this differential length is increased mobility; the smaller postarticular surface has greater potential for anteroposterior movement on the larger prearticular surface than it might were the dimensions of the two surfaces essentially the same.

4. The anterior margin of the prearticular surfaces at L1, T13, and T12 are more ventral than the posterior margin. The resultant convexity, not found elsewhere, is related to the greater degree of flexion at these joints. The curved surface represents an arc, the center of which lies within the nucleus pulposus. During flexion, the articular processes, in the course of "separating," trace an arcuate path which has been incorporated into the form of these prearticular surfaces.

Hypotheses of Arboreal Ancestry

Ever since Huxley (1880) proposed that marsupials descended from an arboreal stock, the possibility of arboreal adaptation in mammalian ancestry has been repeatedly investigated. Did the early primates merely exploit the primitive mammalian mode of life or did they represent an adaptive innovation? In short, were the early placental mammals arboreal or terrestrial in adaptation? Huxley's original idea was based on the fact that members of most families of living marsupials possess an apparently prehensile pes by virtue of an abductable hallux; those species with a vestigial or absent hallux Huxley supposed possessed a "reduced prehensile pes" by virtue of their taxonomic relationships. Dollo (1899) developed the evidence in more detail, particularly in regard to the problem of the atrophied hallux. Dollo argued that in marsupials with tetradactylous feet (the hallux reduced or absent) features such as syndactyly, reduction of toes II and III, and enlargement of toe IV were common to forms with opposable halluces, and therefore indicated an arboreal heritage for all marsupials. Bensley (1901a, b) accepted the hypothesis as well.

Matthew (1904, 1909, 1937), using paleontological evidence, inferred that placental mammals also had an arboreal ancestry. Although Matthew's 1904 paper contained brief reference to such features as a gracile build and "probably prehensile" tail among early Tertiary mammals, the argument was based principally on the interpretation of an opposable or semiopposable digit I. In particular, he regarded the feet in creodonts (early Tertiary carnivores) as representing the basic placental morphology adapted to arboreal life. No other pedal structure or function was given as careful consideration as prehensility, and it is apparent from Matthew's last paper that his interpretation was founded on the identification of this ability. Subsequently, the theory of an arboreal ancestry for mammals gained wide acceptance and has been reinforced to a limited degree by some neontological evidence (see discussion in Haines, 1958, and Martin, 1968b). An early paper by Gregory (1910) probably also contributed to the genesis of the theory.

Only two authors have challenged Matthew's theory. Gidley's (1919) objection that mere pollical or hallical divergence was inadequate evidence of semiopposability was answered by Matthew (1937) with a detailed description of the manus in *Claenodon.* In *Claenodon* (and presumably other creodonts with the presumptively "primitive" mammalian manus), Matthew interpreted the pollical carpometacarpal and carpotrapezium joints as permitting a degree of pollical opposability. Haines (1958), in a careful reexamination of this material, contradicted Matthew's interpretation, and indeed the evidence for opposability appears insubstantial. However, Haines went on to suggest that the feet of the Egyptian mongoose (*Herpestes ichneumon*), a terrestrial carnivore, adequately represent the primitive placental plan and that, consequently, early mammals were probably terrestrial.

Previous theories on primitive mammalian arboreal or terrestrial adaptation have several shortcomings; first, the fossils considered are too late in time (Paleocene-Eocene) or too specialized in form (Creodonta) to represent common ancestral forms, and second, there has been inadequate appreciation of the adaptations among living mammals which bear on the interpretation of fossil forms. Matthew (1937) had available only early Tertiary mammals, rather than Cretaceous forms which he suspected were ancestral. Haines (1958), on the other hand, virtually ignored fossil evidence in his selection of a primitive mammalian analog among living species; without reference to fossil material, *Herpestes* is no more credible an analog of an early mammal than is, for example, a marsupial. Both appropriate fossils and relevant experimental evidence from living mammals are necessary to make interpretations of the anatomical adaptations of early mammals. Although the evidence is far from complete, tree shrews appear to offer at least one viable model for such interpretations on the basis of skeletal similarity to that of early mammals.

The Origins of Primate Arborealism

Discussions of the evolutionary origins of major taxa usually focus on adaptive innovations (e.g., anatomical, physiological, or behavioral features) and new adaptive zones (e.g., in habitat or food source). For the Primates, the adaptive theme is arborealism. Although not all living primates are tree-living, they all are derived phylogenetically from arboreal ancestors. The question naturally follows as to what selective forces initially favored arboreal adaptation and habitat; escape from predation and food source are among the commonly proposed answers. However, the circumstances in which trees became an accessible and viable adaptive zone for early primates remains unknown. Two alternatives seem plausible. The theory that early mammals were basically terrestrial implies that the origin of primates represented an adaptive innovation both in anatomical form and in habitat. Thus, the earliest primates are believed to have developed arboreal specializations which set them apart from other early mammalian groups. Since teeth constitute the major part of the fossil record of the earliest primates, the tendency has been to regard dietary specialization as a primary factor in primate origins (see Szalay, 1969). However, implicit is the recognition that other arboreal features were also acquired which less specialized mammals, both fossil and living, do not possess. This viewpoint, of which Haines (1958) is the most recent proponent, treats "arborealism" and "terrestrialism" as necessarily discrete phenomena. An alternative theory, proposed here, is based upon the implications of tree shrew behavior and habitat in which "terrestrialism" and "arborealism" are not discrete phenomena.

By emphasizing arboreal versus terrestrial behavior, published accounts of tree shrews appear to have misrepresented the true nature of their habitat and adaptations. With the possible exception of a few species, all tree shrews can and do move freely between ground and trees. Even the most "terrestrial" species are known to be good climbers. This ability relates not simply to climbing trees, but more realistically to the fact that the forest offers, to a small mammal, an extremely uneven and disordered substrate for locomotion. The forest floor, traversed by roots and littered with plant debris (including fallen trunks), grades upward through secondary growth (including vines, bushlike growth and small trees) to tree trunks and the lower story. In the forest habitat of tree shrews, the distinction between "arboreal" and "terrestrial" locomotion is artificial in the sense that the substrate everywhere requires basically the same locomotor repertoire for this sized mammal.

The locomotor repertoire of tree shrews is adapted to two major topographical features of their habitat. The first is that most surfaces, in terms of the actual plantar area of tree shrew feet, are not level. The ability to invert–evert the hindfeet, and pronate–supinate the forefeet, is critical in this situation. Combined with movements of plantar flexion and dorsiflexion, the feet are able to appose a nonlevel surface while the animal retains its upright posture. Postural stability is likewise important on uneven surfaces, and may be achieved in part by maintaining a relatively low center of gravity while carrying the body on as wide a base as practicable. In terms of limb posture, this is accomplished by employing the elbow and knee joints in flexed rather than extended positions, and by maintaining the limbs in abducted rather than parasagittal postures. A second feature of the forest habitat is the disordered spacial arrangement of locomotor surfaces. In the perspective of a small mammal, the forest floor, the trees, and all the interconnecting secondary growth are not a smooth continuum but rather a ragged network of substrate possibilities. Irregular spacing of footfalls is as appropriate for moving about among the debris of the forest floor as it is for climbing through forest vegetation. Both situations favor a highly flexible and versatile locomotor pattern, and may account for the predominance of asymmetrical gaits among tupaiids. In tree shrews, such versatility of foot placement is in part a function of the lateral mobility of the limbs (significant abduction at the shoulder and hip is common, particularly when climbing) and the flexibility of the vertebral column. The latter's function in lengthening the stride in some cursorial mammals has often been emphasized. A similar function is served for the tree shrew leap and bounding run. However, in other situations spinal flexion and extension also allows considerable variation in the distance between pectoral and pelvic girdles, and consequently between fore- and hindfeet. In terms of foot placement along an irregularly spaced substrate, the ability to make gross adjustments in the length of the quadrupedal stance pattern is important. Such adjustments during walking have already been discussed (p. 105, Fig. 8). Variations in quadrupedal stance are common in other situations and always involve distinctive vertebral column postures. For example, when positioned transversely on a branch, tree shrews must closely approximate all four feet and to do so use maximal spinal flexion. On steep or vertical surfaces, where, in terms of leverage, it is advantageous to keep the body (center of gravity) close to the surface, the feet are spread widely and the vertebral column is maximally extended.

The locomotor pattern of *Tupaia glis* is significant both in a comparative and phylogenetic context. The basic characteristics of the pattern are not peculiar to tree shrews, but are shared by other relatively generalized, noncursorial mammals (Jenkins, 1971). Although some specialization has undoubtedly occurred among tupaiids, such fundamental features as limb posture and excursion are similar to those in the rat (*Rattus norvegicus*) or the Virginia opossum (*Didelphis marsupialis*). Thus, a common locomotor pattern for noncursorial, relatively generalized mammals is recognizable, and a similar functional pattern—if osteology is a reliable guide—appears to have been present among early mammals.

A new hypothesis on the origin of mammalian posture and locomotion, as well as the origin of primate arborealism, is now possible. Mammalian limb posture traditionally has been interpreted in terms of the presumed biomechanical efficiency of upright, parasagittally oriented limbs; it is now known that this arrangement is not present in noncursorial mammals (Jenkins, 1971). Instead, the primitive mammalian locomotor mode may have evolved as an adaptation to moving on uneven, disordered substrates (which, of course, are not found only in forest habitats). Flexion and extension of the vertebral column, the mobility of the feet, and the flexed, abducted limbs may be interpreted as mechanisms to permit versatility in stance and locomotor pattern. Most Mesozoic mammals were approximately tree shrew sized, and the importance of a versatile locomotor repertoire to a small mammal has already been emphasized. Accordingly, the question of an arboreal or terrestrial ancestry for mammals no longer is relevant. The locomotor niche which ancestral mammals exploited undoubtedly included both "terrestrial" and "arboreal" surfaces. Ancestral primates very likely occupied forest habitats in the manner of some living tree shrews. The adaptive innovation of ancestral primates was therefore not the invasion of the arboreal habitat, but their successful restriction to it.

References

Altner, G. (1971). Histologische und vergleichend-anatomische Untersuchungen zur Ontogenie und Phylogenie des Handskeletts von *Tupaia glis* (Diard 1820) und *Microcebus murinus* (J. F. Miller 1777). *Folia Primatol. Suppl.* **14**, 1–106.

Banks, E. (1931). A popular account of the mammals of Borneo. *J. Malay. Br. Roy. Asiat. Soc.* **9**, 1–139.

Bartels, M., Jr. (1937). Zur Kenntnis der Verbreitung und der Lebensweise javanischer Säugetiere. *Treubia* **16**, 149–164.

Bensley, B. A. (1901a). On the question of an arboreal ancestry of the Marsupialia, and the interrelationships of the mammalian subclasses. *Amer. Natur.* **35**, 117–138.

Bensley, B. A. (1901b). A theory of the origin and evolution of the Australian Marsupialia. *Amer. Natur.* **35**, 245–269.

Bishop, A. (1964). Use of the hand in lower primates. *In* "Evolutionary and Genetic

Biology of Primates" (J. Buettner-Janusch, ed.) Vol. 2, pp. 133–225. Academic Press, New York.
Carlsson, A. (1922). Über die Tupaiidae und ihre Beziehungen zu den Insectivora und den Prosimiae. *Acta Zool.* (Stockholm) 3, 227–270.
Davis, D. D. (1962). Mammals of the lowland rain-forest of North Borneo. *Bull. Raffles Mus.* **31**, 1–129.
Diard, M. (1820). Report of a meeting of the Asiatic Society of Bengal for March 10, 1820. *Asiat. J. Month. Reg.* **10**, 477–478.
Dollo, L. (1899). Les ancêtres des Marsupiaux étaient-ils arboricoles? *Trav. Stat. Zool. Wimereux* **7**, 188–203.
Gambarian, P. P., and Oganesian, R. O. (1970). Biomechanics of the gallop and of the primitive rebounding jump in small mammals. *Proc. Acad. Sci. USSR Biol. Ser.* 3, 441–447. (In Russian.)
Gidley, J. W. (1919). Significance of divergence of the first digit in the primitive mammalian foot. *J. Wash. Acad. Sci.* **9**, 273–280.
Gregory, W. K. (1910). The orders of mammals. *Bull., Amer. Mus. Natur. Hist.* **27**, 1–524.
Gregory, W. K. (1913). Relationship of the Tupaiidae and of Eocene lemurs, especially *Notharctus*. *Bull. Geol. Soc. Amer.* **24**, 247–252.
Haines, R. W. (1955). The anatomy of the hand of certain insectivores. *Proc. Zool. Soc. London* **125**, 761–777.
Haines, R. W. (1958). Arboreal or terrestrial ancestry of placental mammals. *Quart. Rev. Biol.* 33, 1–23.
Hildebrand, M. (1967). Symmetrical gaits of Primates. *Amer. J. Phys. Anthropol.* **26**, 119–130.
Hill, W. C. O. (1953). "Primates. Comparative Anatomy and Taxonomy," Vol. I, Strepsirrhini. University Press, Edinburgh.
Hofer, H. (1957). Über das Spitzhörnchen. *Natur Volk* **87**, 145–155.
Horsfield, T. (1851). "A Catalogue of the Mammalia in the Museum of the Hon. East-India Company." J. and H. Cox, London.
Huxley, T. H. (1880). On the application of the laws of evolution to the arrangement of the Vertebrata, and more particularly of the Mammalia. *Proc. Zool. Soc. London,* pp. 649–663.
Jenkins, F. A., Jr. (1971). Limb posture and locomotion in the Virginia opossum (*Didelphis marsupialis*) and in other non-cursorial mammals. *J. Zool.* **165**, 303–315.
Kaufmann, J. H. (1965). Studies on the behavior of captive tree shrews (*Tupaia glis*). *Folia Primatol.* 3, 50–74.
Kloss, C. B. (1903). "In the Andamans and Nicobars." John Murray, London.
Kloss, C. B. (1911). On a collection of mammals and other vertebrates from the Trengganu Archipelago. *J. Fed. Malay. States.* **4**, 175–212.
Layne, J. N., and Benton, A. H. (1954). Some speeds of small mammals. *J. Mammal.* **35**, 103–104.
Le Gros Clark, W. E. (1924). The myology of the tree-shrew (*Tupaia minor*). *Proc. Zool. Soc. London,* pp. 461–497.
Le Gros Clark, W. E. (1925). On the skull of *Tupaia*. *Proc. Zool. Soc. London,* pp. 559–567.
Le Gros Clark, W. E. (1926). On the anatomy of the pen-tailed tree-shrew (*Ptilocercus lowii*). *Proc. Zool. Soc. London,* pp. 1179–1309.

Le Gros Clark, W. E. (1927). Exhibition of photographs of the tree shrew (*Tupaia minor*). Remarks on the tree shrew, *Tupaia minor,* with photographs. *Proc. Zool. Soc. London,* pp. 254–256.

Le Gros Clark, W. E. (1960). "The Antecedents of Man," Quadrangle, Chicago, Illinois.

Lim, B. L. (1967). Note on the foot habits of *Ptilocercus lowii* Gray (Pentail tree-shrew) and *Echinosorex gymnurus* (Raffles) (Moonrat) in Malaya with remarks on "ecological labelling" by parasite patterns. *J. Zool.* **152**, 375–379.

Lim, B. L. (1969). Distribution of the primates of West Malaysia. *Proc. 2nd Intern. Congr. Primatol.* **2**, 121–130.

Luckett, W. P. (1969). Evidence for the phylogenetic relationships of tree shrews (family Tupaiidae) based on the placenta and foetal membranes. *J. Reprod. Fert. Suppl.* **6**, 419–433.

Lyon, M. W. (1913). Treeshrews: an account of the mammalian family Tupaiidae. *Proc. U.S. Nat. Mus.* **45**, 1–188.

McKenna, M. C. (1966). Paleontology and the origin of the primates. *Folia Primatol.* **4**, 1–25.

Martin, R. D. (1966). Tree shrews: Unique reproductive mechanism of systematic importance. *Science* **152**, 1402–1404.

Martin, R. D. (1968a). Reproduction and ontogeny in tree-shrews (*Tupaia belangeri*), with reference to their general behaviour and taxonomic relationships. *Z. Tierpsychol.* **25**, 409–532.

Martin, R. D. (1968b). Towards a new definition of primates. *Man* **3**, 377–401.

Matthew, W. D. (1904). The arboreal ancestry of the Mammalia. *Amer. Natur.* **38**, 811–818.

Matthew, W. D. (1909). VI.—The Carnivora and Insectivora of the Bridger Basin, Middle Eocene. *Mem. Amer. Mus. Natur. Hist.* **9**(Part 6), 289–567.

Matthew, W. D. (1937). Paleocene faunas of San Juan Basin, New Mexico. *Trans. Amer. Phil. Soc.* **30**, [N.S.], 1–510.

Muul, I., and Lim, B. L. (1971). New locality records for some mammals of West Malaysia. *J. Mammal.* **52**, 430–437.

Napier, J. R. (1961). Prehensility and opposability in the hands of primates. *Symp. Zool. Soc. London* **5**, 115–132.

Napier, J. R., and Napier, P. H. (1967). "A Handbook of Living Primates." Academic Press, New York.

Pfeffer, P. (1969). Considérations sur l'écologie des forêts claires du Cambodge oriental. *Terre Vie* **116**, 3–24.

Polyak, S. (1957). "The Vertebrate Visual System." Univ. Chicago Press, Chicago, Illinois.

Raffles, T. S. (1821). Descriptive catalogue of a zoological collection, made on account of the Honourable East India Company, in the Island of Sumatra and its vicinity, under the direction of Sir Thomas Stamford Raffles, Lieutenant–Governor of Fort Marlborough; with additional notices illustrative of the natural history of those countries. *Trans. Linn. Soc. London* **13**, 239–274.

Ridley, H. N. (1895). The mammals of the Malay Peninsula. *Nat. Sci.* **6**, 23–29.

Robinson, H. C., and Kloss, C. B. (1909). In Thomas, O. and Wroughton, R. C. 1909. On mammals from the Rhio Archipelago and Malay Peninsula collected by Messrs. H. C. Robinson, C. Boden Kloss and E. Seimund, and presented to the

National Museum by the government of the Federated Malay States. *J. Fed. Malay. States* **4**, 99–129.

Schlott, M. (1940). Beobachtungen an Tanas (*Tupaia tana* Raffl.). *Zool. Gart.* **12**, 153–157.

Simpson, G. G. (1965). Long-abandoned views. *Science* **147**, 1397.

Sorenson, M. W. (1970). Behavior of tree shrews. *In* "Primate Behavior: Developments in Field and Laboratory Research" (L. A. Rosenblum, ed.), Vol. 1, pp. 141–193. Academic Press, New York.

Sorenson, M. W., and Conaway, C. H. (1964). Observations of tree shrews in captivity. *J. Sabah Soc.* **2**, 77–91.

Sorenson, M. W., and Conaway, C. H. (1966). Observations on the social behavior of tree shrews in captivity. *Folia Primatol.* **4**, 124–145.

Sprankel, H. (1961). Über Verhaltensweisen und Zucht von *Tupaia glis* (Diard 1820) in Gefangenschaft. *Z. Wiss. Zool.* **165**, 186–220.

Steinbacher, G. (1940). Beobachtungen am Spitzhörnchen und Panda. *Zool. Gart.* **12**, 48–53.

Szalay, F. S. (1968). The beginnings of primates. *Evolution* **22**, 19–36.

Szalay, F. S. (1969). Mixodectidae, Microsyopidae, and the insectivore-primate transition. *Bull. Amer. Mus. Natur. Hist.* **140**, 193–330.

Vandenbergh, J. G. (1963). Feeding, activity and social behavior of the tree shrew, *Tupaia glis,* in a large outdoor enclosure. *Folia Primatol.* **1**, 199–207.

Van Valen, L. (1965). Treeshrews, primates, and fossils. *Evolution* **19**, 137–151.

Verma, K. (1965). Notes on the biology and anatomy of the Indian tree-shrew, *Anathana wroughtoni. Mammalia* **29**, 289–330.

Wharton, C. H. (1950). Notes on the Philippine tree shrew, *Urogale everetti* Thomas. *J. Mammal.* **31**, 352–354.

4

A Cineradiographical Analysis of Leaping in an African Prosimian (Galago alleni)*

F. K. JOUFFROY
and
J. P. GASC

Introduction

The galago's jump—more than 7 feet high and long and nearly fourteen times body length (tail excluded)—is a record for arboreal jumping, and as such may serve as a model for a more general study of mammalian jumping. Indeed, the diverse variations in jumping styles among mammals are related, among other factors, to terrestrial or arboreal modes of life. The galago's record is matched, perhaps even slightly exceeded, by small terrestrial mammals which, although belonging to very different orders, display similar adaptations: kangaroo rats (Zapodinae) are able to bound distances of 10 feet, and rat kangaroos (Potoroinae) jump up to heights of 8 feet. All these animals, like the galago, are of small size. Their relative performances surpass that of the great kangaroo (*Macropus rufus*) which holds the absolute record of high jumps at 9 feet; however, this jump represents only one and a half times body height (tail excluded), about the same ability as in man (Gray, 1953).

Among terrestrial animals, jumping is used either by quadrupedal animals such as the horse in clearing an occasional obstacle, or by bipedal forms such as kangaroos and jerboas in normal progression. However, in

* The authors express their sincere thanks to F. A. Jenkins, Jr., for translating this chapter.

both cases, the accompanying anatomical specializations (such as reduction of the lateral digits, elongation and diametrical increase in the axial metapodials, fusion of the tibia with the fibula and the radius with the ulna) render the feet little suited for other functions. In terrestrial mammals, particularly bipeds, jumping is used to gain distance rather than height; points of departure and landing are on the same level and the jump is gauged by the distance covered. Contrary to terrestrial species, arboreal mammals move about in an essentially three-dimensional environment. In this situation, jumping becomes almost infinitely variable, not only in the direction of movement (horizontal, oblique, or vertical) and in the diversity of support surfaces (ground, trunks, branches, vines, and foliage), but also in anatomical specializations (such as nails, prehensile hands, and feet) used for gripping.

Although arboreal jumping is performed by a number of mammals including carnivores, rodents, and marsupials (the latter two groups have even evolved gliding flight), the best adapted forms are without doubt among the primates. Even in cases where jumping is accompanied by certain anatomical specializations, the foot of jumping primates remains essentially prehensile and without any major limitations of movement. The range of pronation-supination and abduction-adduction, the extraordinary mobility of the ankle, and the opposability of the hallux all make the foot an organ of precise and effective grip which is able to respond immediately to the exigencies of a particularly diverse environment.

Jumping locomotion, used by members of all groups of prosimians and higher primates, varies in modality and degree between genera. The adaptation is characterized in skeletal proportions by a relative elongation of the hind limbs. The intermembral index [(femur + tibia)/(humerus + radius)] varies from about 140 (*Aotus*) to 170 (*Galago*). Among quadrupedal primates, this index is on the order of 110 to 120 in such extant genera as *Papio, Loris,* and *Perodicticus* and in such fossil genera as *Archaeolemur* and *Pachylemur* (Jouffroy, 1956, 1960, 1963).

Among primates, two types of structural adaptations for jumping may be distinguished (see Lessertisseur, 1970, Lessertisseur and Jouffroy 1973, 1974). The first type is characterized by a long thigh and a foot elongated chiefly in the metatarsus. Typically developed among higher primates (e.g., *Hapale, Aotus, Hylobates*), this type is also found among lemurs such as *Indri* or *Propithecus* (the "Brachytarsi" of Fitzinger; fide Brehm, 1868), and is not exclusive of other locomotor modes such as brachiation or bipedalism. The second type comprises the long tarsal jumpers (the "Macrotarsi" of Fitzinger). All are prosimians: *Microcebus, Galago,* and *Tarsius.* They are characterized by short thighs, their long hind limbs being related to the elongation of the calcaneum and navicular (Fig. 1). This type appears to be exclusive of brachiation, but not of bipedalism which the galago in particular uses when on the ground, hopping somewhat in the manner of a jerboa. To this group belong those prosimians which are currently considered to be the most primitive living primates (Napier and Walker, 1967; Walker, 1967; Charles-Dominique and Martin, 1970). It is in this group, comprising only light-bodied forms, that the best mammalian jumpers are found.

Among prosimians, "long-thighed" jumpers are represented by the Madagascan indriids. Although they also belong in Napier and Walker's

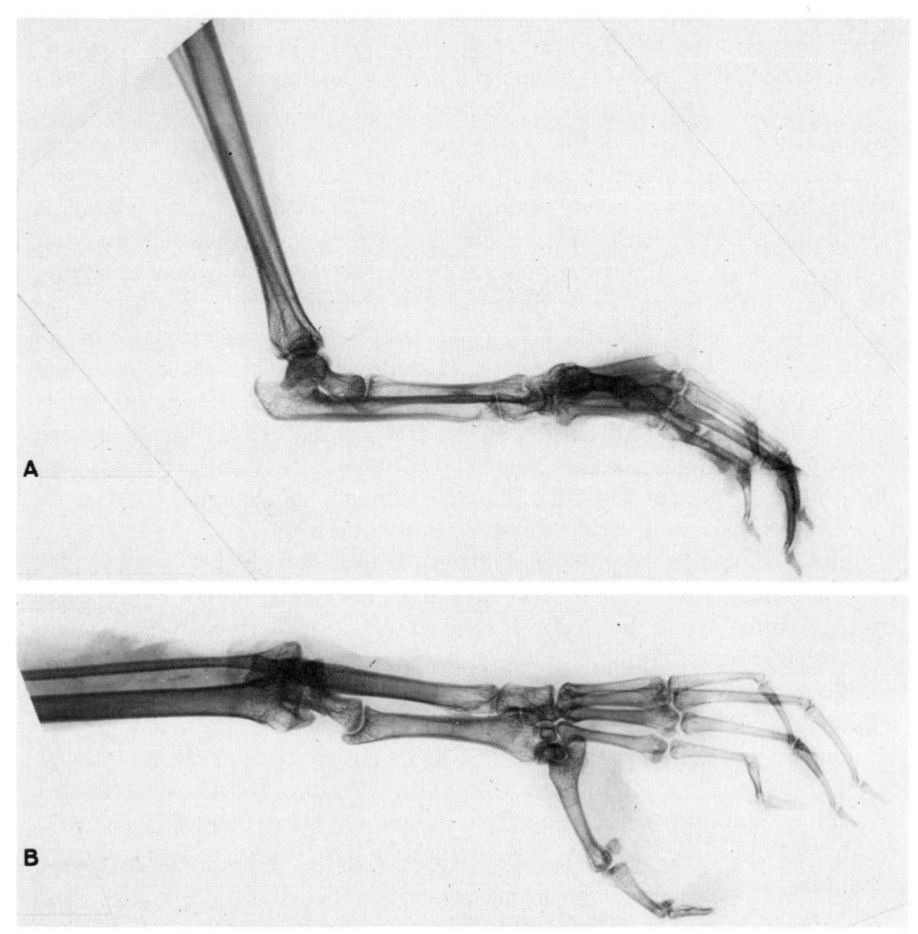

Fig. 1. Radiographs of the left foot of *Galago alleni* in (A) lateral projection, and (B) dorsoventral projection.

(1967) category of "vertical clingers and leapers," they are quite distinct from "long tarsal" jumpers in the sense that they are less adept at quadrupedal locomotion; doubtless this is partly due to the great elongation of the thigh—in the indri, the thigh is 37% of hind limb length and more than half the trunk length. It should be noted that these animals are larger than those in the "long tarsal" category, and perhaps this fact may partly explain known anatomical differences. Although *Lepilemur* (which is of medium size) is placed within this category, it displays several intermediate characters and behaves in certain respects more like a true lemur.

"Long tarsal" jumpers are represented by African Galaginae (*Galago, Euoticus*), Madagascan Cheirogaleinae (*Cheirogaleous, Microcebus*), and the Indonesian tarsier (*Tarsius*). Galaginae and Cheirogaleinae resemble one another in a number of morphological, ecological, and behavioral traits (Charles-Dominque and Martin, 1970); many of these traits are primitive or ancestral in the sense that they occur among primates in general or at least among lemurs (Lorisiformes, Lemuriformes). Jumping from a bipedal posture is one of these traits (Napier and Walker, 1967). A complete spectrum exists from animals that habitually branch run and only rarely leap in a vertical position (e.g., *Microcebus,* which nevertheless easily leaps considerable distances) to animals such as *Galago alleni* that move by leaps and run only occasionally. The tarsier represents an extreme of the latter group; indeed, it stands somewhat apart by virtue of its anatomical specializations, especially the extreme elongation of its hind limbs (250% of trunk length) and tarsus as well as the enormous development of adhesive skin organs (cushions located on the digits and the ventral surface of the tail). Its characteristic movements, whether in trees or on the ground, rather resemble those of a frog.

Among all species of galagos, *Galago alleni* is the most specialized for jumping locomotion (Charles-Dominique, 1971). It moves with great speed, leaping from the vertical support of a small tree trunk or vine to another vertical support as much as $2\frac{1}{2}$ meters away (in the vertical clinging and leaping style defined by Napier and Walker, 1967). The galago also employs a "standing high jump" when, after having descended to ground level to chase insects or gather fruits, it leaps back at the slightest provocation to a trunk or a vine. We have frequently observed the animal's ability to make vertical jumps of 7 feet (about thirteen to fourteen times body length, tail excluded), and have radiographically filmed the jump in a captive animal.

The prosimian jump already has been analyzed behaviorally (Geoffroy-Saint-Hilaire, 1796; Lowther, 1940; Sanderson, 1940; Roberts, 1951; Petter, 1962; Napier and Walker, 1967; Vincent, 1969; Charles-Dominique, 1971), anatomically (Murie and Mivart, 1872; Freudenberg, 1931; Nayak, 1933; Hill, 1953; Le Gros Clark, 1959; Hall-Craggs, 1965a, 1966; Wrobel, 1967; Uhlmann, 1968; Lessertisseur, 1970), cinematographically (Hall-Craggs, 1964, 1965b), and biodynamically (Treff, 1970). Hall-Craggs (1965b, p. 22) in particular has documented the peculiar vertical type of jump by high-speed photography and studied the position of various limb segments during the preparatory phase of the jump: "for it is the velocity with which the animal leaves the ground that determines the height to which it rises." We have attempted to extend this type of analysis by cineradiography. This technique permits precise determination of the relative positions of bones, i.e., mechanical axes (Jenkins, 1970, 1972).

Cineradiographical Techniques

In cineradiography, the image is produced on a small screen which fluoresces when subjected to X radiation and is retained on movie film. Use of an image intensifier reduces the radiation necessary to produce an acceptable image, and consequently reduces the danger to the subject. The quality and amount of information obtained depends not only on radiological factors (e.g., number and speed of electrons, sensitivity of fluorescent screen) and characteristics of the photographical system (e.g., quality of the optics, frames per second), but, in addition, requires an awareness of both the ecology and behavior of the animal to be studied. It is essential to obtain an animal's cooperation in conforming as closely as possible to natural behavior, or, at least for the captive animal, its habitual behavior. In this study we used a species with which we were familiar and an animal which one of us had personally raised.

A Massiot Philips Medio 20 X-ray generator was employed at 60 kv. The milliamperage was automatically regulated during operation, but never exceeded 15 ma. The maximum length of uninterrupted viewing under these conditions was 60 seconds. To prevent the tube from overheating, a pause of equal time was allowed before continuing with a new film sequence. The operator, protected by a lead apron, followed the field on a mirror that reflected the fluoroscope. The cineradiographical apparatus, mounted on a "C" arm, was suspended by a motorized telescopic column which permitted raising and lowering; this apparatus was freely movable along the ceiling on a double system of perpendicular rails. Thus, although the diameter of the image intensifier was only 15 cm, the mobile suspension system made it possible to center the image rapidly when the animal moved.

For movements as rapid and sudden as the galago's jump, the inertia of the apparatus and the lag time between the observation of the event and the start and actual operation of the motors make centering the apparatus a difficult task. Therefore besides rapid reflexes, sufficient familiarity with the tame animal to predict its direction of movement is required. In addition, the sharp noise of the starting generator always startles the animal, interrupts its movement, and usually provokes an escape reaction. For this reason we were unable to obtain any record of one prosimian (*Cheirogaleus* sp.) which crouched on the ground in fright and attempted to hide itself in the least crevice. In the case of the galago, the escape reaction was used to advantage; we were able to study the vertical or semivertical jump that the frightened animal always employed when attempting to reach security in its nest box.

A 16 mm Arriflex camera was run at 64 frames per second with Kodak Plus X T.V. reversible film. When filming small-sized animals, the periphery of the field is the brightest, and this obscures the image definition upon projection. For this reason we found it necessary to stop down the lens (Schneider-Kreznach Arriflex Cine Xenon 1: 2/35) to a setting of 5.6.

The film was initially projected at 16 frames per second to permit observation of the movements in slow motion; the interesting sequences then were examined using a film editor. Two procedures were used to analyze these sequences frame by frame. (1) Printing all the frames on photographical paper with an enlarger, a procedure which is only practical for short sequences on the order of 40 frames. (2) Single frame projection (Bell and Howell Model 163 projector) onto a screen covered with tracing paper. On this surface the axis of various limb bones may be traced, and successive images may be superimposed as required. Sequences of 160 frames (2.5 seconds duration) are readily analyzed by this technique.

General Observations

The subject of this study, a female *Galago alleni*, was a pet of one of the authors who captured her in primary forest in Gabon (May, 1963; Makokou; Mission biologique au Gabon C.N.R.S. France), brought her back to France, and raised her in her house (1963–1968); she was then taken to the laboratory where she lived until her death (April, 1971).

We noted previously that the use of an image intensifier reduces the amount of radiation to the animal. We mention this specifically because the galago used in these experiments died 4 months after the second of two filming sessions. The total dosage is impossible to calculate exactly, for besides a total of 8 minutes of actual filming time, at least 30 minutes was spent in fluoroscopic (nonfilming) examination. The galago showed no signs of radiation sickness other than a slight radiodermatitis (from which she recovered). Death followed rapidly after sudden paralysis of both hindlegs. Since the animal was old, however, it is difficult to conclude that radiation was responsible for her death. The galago had spent 7 years, 11 months in captivity, and the longevity record for this species in zoos is 5 years, 3 months (Jones, 1962). However, a closely related species of similar size, *Galago senegalensis,* is known to have lived for 10 years, 5 months (Jones, 1962).

The galago lived in a large-meshed, metallic cage measuring 30 × 45 cm, and 100 cm high. A box 18 × 16 × 15 cm, fastened 60 cm. above the ground, served as a site for sleeping, eating, and observing, and also as a refuge when frightened. So as not to upset her normal behavior, she was placed during filming sessions in a cage of similar dimensions, but built with a plywood frame and with netting on the sides; the nest box was replaced with a wooden platform. The "C" arm of the cineradiographical apparatus was positioned to deliver a horizontal X-ray beam.

Because of the restrictions of the experimental technique, we analyzed only the vertical or semivertical (within 30° of vertical) jump which the galago invariably used to reach her perch. This jump was employed particularly by the galago when, having just sought food at ground level, she returned to her perch after either having been frightened by a sudden noise or simply to eat at leisure. For angular measurements, we used sequences in which the galago faced the apparatus (frontal view) and also sequences in which the hindlimbs were positioned in lateral view. Frontal views were most common because by natural curiosity this animal tended to watch the investigator and therefore positioned herself frontally to the apparatus. Angular measurements thus recorded are not usually the actual angles between two bones, but represent a projection of the angles onto the frontal or sagittal plane. Comparisons of angular variations are valid only during the time when the animal moves limb segments in the same particular plane (e.g., sagittal or frontal).

Slow-Motion Analysis

Slow-motion analysis at 16 frames per second yielded the following data. Of the seven jump sequences filmed, six were asymmetrical (all the body weight shifted to a single foot—the active foot—before the jump), and one was symmetrical. In *Galago senegalensis,* in contrast, Hall-Craggs (1965b) observed only symmetrical jumps (the center of gravity being in the sagittal plane), and he thus concluded that "each pair of limbs may be considered as a single structure."

In every asymmetrical jump, the active foot was always the *left* foot. First, the right foot was lifted as the body weight shifted onto the left foot, and then at the time of the jump, the different segments of the same left limb became aligned obliquely laterally. In two cases of asymmetrical jumps, the galago started from a quadrupedal posture and was oriented laterally to the apparatus; first she lifted her forelimbs, and, remaining in bipedal posture, turned her body toward the investigator, and then turned her head around in the direction of the jump before pushing off with the hindlimbs. During the preparatory phase of two other asymmetrical jumps, the active (left) foot moved away from the other by a short step sideways. At the same time, the center of gravity shifted to the left as the right foot elevated.

In the case of the symmetrical jump, three successive features are distinguishable in slow motion: (1) straightening of the head and incipient realignment of the trunk in a quadrupedal position; (2) lifting of the forelimbs with accompanying extension of the trunk, lowering of the pelvis and knee flexion; (3) pushing off by both hindlimbs.

Single Frame Analysis

Frame by frame study of sequences and drawings based on individual frames allowed us to make the following chronographical observations.

The Symmetrical Leap

The analysis was made on a sequence of 24 frames (Fig. 2, F1–F24), representing no more than 0.35 seconds and following the propulsive phase up to the launch. The sequence began after a period of relative immobility in which the galago faced the apparatus and kept both forelimbs on the ground.

F1 to F16 (Fig. 2) (About 0.25 Seconds Duration)

In frame 1 the galago's hands are slightly raised but remain touching the ground until frames 9 (left) and 14 (right) when they are lifted rapidly. A movement of the thigh occurs just after the forelimbs are lifted. The movement is characterized by a decrease in the angle of the femur with respect to the ground (Figs. 3A and 4); simultaneously, the femorotibial angle decreases (Figs. 3B and 4), but the tibia remains essentially at the same angle relative to the ground (Figs. 3C and 4). This movement, which actually is flexion followed by extension of the knee, gives the impression of an elastic rebound (the term "elastic" will be used hereafter to denote the rapid reciprocal action of the antagonist muscles without any other physical or physiological implications).

F17 to F24 (Fig. 2) (Approximately 0.10 Seconds Duration)

This sequence represents the actual propulsion and launching by the hindlimbs. Initially, extension at the knee joints carries both femurs through the horizontal to an almost vertical position. There appears to be a slight time lag between the right and left sides; a more marked lowering of the left knee, accompanied by the lifting of the left tarsus slightly ahead of the right tarsus (about 0.03 seconds, from F20 to F22), is compensated by a greater vertical alignment of the femur. Simultaneously, the lower legs (i.e., tibia and fibula) undergo an "elastic" movement (lasting about 0.05 seconds) at the ankle, initially inclining to a position more parallel with the ground, and straightening thereafter (the left in F20, the right in F21). This movement is at its extreme in F20. Both feet [Figs. 2 (F23) and 4] leave the ground together.

The principal features of the symmetrical jump are summarized below.

1. The symmetrical jump involves voluntary, flexion–extension movements at the knees and ankles.

2. The launch, strictly speaking, is relatively rapid; whereas the preparatory knee movements begin about 0.25 seconds before the launch, the actual launching action takes no more than 0.10 seconds.

3. The orientation of the distal bony segments, which are the only ones visible at the time that the feet leave the ground, is almost vertical (approximately 80°).

4. There is almost perfect synchrony between the right and left hindlimbs. This synchronism is related to the galago's ability to jump with both feet bound together (Lowther, 1940) although, as will be evident below, other modes of jumping are usual.

The Asymmetrical Leap

Asymmetrical jumps are defined as jumps from a single foot (the "active foot") which in all recorded cases was the left foot. Elevation of the right foot (the "free foot") places all the body weight on the active foot.

Analysis of asymmetrical jumps is more complex than for symmetrical jumps. First, the movements of the different segments of both limbs, not being synchronized or symmetrically arranged, tend to make image anal-

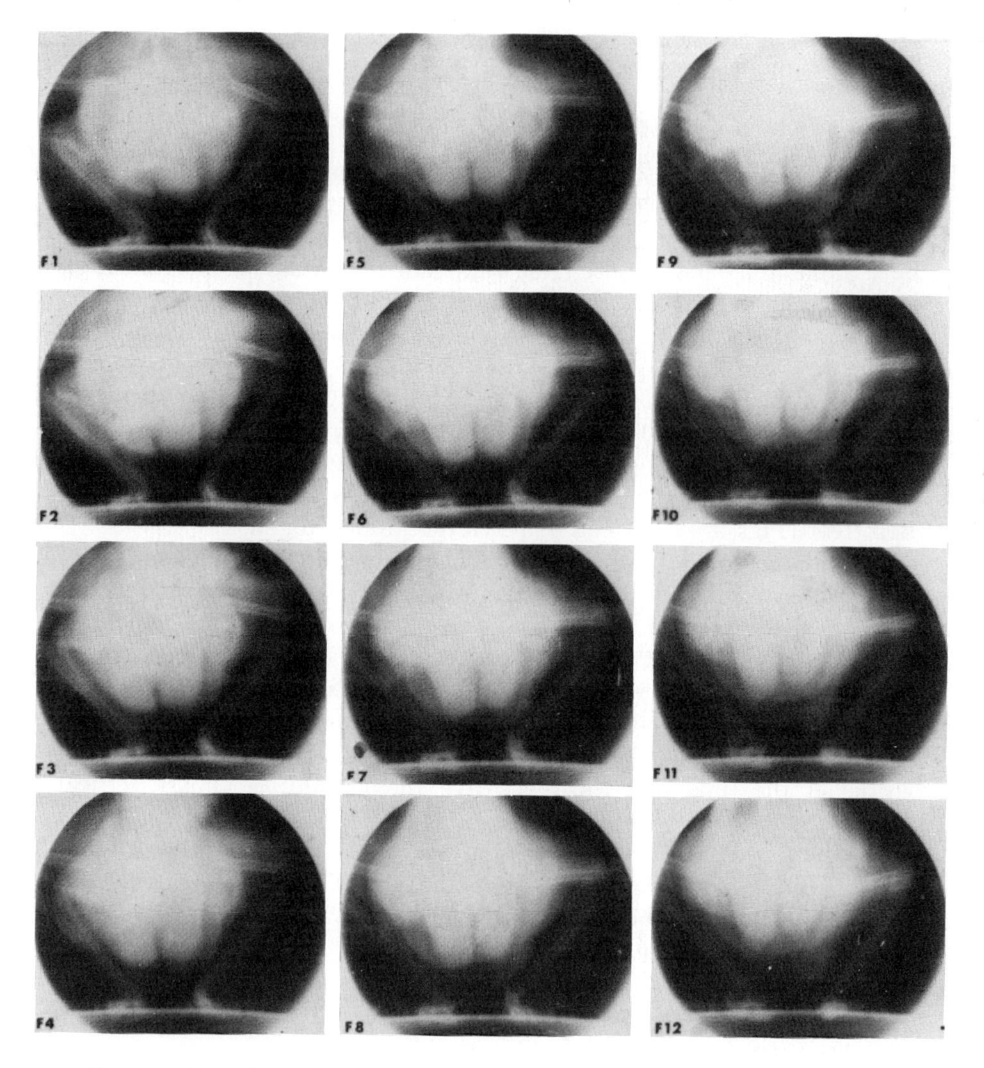

Fig. 2. Cineradiographical sequence of the symmetrical jump in *Galago alleni.* F numbers are frame numbers.

ysis difficult. Second, the direction of movement is not vertical but about 30° from vertical in the frontal plane. The different segments of the active limb also become oriented in this direction.

Various phases of the asymmetrical jump were studied from seven sequences. The following analysis is based on a 36-frame sequence which

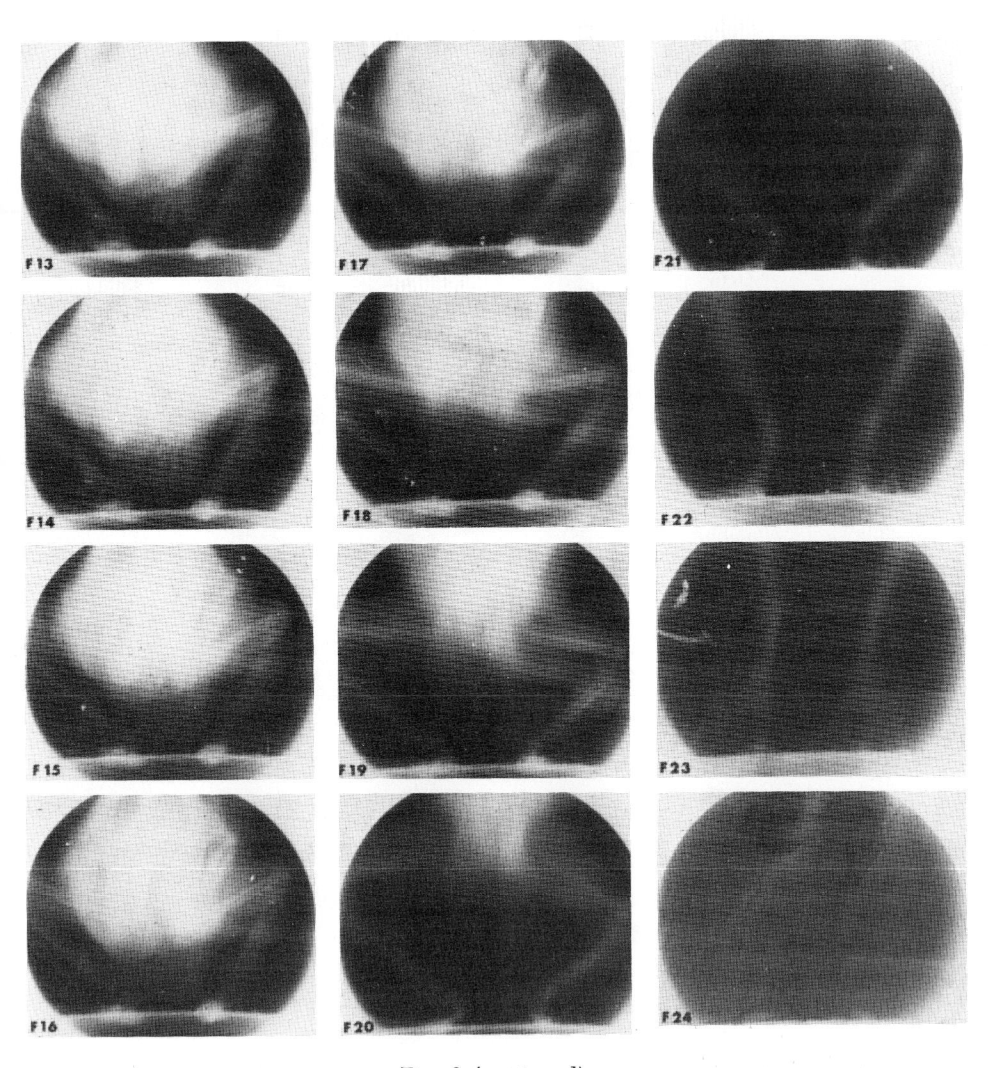

Fig. 2 (continued)

is illustrated in Fig. 5 (the first 12 frames are omitted because the degree of movement is relatively small).

F1 to F15 (Fig. 5)

As in the symmetrical jump, a preparatory flexion-extension ("elastic") movement (not illustrated) occurs before pushing off on the active foot.

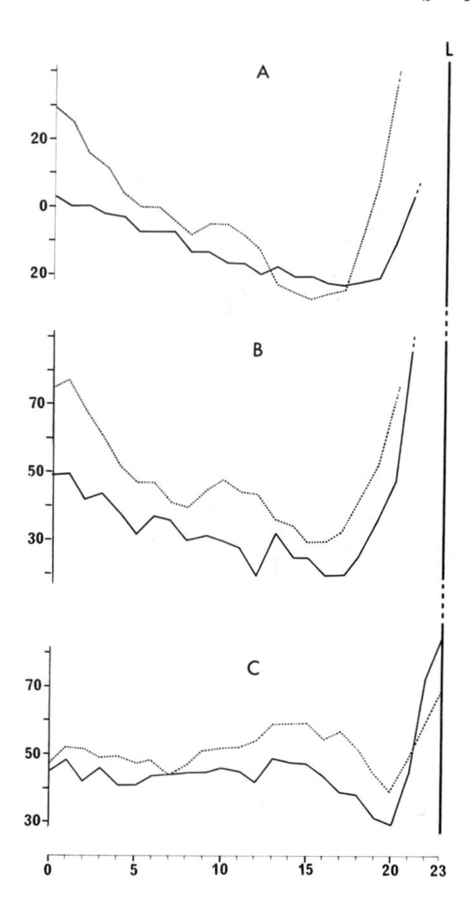

Fig. 3. Angular variations of the femur and lower leg during a symmetrical jump, interpreted by cineradiography. A, Angle of the femur relative to horizontal. B, Femorotibial angle. C, Angle of the tibia relative to horizontal. The right limb is represented by an unbroken line, the left by a dotted line. The abscissa is the frame sequence; the ordinate is angular value in degrees. L is the instant of the launch.

At the start of the sequence, the center of gravity is shifted; while the left tarsus is kept flat on the ground, the right tarsus is raised and dorsiflexed at the tarsometatarsal joint. This movement takes place in less than 0.20 seconds. During this entire time, the pelvis remains stationary and at a minimal height above the ground; the weight-bearing limb is immobile. The right tarsus oscillates relative to the ground (Fig. 7B) as the tibiotarsal joint flexes and extends (Fig. 7C). The inclination of the right tibia and fibula relative to the ground remains relatively constant (Figs. 7A and 9). Oscillation of the right tarsus, with an amplitude on

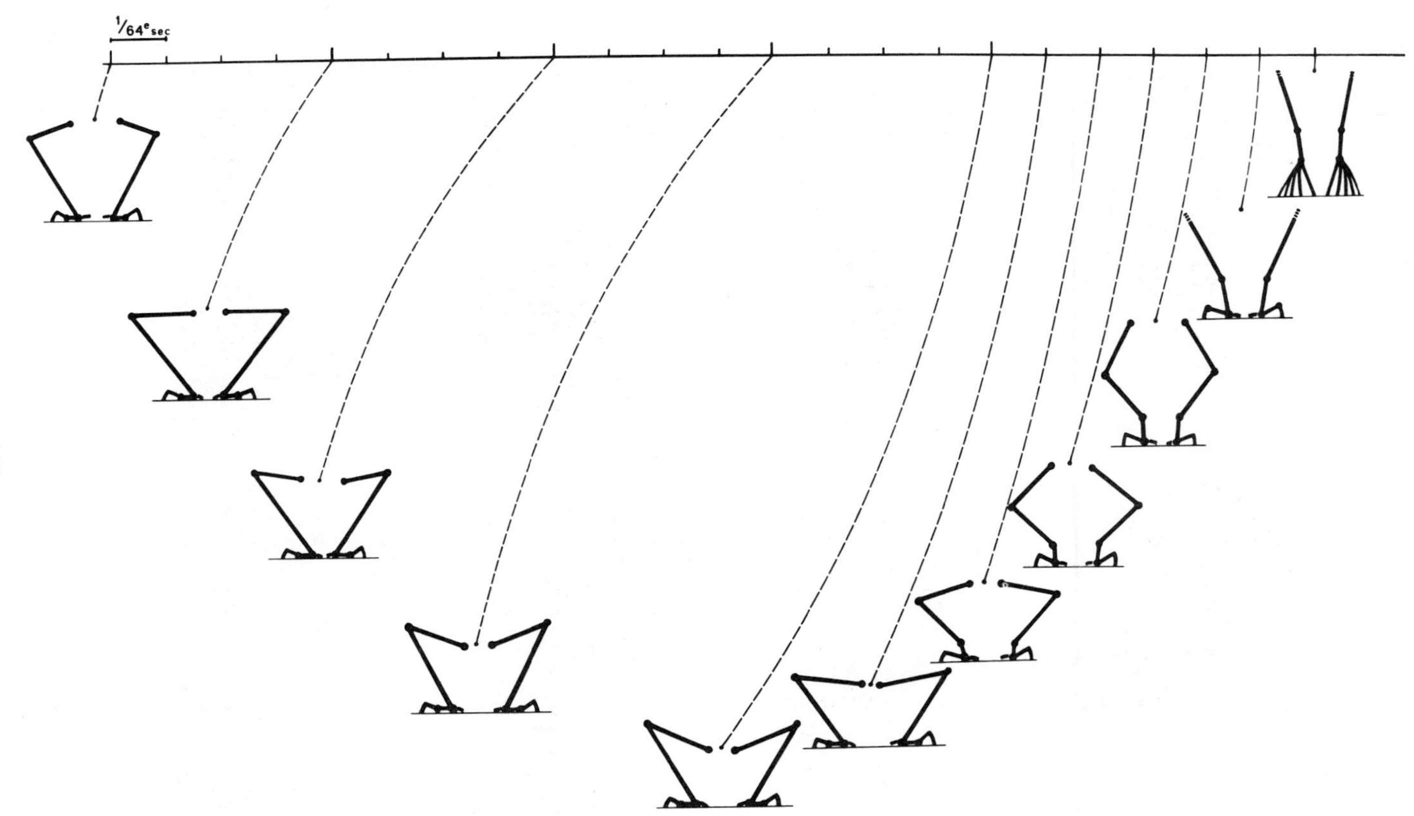

Fig. 4. Schematic representation of the principal phases of the symmetrical jump (based on cineradiography). The scale represents 1/64 second intervals.

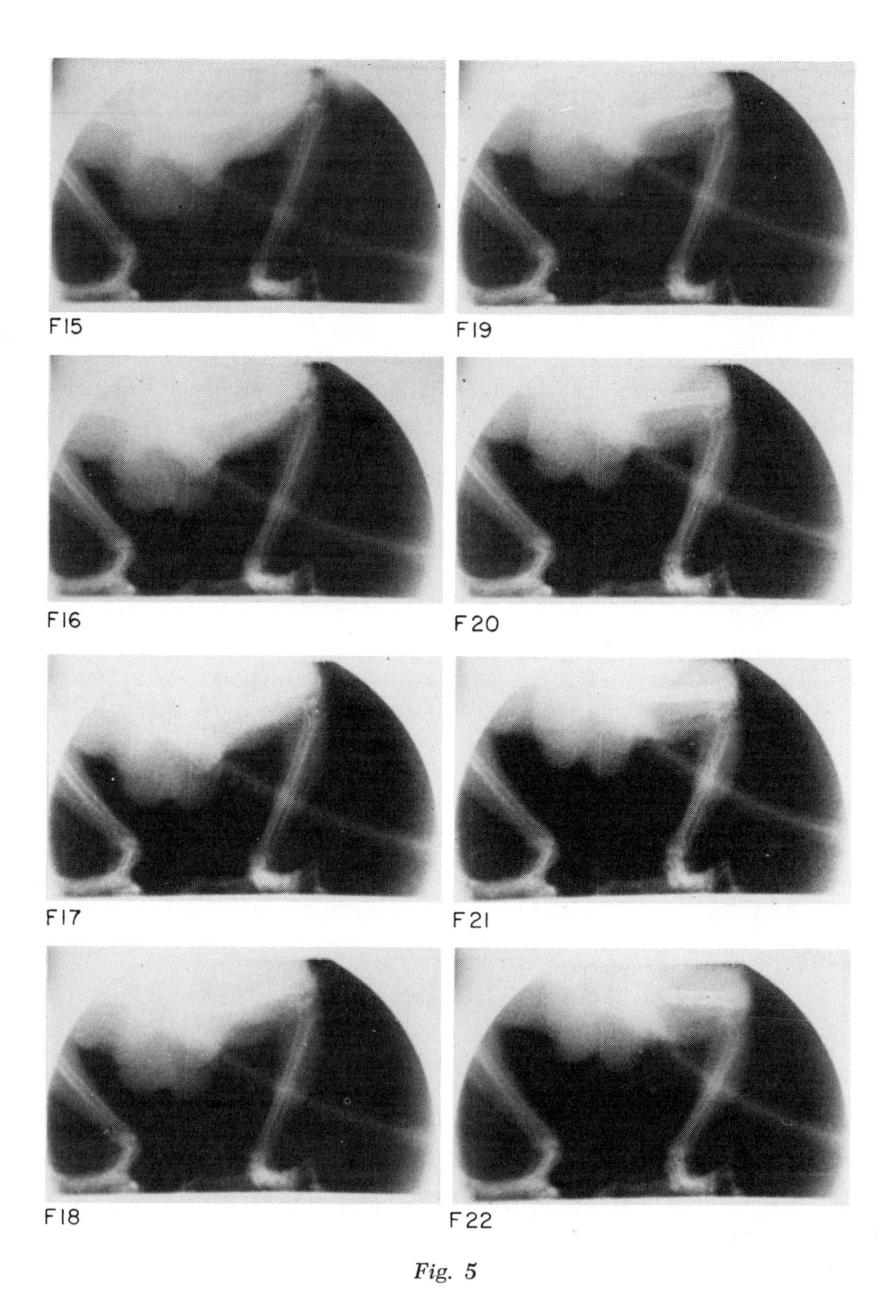

Fig. 5

Fig. 5. Cineradiographical sequence of the asymmetrical jump in *Galago alleni*. F numbers are frame numbers.

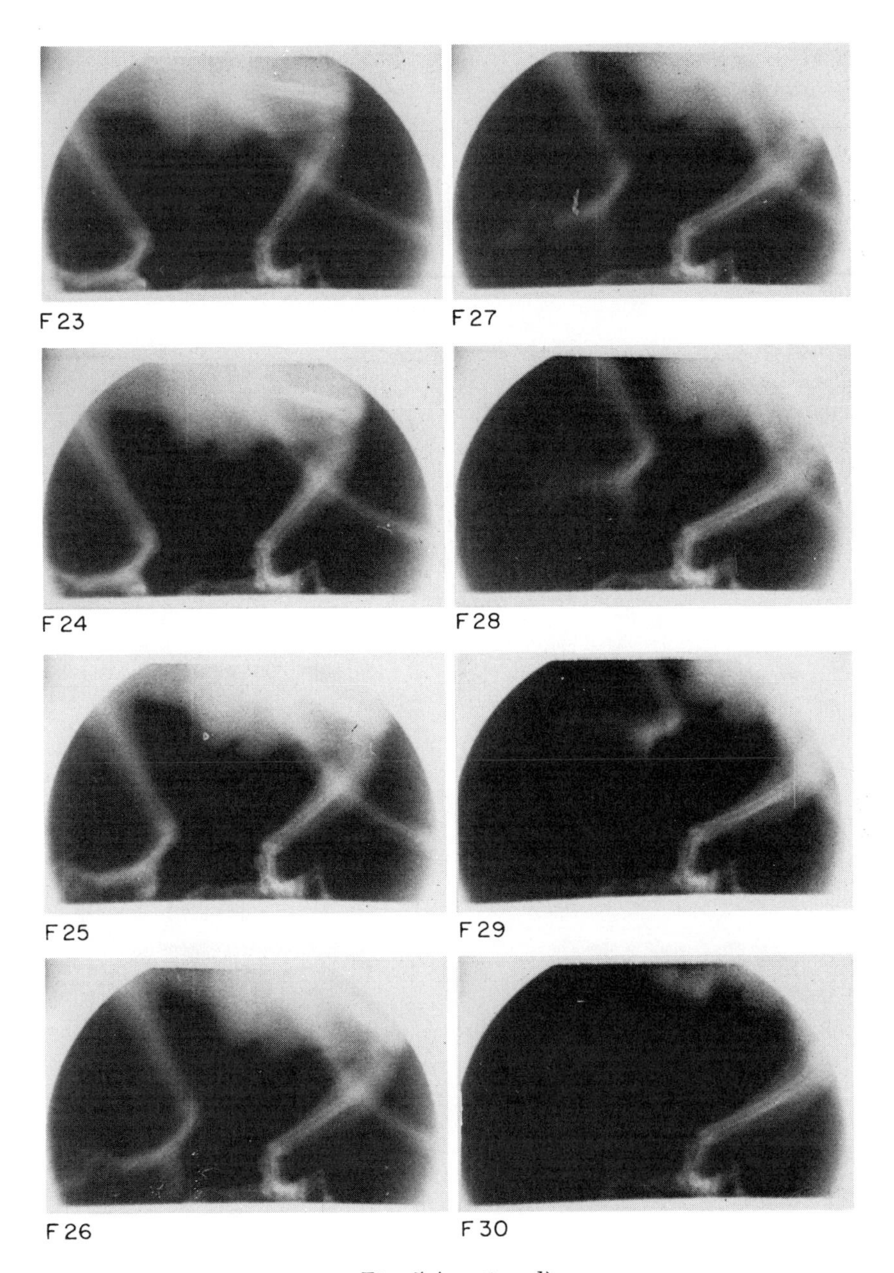

Fig. 5 (continued)

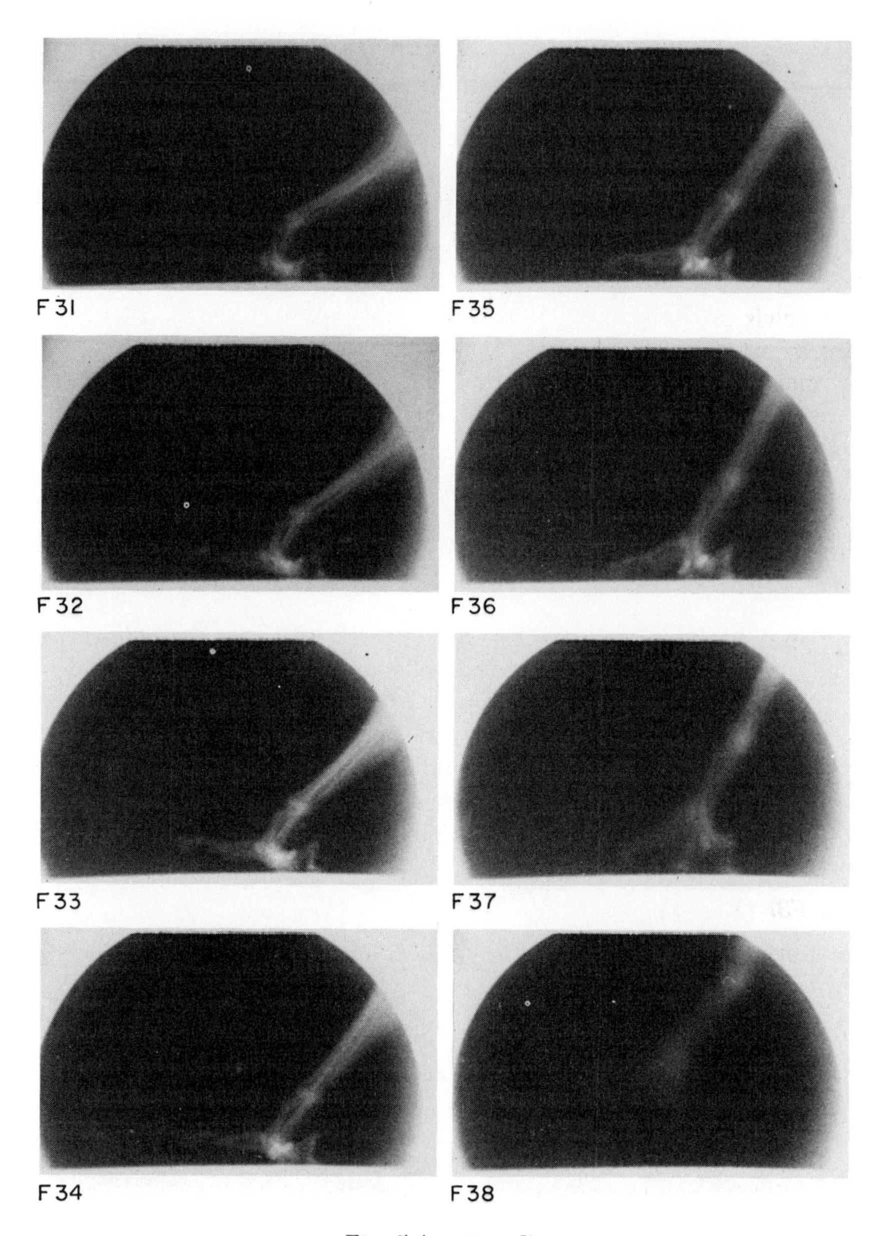

Fig. 5 (continued)

the order of 15 to 20° and a period of about 0.04 to 0.08 seconds, continues until the foot is lifted (Fig. 7).

The left hindlimb—the weight-bearing limb—begins the launch in F15 (Figs. 5 and 8). Launching takes place with three movements: the femur extends and thus becomes oriented vertically, the tibia undergoes the flexion-extension ("elastic") rebounding movement described previously and the tarsus dorsiflexes.

F15 to F19 (Fig. 5)

The pelvis is raised and is tilted to the left (Fig. 5). The left femur begins to extend, and the thigh becomes more vertical (Fig. 6A). The tibia remains almost immobile (Figs. 6C and 9). The tarsus, however, also begins to assume a vertical orientation; having been flat on the ground, the tarsus is elevated by dorsiflexion at the tarsometatarsal articulations. However, as the tarsus is almost perpendicular to the frontal plane, the actual angle of tarsus relative to the ground is impossible to determine.

F19 to F24 (Fig. 5)

The right tibia assumes a slightly more vertical orientation (Figs. 7A and 9). Just before the foot is lifted, tarsal oscillations become more rapid (less than 0.02 seconds). After the foot is off the ground, it dorsiflexes to a position almost at a right angle to the leg (Figs. 5 and 8); this movement is very different from that of the active foot in which the tarsus and tibia remain aligned after the launch. Throughout the jump, the right foot is nonweight bearing; it is lifted almost passively with the limb and exerts no propulsive force on the ground.

F19 to F37 (Fig. 5)

Movement is primarily restricted to the knee joint. The femur, which disappears out of the field in F27, continues to become more vertically oriented.

This part of the jump may be broken down into three phases.

F19 to F27 (about 0.12 seconds)

The active tibia is lowered about 40° from its previous position (Figs. 6C and 9).

F27 to F31

A pause of 0.06 seconds occurs in this extreme position. At the same time, the tarsus, which previously was elevated to a vertical position by tarsometatarsal dorsiflexion, remains perpendicular to the frontal plane

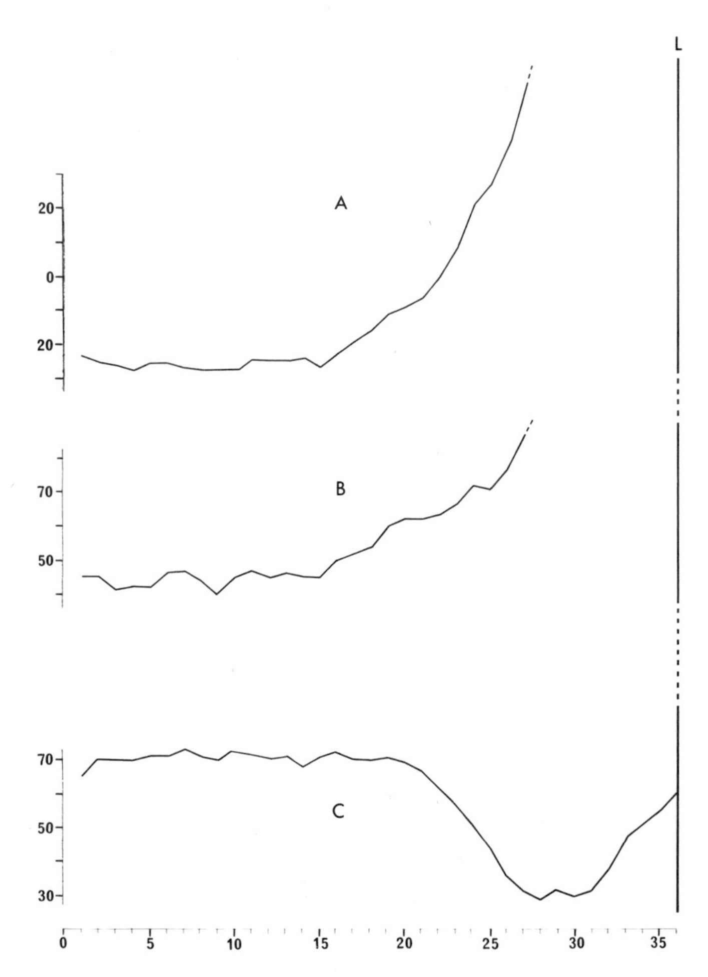

Fig. 6. Angular variations in the femur and lower leg of the active left limb during an asymmetrical jump, interpreted from cineradiography. A, Angle of the femur relative to horizontal. B, Femorotibial angle. C, Angle of the tibia relative to horizontal. Other abbreviations as in Fig. 3.

and begins to become oriented laterally; thus the tarsus becomes aligned with the other segments in the direction of the jump which is approximately 30° from vertical (Figs. 5 and 8).

F31 to F37

The tibia moves rapidly through a 30° angle so that it comes to the launching position (Figs. 6C and 9). Finally, by a slight movement the tip of the hallux is lifted before resuming contact with the ground just as the foot leaves the ground.

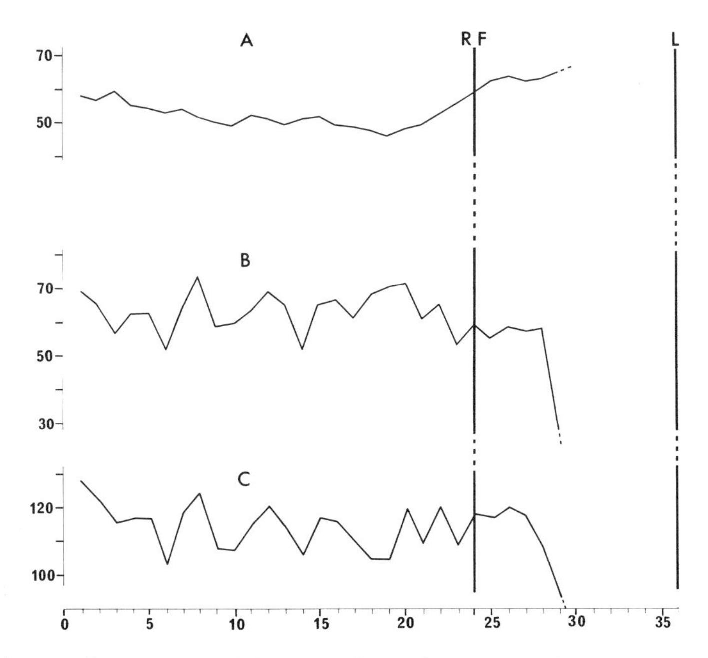

Fig. 7. Angular variations of the right tibia and calcaneum during an asymmetrical jump. A, Angle of the tibia relative to horizontal. B, Angle of the calcaneum relative to horizontal. C, Calcaneotibial angle. RF, point at which right foot is lifted. L, point of launch. Abscissa and ordinate as in Fig. 3.

The principal features of the asymmetrical jump are summarized below.

1. The alignment of the distal segments of the active limb is approximately 30° from vertical in or near the frontal plane.
2. Propulsion by the hind limbs begins with straightening of the femurs and elevation of the trunk.
3. Before the free foot is lifted, its tarsus is actively engaged in an oscillating movement.
4. In the active limb, flexion-extension propulsive movements occur principally at the knee and to a lesser degree at the ankle (tibiotarsal) joint; however, the bones providing support on the ground, the distal extremity of the tarsus and terminal phalanges, and especially those of the hallux, participate in the final phase of propulsion.

Discussion

The jumps of the galago are too variable to permit definitive conclusions on the basis of this limited cineradiographical study. Nevertheless, besides confirming Hall-Cragg's observations on the lack of forelimb par-

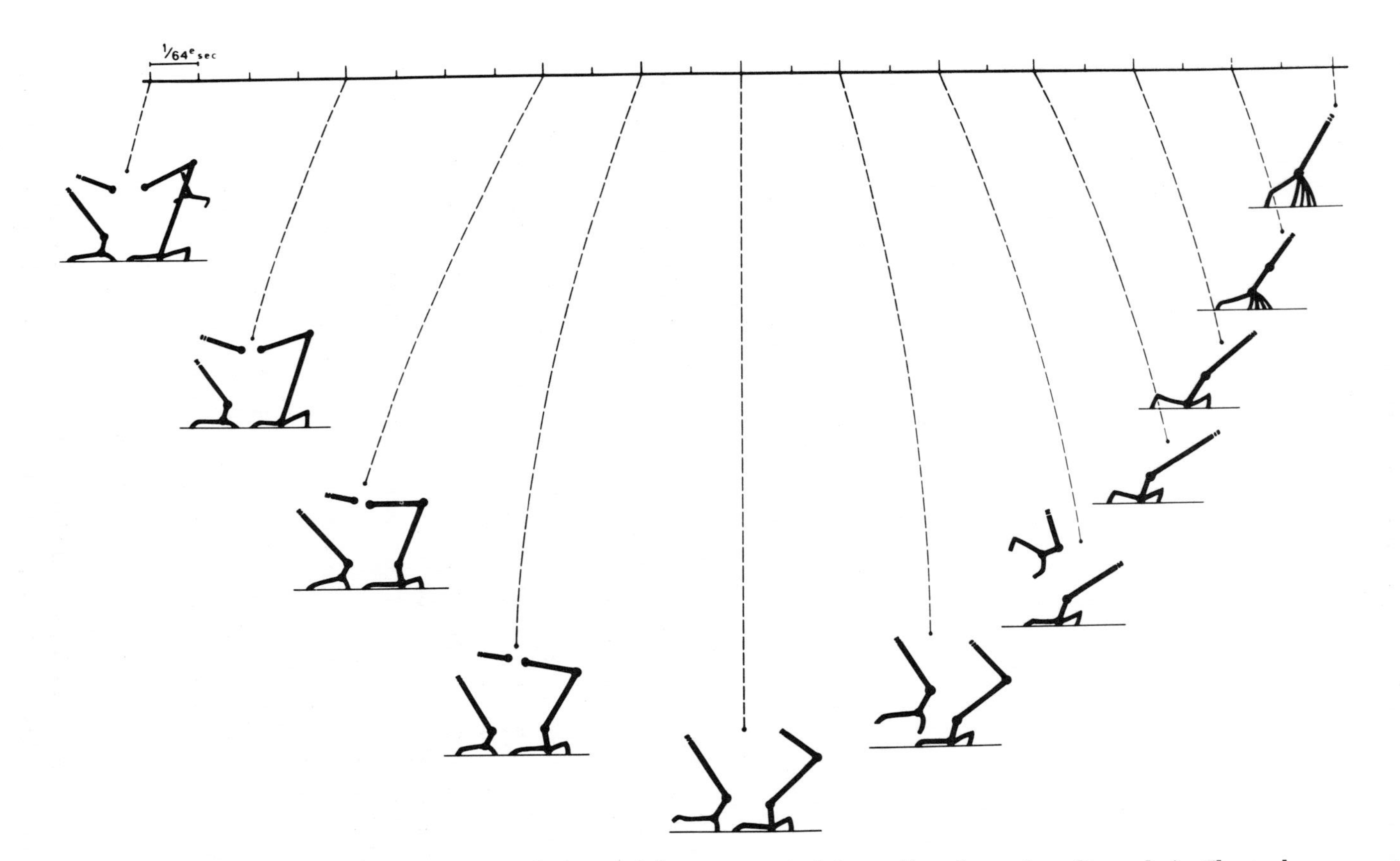

Fig. 8. Schematic representation of the principal phases of the asymmetrical jump (based on cineradiography). The scale represents 1/64 second intervals.

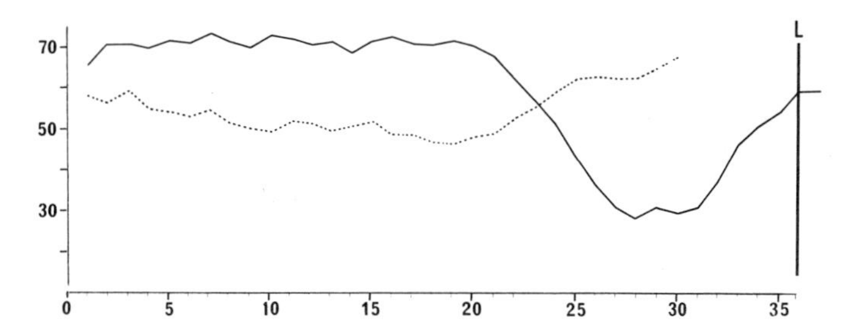

Fig. 9. Comparison between the angle of the tibia of the active limb (solid line) and the free limb (dotted line) relative to horizontal during an asymmetrical jump. Abbreviations as in Fig. 3.

ticipation in launch, and on the location of the contact area for the take-off in the region of the tarsometatarsal joints, we have documented in addition that there are at least two types of vertical jumps from the ground and that these are clearly distinct. The symmetrical jump does not represent a restricted asymmetrical jump in which the launching movements of both feet happen to coincide (Fig. 10). In both types of jumps, extension of the hindlimb joints is preceded by a slight flexion. This movement, which we have referred to as "elastic," is the result of antagonistic muscle action. In the asymmetrical jump, the body weight is placed completely on the active foot and the other foot is completely free of weight throughout. A flexion–extension propulsive action occurs only at the ankle. In the symmetrical jump, both limbs function simultaneously at every level; the push-off by both feet is exactly simultaneous and the alignment of the distal segments shows that the forces are evenly distributed between them. In this jump, a flexion-extension propulsive action occurs not only at the ankle, but at the knee as well. In addition, the two types of jump have different trajectories: in the sagittal plane in the case of symmetrical jump, and more or less oblique to the sagittal plane in the asymmetrical jump.

As to the question of the efficiency of the jump, only the preparatory phase was studied; the relative performances were not studied. In any case, efficiency is measured above all by the height that the animal can achieve, and the experimental conditions did not allow us to judge the maximum height. However, a comparison of the speed of different phases of the spring in the two types of jump yields some information.

Since the total duration of the spring and launch were of the same order in the two cases studied (about 0.33 seconds), a comparison of the speed of different phases of the launch will be made:

1. In the case of the symmetrical jump, the launch is very rapid; it

does not last more than 8 frames (about 0.12 seconds), in contrast to 21 frames (about 0.33 seconds) for the asymmetrical jump.

2. If only the distal elements—lower leg and foot—are considered, the movements are much more rapid, and therefore more closely grouped and associated with the launch in the symmetrical jump (Fig. 10). In

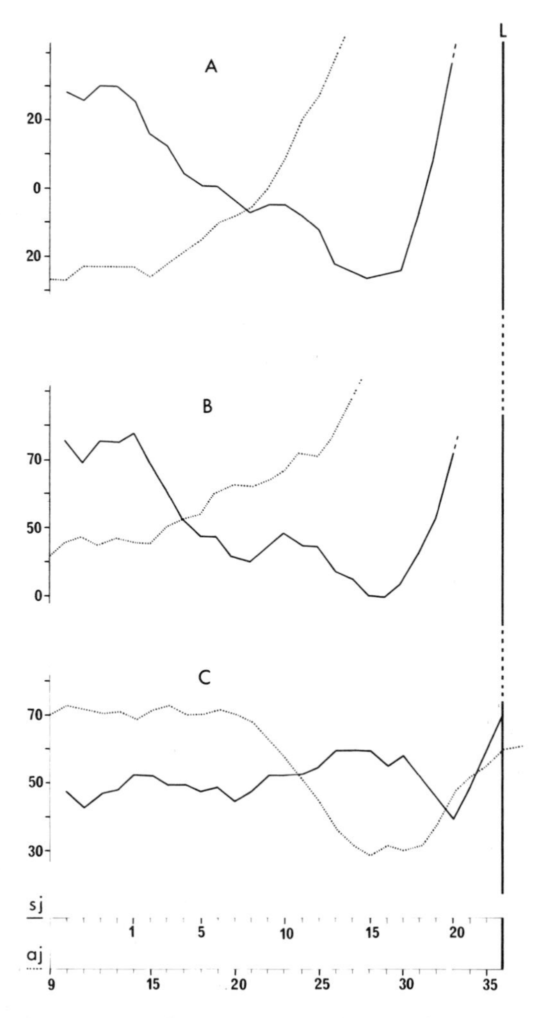

Fig. 10. Angular variations between A, the femur relative to horizontal; B, the femur and tibia; C, the tibia relative to horizontal in symmetrical (solid line) and asymmetrical (dashed line) jumps. In the case of the asymmetrical jump, these data represent only the active limb. The scales at the bottom represent the frame numbers of the symmetrical (sj) and asymmetrical (aj) jumps.

this case, in effect, both tarsi are lifted and both tibiae straightened in a single sudden movement only about 0.04 or 0.05 seconds before the launch. Until that time, the feet remain on the ground in a plantigrade position. In the asymmetrical jump, on the contrary, the movements of the tibia and tarsus of the active leg are sequential rather than simultaneous. First, the tarsus assumes a semidigitigrade position approximately 0.25 seconds before the launch; from this position the tarsus then moves out of the parasagittal plane to the frontal plane in which it becomes reoriented in the direction of movement at about 30° from the vertical.

The distinction between the symmetrical and asymmetrical types of jumps should be interpreted as an adaptation to diverse environmental conditions. The highest jumps of which Allen's galago is capable (on the order of 7 feet) are symmetrical. In reaching lesser heights, the galago usually uses an asymmetrical jump.

We have already noted that the anatomical specializations which characterize this adaptation do not entail any limitation in other locomotor abilities. Allen's galago, even more than other galagos, moves almost exclusively by jumping from vertical support to vertical support, even in the same tree (Charles-Dominique, 1971). Statistically, Allen's galago only rarely runs along branches although it is perfectly capable of doing so. On the ground, where it goes frequently to feed, the galago moves quadrupedally or bipedally in small hops; finally it returns to the trees by means of those high jumps which we have analyzed cineradiographically.

The symmetrical jump is not used exclusively on the ground, for it appears probable that leaping from vertical to vertical support also requires simultaneous extension of the hind limbs. Whether symmetrical or not, these jumps are properly classified as "vertical leaping" in the sense of Napier and Walker (1967), since the animal's body is vertically positioned at the start of the jump. The direction of movement at the start invariably is approximately perpendicular to the support surface; from the ground, movement is vertical, and from a vertical trunk, movement is horizontal. When jumping from the ground, the animal always assumes an erect posture on his hind limbs prior to jumping.

The symmetrical jump from ground level requires by definition a landing surface above the animal. The asymmetrical jump, on the contrary, permits greater flexibility in the choice of the landing surface. The fact that we observed only the left limb used in the asymmetrical jump perhaps is an artifact of the experimental situation in which the galago and its perch always occupied the same relative position. Further cinematographical observations perhaps would have enabled us to confirm whether the same animal would be able to jump equally well with either foot under these circumstances.

In contrast to the symmetrical jump, the asymmetrical jump (at least the type which we observed) is properly speaking a leap from the ground; indeed, the way in which the center of gravity is shifted at the beginning of the jump, thus placing all the weight on one foot, would appear to be unsuitable for support on vertical surfaces.

More observations and analysis will be required to achieve a satisfactory understanding and classification of the locomotory modes and adaptations among prosimians. The locomotion of galago is precise and well adapted as much from the point of view of anatomical features as behavioral variation which allows them to adapt to ecological variations of the equatorial forest. For this reason, continued study of these animals' movements is important not only because it will contribute to our understanding of this particular group, but also because it will clarify our general ideas on the evolution of primate locomotor behavior.

Acknowledgment

We wish to thank Professor J. Anthony, Director of the Comparative Anatomy Laboratory of the National Museum of Natural History (Paris), for putting at our disposal the cineradiographical equipment.

References

Brehm, A. E. (1868). "La Vie des Animaux Illustrée, Les Mammifères," Vol. 1, pp. 131 and 140. Baillière et fils, Paris (French trans. by Z. Gerbe.).

Charles-Dominique, P. (1971). Eco-éthologie des Prosimiens du Gabon. *Biol. Gabon.* **7**, 121–228.

Charles-Dominique, P., and Martin, R. D. (1970). Evolution of Lorises and Lemurs. *Nature (London)* **227**, 257–260.

Clark, W. E. Le Gros (1959). "The Antecedents of Man." University Press, Edinburgh.

Freudenberg, W. (1931). Bemerkungen zum Fussgerüst der Gattung Galago. Z. *Säugetier-Kunde* **6**, 233–235.

Geoffroy-Saint-Hilaire, E. (1796). Mémoire sur les rapports des makis Lemur L. et description d'une espèce nouvelle de mammifère. *Magasin Encyclop. J. Sci. Lett. Arts* **1**, 20–50.

Gray, J. (1953). "How Animals Move." University Press, Cambridge.

Hall-Craggs, E. C. B. (1964). The jump of the Bush-baby, a photographic analysis. *Med. Biol. Ill.* **14**, 170–174.

Hall-Craggs, E. C. B. (1965a). An osteometric study of the hind limb of the Galagidae. *J. Anat.* **99**, 119–126.

Hall-Craggs, E. C. B. (1965b). An analysis of the jump of the lesser Galago (Galago senegalensis). *J. Zool.* **147**, 20–29.

Hall-Craggs, E. C. B. (1966). Rotational movements in the foot of *Galago senegalensis*. *Anat. Rec.* **154**, 287–293.

Hill, W. C. O. (1953). "Primates. Comparative Anatomy and Taxonomy," Vol. 1; Strepsirrhini. University Press, Edinburgh.

Jenkins, F. A., Jr. (1970). Limb movement in a monotreme *Tachyglossus aculeatus:* a cineradiographic analysis. *Science* **168**, 1473–1475.

Jenkins, F. A., Jr., (1972). Chimpanzee bipedalism: cineradiographic analysis and implications for the evolution and gait. *Science* **178**, 877–879.

Jones, M. L. (1962). Mammals in captivity—Primate longevity. *Lab. Primate Newsletter* **1**, 3–13.

Jouffroy, F. K. (1956). Le membre antérieur d'Archaeolemur. *Mém. Inst. Sci. Madagascar,* A,**11**, 189–198.

Jouffroy, F. K. (1960). Caractères adaptatifs dans les proportions des membres chez les Lémurs fossiles. *C. R. Acad. Sci.* (*Paris*) **251**, 2756–2757.

Jouffroy, F. K. (1963). Contribution à la connaissance du genre Archaeolemur, Filhol 1895. *Ann. Paleontol.* **49**, 129–155.

Jouffroy, F. K. (1971). Musculature des membres. *In* Grassé, P.-P., *Traite Zool.* **16**, Part 3, 1–475.

Lessertisseur, J. (1970). Les proportions du membre postérieur de l'Homme comparées à celles des autres Primates. Leur signification dans l'adaptation à la bipédie érigée. *Bull. Mem. Soc. Anthropol.* **6**, 227–241.

Lessertisseur, J., and Jouffroy, F. K. (1973). Ostéométrie comparée du pied de L'Homme et des Singes bipèdes occasionnels. *Commun. Int. Cong. Anthro. Ethno. Sci. 9th, 1973,* 16 pp.

Lessertisseur, J., and Jouffroy, F. K. (1974). Tendances locomotrices traduites par les proportions du pied. L'adaptation à la bipédie. *Primatologia* (in press).

Lowther, F. (1940). A study of the activities of a pair of Galago senegalensis moholi in captivity. *Zoologica* **25**, 433–462.

Murie, J., and Mivart, St. G. (1872). On the anatomy of the Lemuroidea. *Trans. Zool. Soc. London* **7**, 1–114.

Napier, J. R., and Napier, P. H. (1967). "A Handbook of Living Primates. Morphology, Ecology, and Behaviour of Nonhuman Primates," 456 pp. Academic Press, London and New York.

Napier, J. R., and Walker, A. C. (1967). Vertical clinging and leaping, a newly recognized category of locomotor behaviour among primates. *Folia Primatol.* **6**, 204–219.

Nayak, U. V. (1933). A comparative study of the Lorisinae and Galaginae. Ph. D. Thesis, London University, London.

Petter, J. J. (1962). Recherche sur l'écologie et l'éthologie des Lémuriens malgaches. *Mem. Mus. Natur. Hist. Natur.* **27**, 1–14.

Roberts, A. (1951). "Mammals of South Africa." Johannesburgh.

Sanderson, I. T. (1940). The mammals of the north Cameroons forest area. *Trans. Zool. Soc. London* **24**, 623–725.

Treff, H. A. (1970). Der Absprungwinkel beim schrägen Sprung des Galago (*Galago senegalensis*). *Z. Vergl. Physiol.* **67**, 120–132.

Uhlmann, K. (1968). Hüft-und Oberschenkelmuskulatur. Systematische und vergleichende Anatomie. *Primatologia* **4**, No. 10, 1–442.

Vincent, F. (1969). Contribution à l'étude des Prosimiens africains; le Galago de Demidoff. Reproduction (biologie, anatomie, physiologie) et comportement *Copedith (Paris)* **1** and **2**, 1–460.
Walker, A. (1967). Locomotor adaptation in recent and fossil Madagascar Lemurs. Ph. D. Thesis, 535 pp. Univ. London, London.
Wrobel, K. H. (1967). Zur Fussmuskulatur bipedspringender Halbaffen. *Morphol. Jahrb.* **110**, 579–602.

5

The Wrist Articulations of the Anthropoidea

O. J. LEWIS

Introduction

The key role which locomotor experimentation must have played in the adaptive radiation of the Primates has been repeatedly emphasized. Few, for instance, would quarrel with the idea that a trend toward the perfection of forelimb suspension was of major significance in the evolution of the Pongidae; some would go further and stress the importance of such an apprenticeship in the emergence of the Hominidae. However, a dearth of reliable osteological criteria clearly identified with the various locomotor modes has hindered satisfactory interpretation of this aspect of the fossil record. For instance, judgment in favor of brachiation has usually been given solely on the basis of a raised intermembral index. On this basis alone it has been suggested that brachiation has evolved in parallel three to six times (Simons, 1962).

Recent work (Lewis, 1965, 1969; Lewis *et al.*, 1970), however, has demonstrated that the Hominoidea share a hitherto unsuspected major reconstruction of the wrist articulations which clearly distinguishes them from the monkeys. Gregory (1916) perceptively realized the necessity for an extreme range of supination during brachiating locomotion and a behavioral study by Avis (1962) has confirmed this. The remodeling of the hominoid wrist fulfils this requirement and is thus clearly correlated with the change from a quadrupedal posture to one of forelimb suspension. Surprisingly these changes had been overlooked yet they are considerable in scale, rivaling the most extreme examples of joint evolution recorded in other mammalian orders. A new diarthrosis is fashioned (inferior radio-ulnar joint); an evolutionary novelty, a large intra-articular meniscus, appears which excludes the pisiform from the wrist joint, relegating it to a new function; the meniscus progressively becomes an integral part of the proximal articular surface thereby excluding the ulna from what has become simply a radiocarpal joint; a new bony carpal element occurs constantly in gibbons (os Daubentonii) and sporadically in man. These changes fashion a joint quite unlike that of monkeys, and the progressive modifications are clearly mirrored in the dry bones, recent or fossil. Wrist specialization, then, has undoubtedly played a key role in the locomotor evolution of the Hominoidea.

Cercopithecoidea and Ceboidea

Wrist Morphology

Despite the widely differing locomotor habits of monkeys (Tuttle, 1969a) their wrist architecture is surprisingly uniform, preserving the basic mammalian quadrupedal arrangement characterized by direct articulation of the ulna with the triquetral and pisiform (Lewis, 1965). Even the so-called semibrachiators, both the New World (Fig. 1) and the Old World varieties, retain this structure and no hint is seen of any trend

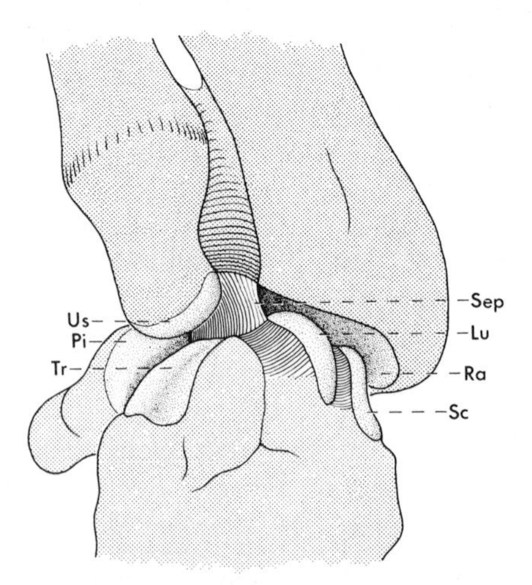

Fig. 1. The right wrist joint of *Ateles geoffroyi* with the dorsal capsule removed. The stippled articular surfaces of the bones participating in the joint are shown: Lu, lunate; Pi, pisiform; Ra, radius; Sc, scaphoid; Tr, triquetral; Us, ulnar styloid process. The synovial septum (Sep) partitioning the joint is also indicated. [From Lewis (1971b) *Folia Primatol.* **16**, 248–256.

toward the hominoid type of specialization (Lewis, 1971b). Only baboons show any departure from the basic pattern and this is trivial when compared with the hominoid modifications, and quite different in nature.

The inferior radio-ulnar joint in monkeys may retain the primitive mammalian character (as seen, for example, in the marsupial *Trichosurus vulpecula*) and be a syndesmosis—a joint consisting of a firm ligamentous bond without elaboration of a synovial cavity. There is, however, an obvious trend toward the refashioning of this joint into a diarthrosis, but this is apparently always imperfect and falls short of the fully established classical type of synovial joint constantly found at this situation in the Hominoidea. A simple syndesmosis has been observed in *Cebus nigrivittatus* and *Procolobus verus*. An incipient synovial cavity—a mere bursa deep within the ligamentous union between radius and ulna—has been recorded in *Ateles geoffroyi, Lagothrix lagotricha,* and *Colobus polykomos.* A considerable synovial cavity intervening between a cartilage-clothed ulnar head and a less perfectly elaborated articular surface on the radius has been encountered in *Cercopithecus nictitans.* In general, the Cercopithecoidea appear more likely to possess a fairly well-formed diarthrosis.

The distal part of the ligament uniting radius and ulna forms an intermediate segment of the proximal articular surface of the wrist joint (Lig, Fig. 9) and when a well-developed inferior radio-ulnar synovial cavity is present the ligamentous sheet is identifiable as the homolog of the hominoid triangular articular disk. Apparently in the primitive condition the wrist joint was subdivided at this situation by an anteroposterior synovial septum. Such partitioning of the wrist joint into separate ulnar and radial compartments has been described in certain marsupials, carnivores, and lemurs (Parsons, 1899; Nayak, 1933) and has been observed by the author in *Lemur fulvus, Trichosurus vulpecula,* and even in a specimen of *Hylobates lar;* it is also a transient developmental feature in the human fetus. In monkeys, a complete septum has been recorded only in *Ateles geoffroyi* (Fig. 1).

In the ulnar part of the wrist joint, the ulnar styloid process articulates with a concave cup formed by the pisiform and triquetral. The facet on the ulnar styloid process is borne upon its convex distal extremity and on the surface facing the interior of the joint. By contrast, the exposed surface, facing laterally in the habitually pronated quadrupedal posture, is nonarticular. The disposition of the styloid process is analogous to that of the lateral malleolus of the hindlimb. In the dorsiflexed and pronated palmigrade attitude, the pisiform projects back into the heel of the hand, acting as a biomechanical counterpart to the heel in the hindlimb, and forming the bony basis underlying the hypothenar skin pad, which is thus related to the lower extremity of the ulna, extending in effect onto the forearm; this is in contrast to the more distal location of the pad in the Hominoidea (Midlo, 1934).

In the radial part of the joint, the convex surfaces of lunate and scaphoid articulate with the radius. There is, however, a posterolateral concave prolongation of the scaphoid articular surface which locks onto the convex dorsal border of the radius in full extension (Fig. 9).

The essential mammalian ligamentous apparatus is retained at the wrist joint in all monkeys; intracapsular palmar ulnocarpal and radiocarpal ligaments converge on the lunate and dorsally there is a thinner radio-triquetral ligament.

Variations from this basic joint pattern are only rarely encountered among either New or Old World monkeys. Reports in the literature (Parsons, 1899; Schwarz, 1938) that in *Ateles* the pisiform is distally displaced and takes no part in the wrist joint are without foundation. Neither has it been found (Lewis, 1971b) that *Lagothrix* has a more complex ligamentous apparatus than that of other monkeys, as one might gather from the description given by Robertson (1944). This form does, however, present a minor modification in that the ligamentous bond uniting radius and ulna, the homolog of the triangular articular disk, has its lateral attachment prolonged along the posterior margin of the radial articular surface, here effectively forming a labrum. Thus, no striking specializations are featured in the wrist morphology of those monkeys exhibiting at least a fairly substantial change in emphasis from quadrupedal toward suspensory locomotion. Conversely, invasion of a terrestrial habitat has been accompanied by some change in structure. Jones (1967) has observed that the dorsolateral aspect of the wrist joint of *Papio ursinus* contains a semilunar meniscus. I have noticed what is doubtless the same morphological entity in *Papio papio:* the lateral part of the dorsal radiotriquetral ligament (which is additionally here tethered to the scaphoid) becomes interposed between the dorsal concavity of the scaphoid and the radius during extension and is here modified to form a somewhat rudimentary meniscus. No new morphological structure has appeared and the remainder of the joint shows no especially noteworthy features.

The dry bones of monkeys bear the clear imprint of the type of wrist architecture outlined above (Fig. 2). The ulnar styloid process bears a facet on its carpal aspect (facing the interior of the joint) for the articular cup formed by pisiform and triquetral; the opposite peripheral aspect of the process is nonarticular. This distinction between articular and non-articular areas is particularly apparent in the Ceboidea, but is rather ill-defined in a few species of the Cercopithecoidea.

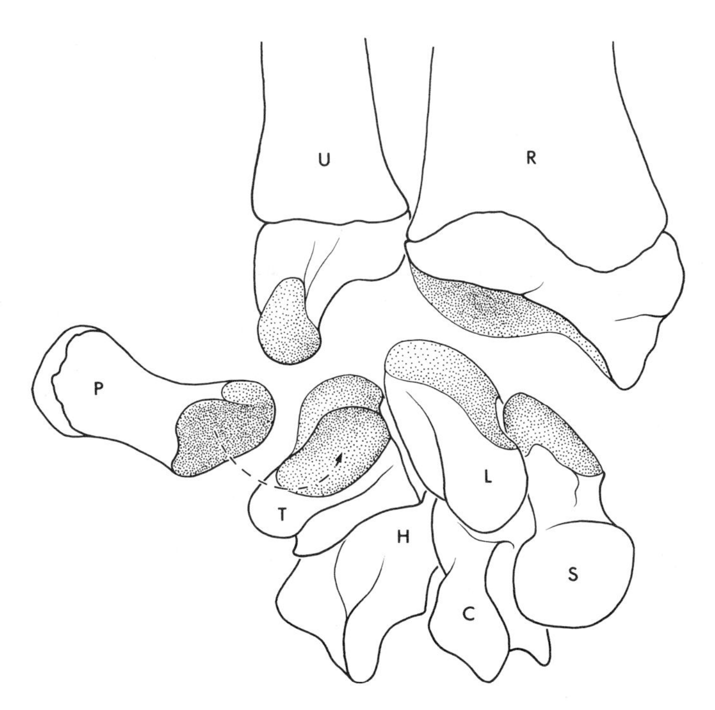

Fig. 2. The palmar aspects of the bones entering into the left wrist joint (with the articular surfaces participating in the joint stippled in each case), together with the capitate and hamate, in *Procolbus verus.* C, capitate; H, hamate; T, triquetral; L, lunate; S, scaphoid; P, pisiform (rolled away from its articulation with the remainder of the carpus); R, radius; U, ulna. [From Lewis (1971a) *Nature* (*London*) **230**, 577–579.]

The triquetral is a relatively massive cuboidal bone bearing a non-articular tuberosity at its medial end. Its proximal surface exhibits a shallow concave facet which is confluent with a second facet on the anterior surface for the articulation of the pisiform, completing the receptive cup for the ulnar styloid process. The elongated pisiform is vertically compressed at its carpal extremity where it presents a terminal facet for the triquetral and a small facet, usually restricted to the medial part of the upper surface, for the ulnar styloid process. In the Atelinae this latter facet is more extensive.

The scaphoid is a squat bone with anterior convex and posterior concave areas identifiable on the proximal surface of its body which is only slightly constricted off from the tuberosity. Between the two parts is a shallow groove for the intracapsularly situated palmar radiocarpal ligament.

The hamate is a broad wedge of bone, obliquely set in the carpus, and often virtually excluding the capitate from contact with the lunate. Its midcarpal articular surface which supports the triquetral faces somewhat proximally. The facet here may be restricted to the proximal part of the bone and is then slightly convex; in some monkeys it is more extensive, concavoconvex and gently spiral, and may reach the distal border of the bone.

The head of the capitate is generally small and maintains only limited contact with the lunate, but is snugly enveloped by the free centrale. In the terrestrial and habitually digitigrade members of the Cercopithecinae —*Papio cynocephalus, Mandrillus sphinx, Erythrocebus patas*—the head of the capitate is enlarged medially, producing a more extensive contact with the lunate, and imparting an obliquely truncated form to the apex of the hamate.

The monkey centrale is an independent bone of somewhat wedge-shaped form, with the thickened base distally located and bearing the trapezoid articulation, and the thin edge interposed between the capitate and scaphoid. The bone is generally free to move with either the proximal or distal carpal rows. However, a firm ligamentous union between the scaphoid and centrale has been recorded in *Papio ursinus* (Jones, 1967) and the present author has observed this in *P. papio.*

Functional Anatomy

The biomechanics of the monkey wrist conform in general to a uniform pattern adapted to quadrupedal progression; only in the digitigrade baboons is this somewhat modified.

In the palmigrade weight-bearing position the monkey limb is pronated and the wrist moves into full dorsiflexion bringing the rolled-over dorsal border of the radius into contact with the concave posterior articular tongue on the scaphoid (Fig. 9). The articular surfaces of the wrist joint proper are then in the position of maximum congruence, the close-packed position. The form of the midcarpal joint reinforces this effect. The distal carpal bones, together with the centrale, form a broad wedge underlying the proximal row. Dorsiflexion exerts a concertina effect on the dorsal aspect of the carpus causing the wedge-shaped centrale to become impacted between capitate and scaphoid, splaying the proximal carpal row and accentuating the close packing of the carpus which becomes welded into a firm unit. This is an excellent example of the basic biomechanical principle (MacConaill, 1950) that joint surfaces are truly congruous in only one position which is commonly the weight-bearing attitude.

In baboons (at least in *Papio ursinus* and *P. papio*), the close-packed position is brought about at an earlier point in the movement of extension by the presence of the wedge-shaped meniscus interposed between the approximating dorsal parts of radius and scaphoid. The firm ligamentous union of the centrale to the scaphoid also appears to limit midcarpal dorsiflexion. These appear to be the factors responsible for the limitation of dorsiflexion in baboons which has been recorded by Tuttle (1969a).

As in all nonpivot joints the swing movement of extension at the monkey wrist appears to involve a small element of rotation. As will be seen later this twisting action is greatly accentuated in the Hominoidea.

Hominoidea

Introduction

All living hominoids possess a fully elaborated diarthrodial inferior radio-ulnar joint which incorporates a neomorphical ulnar head and which is closed inferiorly by the triangular articular disk, derived from the ligamentous union primitively occupying this site. The original carpal extremity of the ulna is relegated to a minor role, becoming the descriptive ulnar styloid process.

For a century it has been recognized that the Hominoidea differ from other Primates in possessing an ulna withdrawn from its primitive articulation with pisiform and triquetral. Strangely, this observation excited no further enquiry. However, an odd assemblage of apparently unrelated facts from human and comparative anatomy have recently furnished the clue as to the nature of the phylogenetic changes involved. It has been on record since the original observation by Daubenton (1766) that gibbons possess a supernumerary carpal element (os Daubentonii) which happens to be located in the interval created by retreat of the ulna from the carpus; it has also been recorded that at a comparable site in the marsupial ankle lies the so-called "os intermedium tarsi." In the course of a study on tarsal homologies, this latter bone was shown (Lewis, 1964) to be no more than a lunula, a sesamoidlike ossification within a meniscus in the marsupial ankle joint. It has also been repeatedly noted that bony elements comparable in situation to the os Daubentonii can occur in the wrist of man and that a pea-sized diverticulum of the human radiocarpal synovial cavity is normally found in close relationship to the ulnar styloid process. Stemming from these observations an investigation then revealed (Lewis, 1965) that the hominoid ulnar styloid process is excluded from the carpus by an intra-articular meniscus (containing a lunula in gibbons) which shows a progressive trend toward incorporation as an integral part of the proximal articular surface of the hominoid wrist joint, leaving the ulnar styloid process isolated within a proximal diverticulum of the joint. These changes, though of considerable scope, even involving the appearance of a true evolutionary novelty, and of obvious functional importance, had suprisingly gone unrecorded.

Pongid Wrist Morphology

Hylobates

The ulnar part of the wrist joint cavity includes a large fibrocartilaginous meniscus, penetrating into the joint with a radially directed, concave, free margin and thus partially excluding the ulnar styloid process from contact with the triquetral (Fig. 3). The meniscus attaches by horns to the radius behind and the lunate in front. This neomorphical structure completely isolates the ulnar styloid process from contact with the pisiform; instead, the base of the pisiform now articulates with the periphery of the

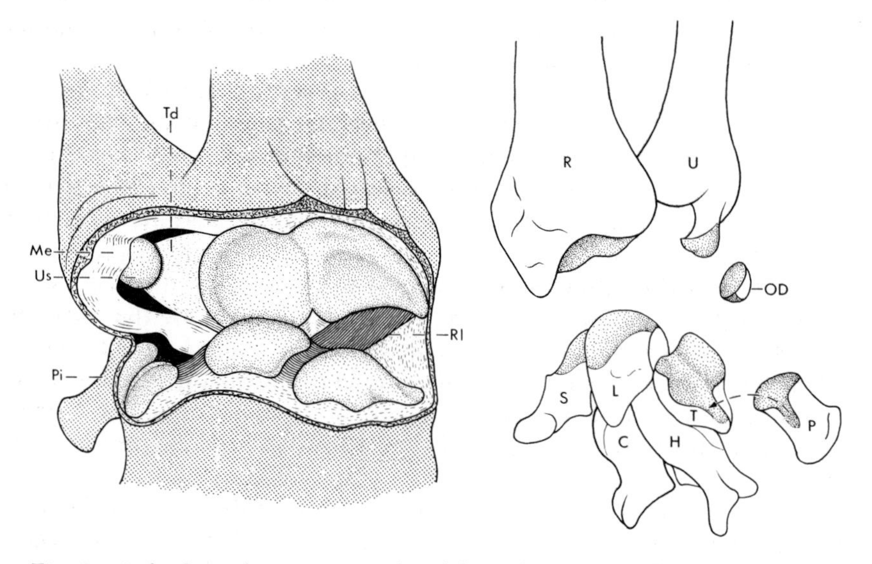

Fig. 3. Left, the right wrist joint of *Hylobates lar* with the dorsal capsule incised and the wrist flexed. Me, meniscus, here containing the os Daubentonii; Pi, pisiform; Rl, palmar radiocarpal ligament; Td, triangular articular disk; Us, ulnar styloid process. Right, the macerated bones of the same specimen viewed from the palmar aspect, with the articular surfaces participating in the wrist joint (stippled). OD, os Daubentonii; other labeling as in Fig. 2. [The left-hand figure after Lewis (1970) *Anat. Rec.* **166**, 499–516.]

meniscus, here strengthened by a bony lunula, the os Daubentonii. The pisotriquetral joint cavity, ballooning up between pisiform and meniscus, is in free communication beneath the meniscus with the wrist joint cavity. In contrast with the situation in monkeys the pisiform is reorientated distally into the palm. This is further accentuated in other hominoids and is the causative factor underlying the apparent distal migration of the hominoid hypothenar pad, noted by Midlo (1934).

In the radial part of the joint cavity the suggestion of a dorsal concavity on the scaphoid (constant in monkeys), locking in extension with a corresponding convexity on the radius, is retained. Persistence of the primitive synovial septum separating ulnar and radial parts of the joint cavity has been observed in a specimen of *Hylobates lar*.

The ligamentous apparatus is clearly derived from a pattern like that of monkeys. The palmar ulnocarpal ligament is, however, united with the anterior border of the triangular articular disk, attaching to the lunate in common with the anterior horn of the meniscus. The thick palmar radiocarpal ligament is essentially intracapsular where it grooves the scaphoid, and becoming bifascicular (in contradistinction to its undivided form in monkeys) attaches not only to the lunate but also to the capitate. A dorsal radiotriquetral ligament is also present.

The dry bones of the gibbon wrist present several specialized features correlated with the acquisition of the meniscus, together with other aspects clearly recalling the more primitive pattern found in monkeys. In this amalgam of primitive and progressive carpal characteristics, the gibbons occupy an intermediate position between monkeys and the remaining hominoids.

The ulnar styloid process is remodeled to provide an articular surface on its peripheral aspect for the concave upper surface of the meniscus, but also retains a more or less demarcated facet at the tip for the triquetral. The process has a hooklike form (Fig. 3). In monkeys, the major articular (carpal) surface faces the interior of the joint; gibbons present a strong contrast, with the major (meniscal) facet on the opposite aspect of the process.

The triquetral, so-called, retains the essentially cuboidal form found in monkeys but is compressed into a flatter plaque. The pisiform facet is displaced distally and medially onto the tuberosity and here the pisiform articulates by a small concave groove on its dorsal surface. The changed disposition of this articulation, when compared with that of monkeys, results in a distal angulation and displacement of the gibbon pisiform into the palm. The broad proximal surface of the bone is articular for the periphery of the meniscus.

The scaphoid is more deeply waisted and more sickle-shaped than that of monkeys due to a widening of the groove for the bifascicular palmar radiocarpal ligament. It has a firm ligamentous union with the centrale, the two bones becoming fused in old individuals (Schultz, 1936, 1944).

The hamate retains an extensive contact with the lunate, as in monkeys, and presents a convex facet, limited to its apical portion, for the triquetral. The adjacent heads of the capitate and hamate entering into the midcarpal joint form an articular ball—rather more than a hemisphere—which is firmly embraced by the proximal carpal row, in particular by the centrale, during wrist extension.

Pan

A meniscus similar in form to that of gibbons and having similar attachments, but lacking a bony lunula, is also a characteristic and striking feature of the ulnar part of the chimpanzee wrist joint (Fig. 4). As in gibbons the pisotriquetral joint cavity is in communication with that of the wrist joint beneath the indented anterior aspect of the meniscus; adjacent to this the pisiform retains a restricted contact with the periphery of the meniscus. The aperture encircled by the meniscus is in some specimens quite large and through it the ulnar styloid process may retain a quite considerable direct articulation with the triquetral. In other specimens (Fig. 5), the meniscus is well integrated into the proximal articular surface of the wrist joint, merging with the triangular articular disk, and thus more effectively segregating the ulnar styloid process within its own proximal synovial compartment. The ligamentous apparatus is similar to that of *Hylobates*.

Clearly, the specialization of the joint is more advanced than that of gibbons, and this is reflected in the skeletal features (Fig. 6). Again the ulnar styloid process presents an articular surface on its periphery for the joint meniscus, but retains a more or less distinguishable facet at the tip, where limited contact with the triquetral persists. The process frequently has a hooklike form similar to that of gibbons.

The triquetral is quite unlike the cuboidal bone characteristically found in monkeys and gibbons. It is remodeled into the shape of a triangular pyramid with the edge between palmar and dorsal surfaces rounded off proximally to form a smooth facet for the inferior meniscal surface and that exposed part of the ulnar styloid process encircled by it. This facet is confluent with a shallow, concave pisiform articular area occupying most of the palmar surface of the bone.

The pisiform articulates with this palmar cup on the triquetral by a large, dorsally located, convex articular facet and bears a second narrow facet at its proximal extremity where it contacts the periphery of the

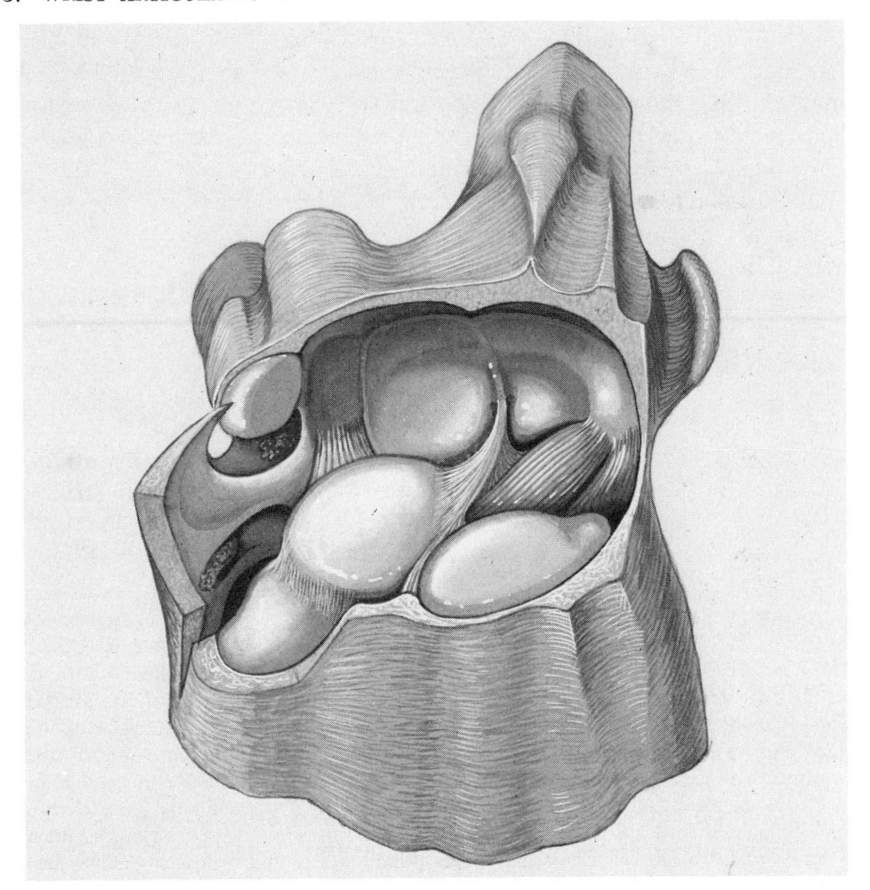

Fig. 4. The right wrist joint of *Pan troglodytes* opened dorsally and flexed. The semilunar meniscus has been detached posteriorly from the radius and pulled forward from its contact with the cartilage-clothed ulnar styloid process, thus revealing the proximal compartment of the joint cavity; this diverticulum extends mainly anterior to the ulnar styloid process (where it contains an aggregation of synovial villi). The anterior margin of the meniscus exhibits a free, crescentic indentation; here the wrist joint cavity communicates distally with the pisotriquetral cavity which extends as a short, blind proximal recess anterior to the meniscus. Also illustrated are the synovial-covered palmar ulnocarpal ligament anterior to the triangular articular disk, the thick intracapsular palmar radiocarpal ligament, and a synovial fold attaching to the radius between its facets for the scaphoid and lunate. [From Lewis (1969) *Amer. J. Phys. Anthropol.* **30**, 251–268.]

meniscus. This drastic reorganization of the pisotriquetral joint surfaces results in a striking reorientation of the pisiform, bringing it into a position where it projects distally into the palm, almost reaching the hook of the hamate.

The hamate is more narrowly wedge-shaped than that of monkeys (or

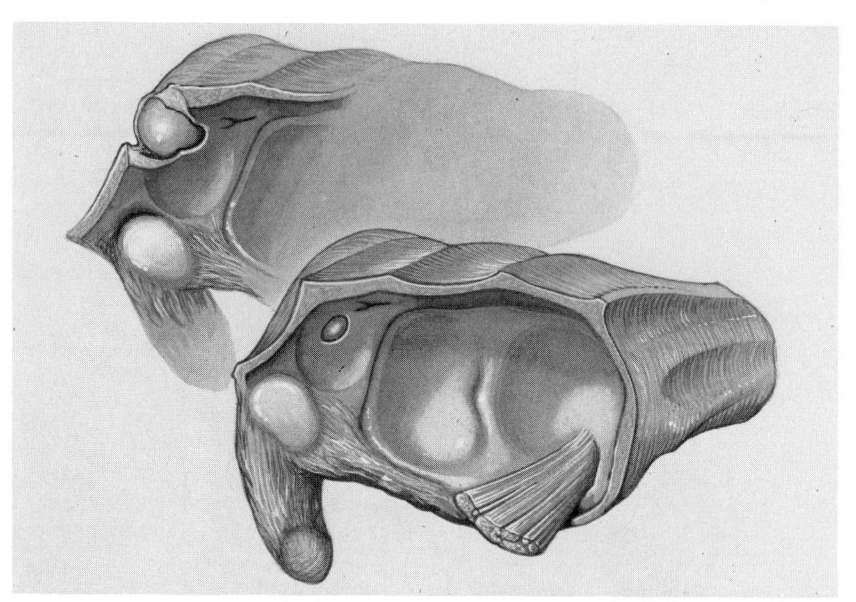

Fig. 5. The proximal articular surface of the right wrist joint (together with the pisiform) of *Pan troglodytes.* The meniscus is merged with the triangular articular disk, the two structures contributing to the concave ulnar component of the articular surface. The aperture leading to the proximal joint compartment is relatively restricted, and through this opening a small portion of the cartilage-clothed ulnar styloid process is exposed. The elongated pisiform with its articular surface for the triquetral is apparent, demonstrating again continuity between pisotriquetral and wrist joint cavities. The thick palmar radiocarpal ligament has been cut from its attachment to the lunate and capitate. In the detail shown above the proximal synovial compartment has been opened (by incising that part of the articular surface derived from the meniscus), thus revealing the cartilage-clothed ulnar styloid process. [From Lewis (1969) *Amer. J. Phys. Anthropol.* **30**, 251–268.]

gibbons) and lies more vertically within the carpus. This restricts its area of contact with the lunate; conversely, the capitate has a more extensive articulation with that bone. The proximal part of the triquetral facet on the hamate is thus almost vertically disposed and its distal termination is inflected toward the root of the hamate hook, giving the facet a spiral form.

The scaphoid, incorporating the centrale, is deeply grooved for the bifascicular palmar radiocarpal ligament. Its radiocarpal articular surface may present a residual indication of the dorsal concavity found in monkeys but conspicuous dorsal ridges, which could plausibly be considered as effective articular stops associated with knuckle walking (Tuttle, 1967), have not been found to be normal features of the bone.

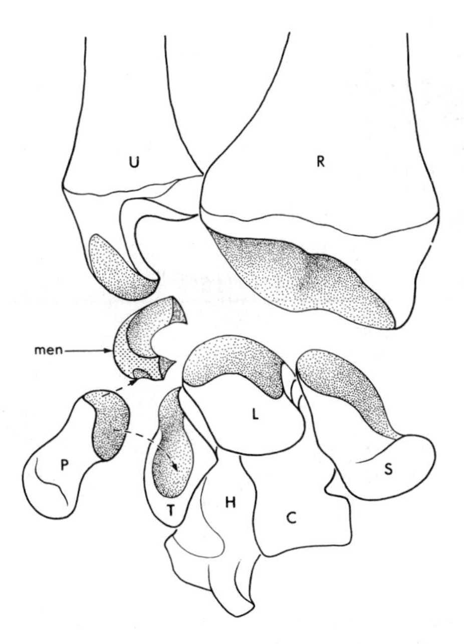

Fig. 6. The palmar aspects of the bones entering into the left wrist joint of the same specimen of *Pan troglodytes* as illustrated in Fig. 4 (with the articular surfaces participating in this joint stippled in each case), together with the capitate and hamate. Part of the meniscus (men) is shown, its horns of attachment to radius and lunate having been removed. The labeling is as in Fig. 2. [From Lewis (1971a) *Nature* (*London*) **230**, 577–579.]

The midcarpal joint surfaces are strikingly remodeled. As noted above the triquetral facet on the hamate is spirally concavoconvex and its almost vertical alignment accentuates the separation of the triquetral from the ulnar styloid process. The head of the adjoining capitate is markedly swollen laterally (Fig. 9). This expansion gives the bone a "waisted" appearance, either when viewed from in front or from behind. Posteriorly, this waist, or neck, is articular and is confluent below with a facet for the trapezoid; anteriorly it forms a deeply undercut nonarticular groove. As will be seen later the waist serves to lodge the centrale element of the scaphoid during extension and thus forms an important part of a uniquely hominoid midcarpal specialization; a thin membranous sheet stretches from the anterior nonarticular part of the waist to the scaphoid.

Gorilla

Incorporation of the meniscus with the triangular articular disk to form a smoothly concave proximal articular surface is the normal state of affairs in gorillas; in chimpanzees this trend may already be apparent, as noted above. The meniscus thus loses its separate identity and the cartilage-clad ulnar styloid process becomes totally excluded from direct participation in the radiocarpal joint. Instead it is lodged in its own capacious synovial cavity communicating with the wrist joint by a small irregular opening. This arrangement has been observed in *Gorilla gorilla gorilla* and in *Gorilla gorilla beringei;* in one example of the latter subspecies (Fig. 7), however, the neck of the proximal synovial cavity lodging the ulnar styloid process was found to be completely sealed off from the wrist joint proper.

Not surprisingly, the carpal skeleton shows specializations similar to those noted in *Pan,* but the capitate is less obviously "waisted." The ulnar styloid process is shorter than that of the chimpanzee, lacks a hooklike character, and is articular at its truncated termination; these features are in accord with its greater degree of withdrawal from the carpus.

Pongo

The specialized structure of the orangutan wrist represents the culmination of trends similar to those noted above, which have been observed in a single mountain gorilla specimen, where the ulnar styloid process was totally excluded from the wrist joint and contained within its own closed proximal synovial compartment.

An orangutan specimen has been described (Lewis, 1971b) in which the cartilage-covered tip of the ulnar styloid process was similarly lodged in a closed synovial compartment which demarcated the upper surface of what was clearly the homolog of the typical hominoid meniscus (Fig. 8). Within the wrist joint a blind pit was located at the apex of the triangular articular disk; this is precisely the situation occupied in all other hominoids by the communicating aperture leading to the synovial compartment proximal to the meniscus. In this specimen, also, a synovial-lined channel led from the radiocarpal cavity to the distally located pisotriquetral joint. In the only other specimen described (Lewis, 1969), this communication was sealed off as was the proximal synovial cavity, leaving the nonarticular ulnar styloid process embedded in a thick wedge of tissue—the meniscus homolog—at the ulnar side of the joint. The true nature of these specialized morphological arrangements only emerges when comparison is made with other hominoids.

The bones bear the clear imprint of this specialized morphology. The ulnar styloid process is typically very short and conical. The hamate preserves some degree of the oblique alignment and rather large articulation with the lunate found in monkeys and gibbons and its facet for the triquetral is less markedly spiral than in the other great apes. The reduced triquetral bears a small facet at its distal extremity for the small pisiform which in some orangutans has migrated so far distally that it also articulates with the hook of the hamate. The scaphoid is deeply grooved for the

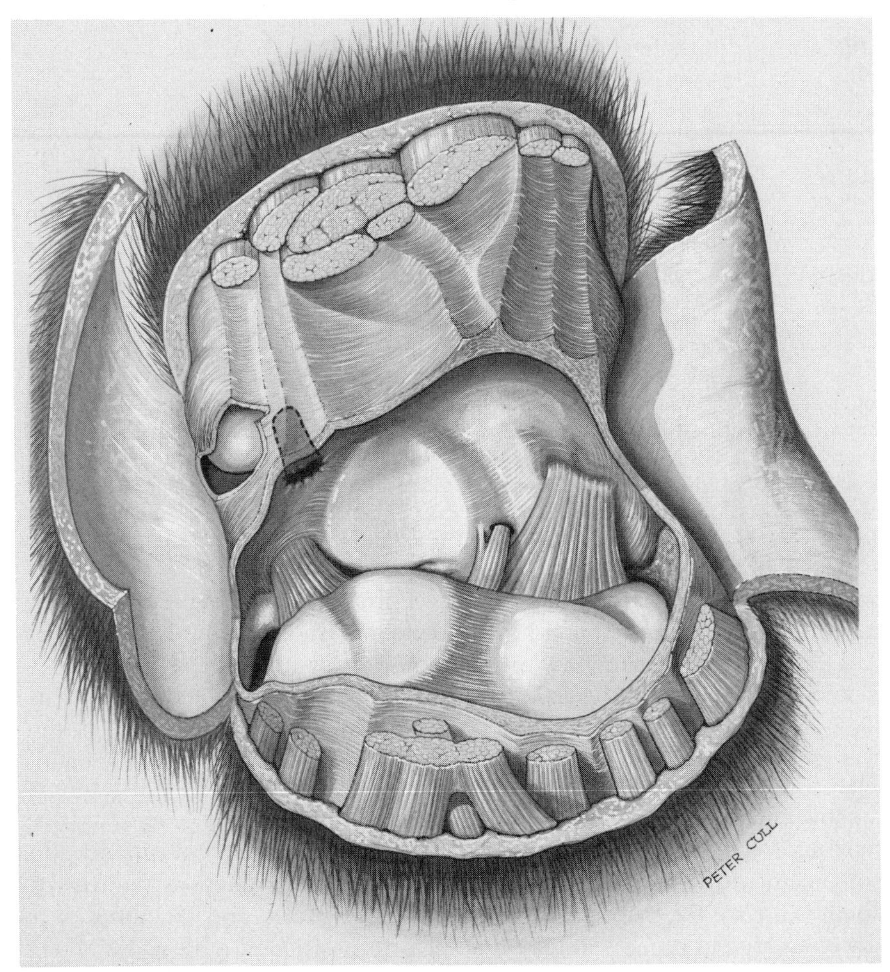

Fig. 7. The right wrist joint of *Gorilla gorilla beringei* opened dorsally. In this specimen the proximal compartment of the joint is sealed off to form a closed cavity which is shown opened in the illustration, revealing the contained ulnar styloid process. The original opening of this compartment into the wrist joint cavity is shown, leading into a blind synovial diverticulum (indicated by the broken line). The continuity of the pisotriquetral articulation with the wrist joint cavity is clearly apparent, and the prominent palmar ulnocarpal and radiocarpal ligaments are shown. [From Lewis (1969) *Amer. J. Phys. Anthropol.* **30**, 251–268.]

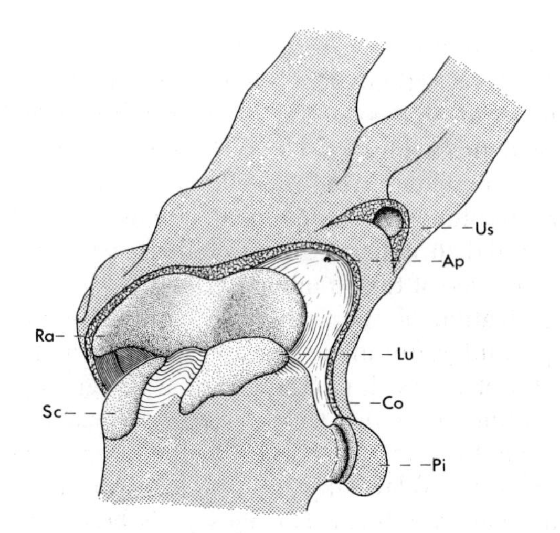

Fig. 8. The left wrist joint of *Pongo pygmaeus* opened dorsally to display the following features: Ra, articular surface of radius; Sc, articular surface of scaphoid; Lu, articular surface of lunate; Ap, aperture leading into a blind pit at the apex of the triangular articular disk; Co, synovial lined communication leading to the articulation of pisiform (Pi) with the triquetral. An incision is shown opening into the closed synovial cavity lodging the ulnar styloid process (Us) and effectively demarcating the meniscus homolog. [From Lewis (1971) *Folia Primatol.* **16**, 248–256.]

palmar radiocarpal ligament and the centrale is independent in young specimens, but is firmly bound to the scaphoid and moves in concert with it. The capitate presents only an ill-defined neck.

Pongid Functional Anatomy

The extent of the specialization of the pongid wrist must be indicative of drastic changes in functional requirements. It is clearly apparent that retreat of the ulna from the carpus is an essential prerequisite for the enhancement of the range of supination which in monkeys is limited (to about 90°) by the restricting articulation of pisiform and triquetral with the ulnar styloid process. In pongids the interposition of a meniscus allows relative displacement between the carpus and the ulnar styloid process, increasing this rotatory range to about 180°.

It is known from behavioral studies (Avis, 1962) that the suspensory locomotion favored by pongids involves swinging from hand to hand with the body rotating a half circle about the grasping hand. This is achieved by 180° of supination. Monkeys, even those most adept at suspensory locomotion, never exhibit this particular type of activity which is properly considered to be brachiation. The rotatory capacity of the forelimbs essential for this type of locomotion of course brings with it also the capacity for a widened range of climbing and feeding activities. This versatile forelimb-dominated arboreal activity is dependent upon modification of a key morphological component, the wrist joint, and may be presumed to have led to enhanced exploitation of the arboreal environment and the emergence of the Hominoidea.

Drastic reorganization of the wrist joint proper seems then to have created the opportunity for maneuverable suspensory posturing. Other correlated changes of a less dramatic nature are engendered in the midcarpal joint, adapting it to the newly imposed tensional forces. These changes are shown to best effect in the chimpanzee wrist where the capitate is deeply indented laterally and the triquetral facet on the hamate has a spiral concavoconvex form. Dorsiflexion is here accompanied by a rotatory movement at the midcarpal joint so that in full extension the scaphoid fits firmly onto the articular neck of the capitate, its incorporated centrale element (bearing the articulation for the trapezoid) becoming closely locked underneath the bulbous capitate head (Fig. 9). In this attitude the triquetral also sits snugly upon the posterior part of the spiral hamate facet—the midcarpal joint is then in the close-packed, maximally congruent position. This is, in fact, an excellent example of a basic principle in joint mechanics which has been discussed by MacConaill (1953) and by MacConaill and Basmajian (1969): at all nonpivot joints the characteristic or habitual movement is a composite one involving some degree of rotation (conjunct rotation) which is most marked in joints with concavoconvex surfaces and is effective in screwing the bones together into a locked position. The advantage of this in stabilizing the extended chimpanzee wrist against tensional forces is obvious. It has been suggested by Jouffroy and Lessertisseur (1960) that fusion of the centrale to the scaphoid may be associated, in some unspecified manner, with limitation of midcarpal mobility. It is now clear that the wrist bones will be most effectively welded together by a screwing action, as the hand is brought into line with the forearm, if the centrale is fused to the scaphoid. This mechanism involving conjunct rotation reaches its highest expression in the African great apes; in gibbons it is rudimentary but it is more effective in orangutans. There is here an apparent paradox in that the least arboreal of extant apes (*Pan* and *Gorilla*) are the possessors of midcarpal joints which seem best suited to forelimb suspension. There appears to be little doubt, however, that the Asiatic apes diverged early from the

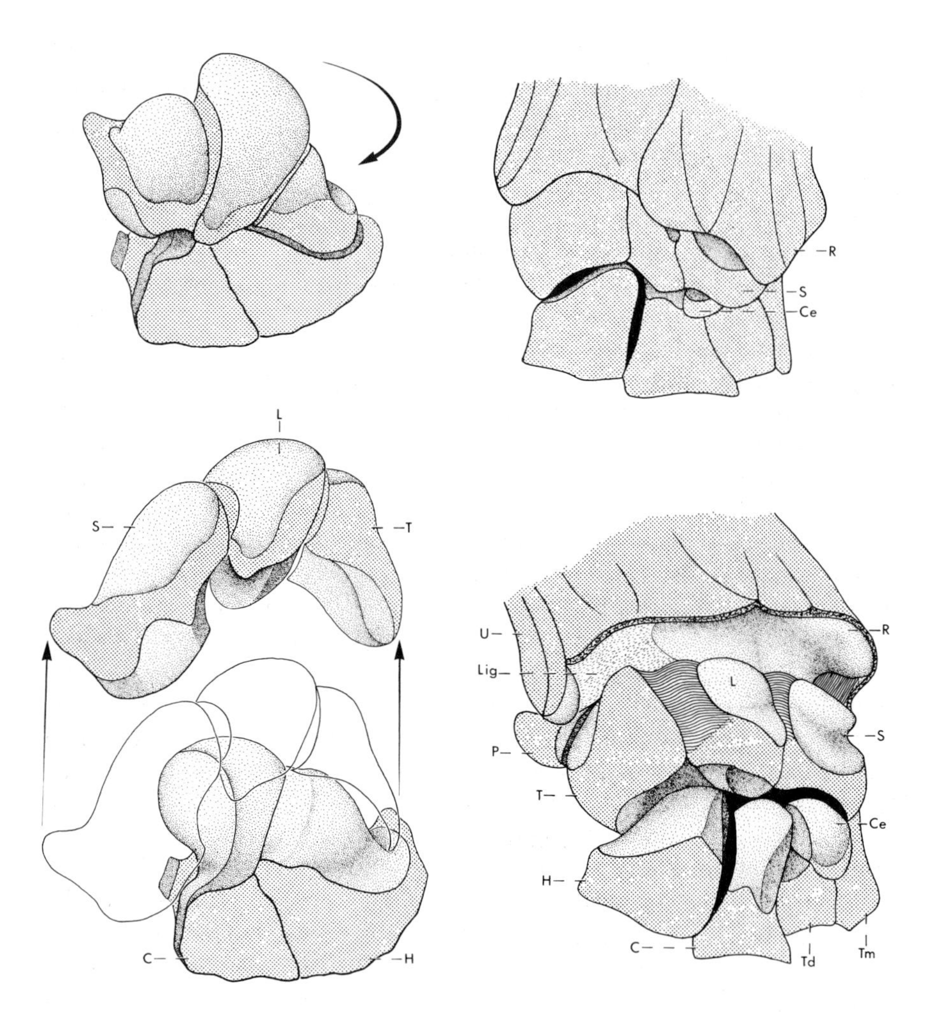

Fig. 9. Left, the midcarpal mechanism in ***Pan troglodytes,*** drawn from the cartilage-clad bones of the left wrist of a wet specimen; below, the capitate and hamate are viewed from behind with the proximal carpal row (middle) shown superimposed, in line, in the position in flexion; above, the bones are shown in the position occupied upon moving into extension, with the arrow indicating the direction of conjunct rotation of the proximal row. Right, a dorsal view of the right wrist of a typical monkey, ***Colobus polykomos;*** below are shown the positions occupied by the bones (slightly separated) in flexion; above is shown the close-packed position achieved in dorsiflexion. Ce, centrale; Lig, interosseous radio-ulnar ligament; Td, trapezoid; Tm, trapezium; other labeling as in Fig. 2.

brachiating lineage leading to *Pan* and *Gorilla,* and that their locomotor behavior developed its own special characteristics; gibbons commonly complete their bodily rotation while airborne (Avis, 1962) and the orang-utan is a modified brachiator dependent for much of its support upon the grasping power of its feet.

The midcarpal mechanism is clearly one factor responsible for that restriction in the range of wrist extension which has been recorded in *Pan* and *Gorilla* (Tuttle, 1969b). The changed orientation of the hominoid pisiform, which comes to act as a bony splint spanning the midcarpal joint, may reinforce this effect.

It could perhaps be argued that these mechanisms limiting wrist dorsiflexion are adaptations to knuckle walking locomotion. It might be argued that the firm bonding of the centrale to the scaphoid in the digitigrade baboons (where stabilization of the extended wrist must also be advantageous) is a point in favor of this view. The baboon mechanism, however, is fundamentally different, lacking the conjunct rotation whereby the scaphoid is locked about the constricted neck of the capitate, thus stabilizing the wrist against disruptive tensional forces. Of course, those factors tending to limit dorsiflexion of the ape wrist are of presumptive benefit during knuckle walking and are in a sense preadaptive for this activity. Suggestions that the wrists of knuckle walkers possess other unique morphological specializations (Tuttle, 1967, 1969b, 1970) seem without sound foundation: thick palmar ligaments are not a unique attribute of chimpanzees and gorillas but represent a primitive mammalian heritage; the belief that the prominent scaphoid tubercle of African great apes is associated with elaboration of a massive radial collateral ligament is the result of misinterpretation of the attachments of the ligamentous apparatus of the wrist; dorsal scaphoid ridges are inconstant features and cannot have been significant factors in the evolution of knuckle walking.

HOMO SAPIENS

Traditional anatomical descriptions might appear to indicate that the human radiocarpal joint has little in common with the corresponding pongid articulations described above. However, two well-documented features of the human wrist seldom appear in these textbook descriptions, yet they are features which have considerable import when viewed in a comparative context. A radiographical opacity is sometimes ($\frac{1}{2}$ to 1%) demonstrable in the human wrist at the site occupied by the gibbon os Daubentonii and a synovial cul-de-sac (the prestyloid recess) regularly extends from the wrist joint cavity into relationship with the ulnar styloid process.

Arthrograms reveal that the human prestyloid recess is similar in size and position to that compartment of the chimpanzee wrist joint found proximal to the meniscus. This technique also clearly suggests that the most ulnar portion of the proximal articular surface in the human joint is the homolog of the often free meniscus in *Pan*. A similar free meniscus has now been demonstrated as an occasional (2%) attribute of the human joint; usually, however, the meniscus is well merged with the triangular articular disk, thus constricting the opening into the prestyloid recess which may, nevertheless, be easily located adjacent to the apex of the triangular articular disk. This latter arrangement is comparable to the pattern sometimes seen in *Pan* and normally in *Gorilla,* the human joint being distinguished by the commonly more extreme withdrawal of the ulnar styloid process beyond the confines of the synovial prestyloid recess with consequent loss of articular character. The human joint has now been shown (Lewis *et al.*, 1970) to present a considerable range of variations, all of which are clearly derived from a basic hominoid type of structure, and which include varieties comparable to those normally found in chimpanzees and gorillas. Embryological studies further emphasize these structural affinities with the Pongidae, for the developmental history of the human joint (Lewis, 1970) recapitulates in a quite striking fashion the phylogenetic stages represented among the Anthropoidea.

The osteology of the human wrist is consistent with this structure and presents certain progressive features when compared with the pattern found in *Pan* and *Gorilla.* Again the hamate is vertically aligned, retaining only a minimal articulation with the lunate, and bears the obliquely disposed triquetral upon a markedly spiral facet. The triquetral itself presents a convex facet proximally for the meniscal component of the proximal articular surface of the radiocarpal joint and a distally located pisiform facet. These two facets on the triquetral may be confluent as in the African great apes; more often they are separated by a nonarticular area representing the site of obliteration of the communication between radiocarpal and pisotriquetral joints. The pisiform itself is, of course, a squat, shortened bone. The ulnar styloid process varies considerably in length. The capitate sometimes preserves in minor degree the "waisted" appearance noted in *Pan* but the lateral swelling of the head (when present) is less marked, being more comparable in appearance to that of *Gorilla.* Clearly, however, conjunct rotation at the midcarpal joint also accompanies extension of the human wrist, even if locking of the scaphoid about an undercut capitate head is less apparent than in *Pan.*

Fossil Hominoidea

Dryopithecus (*Proconsul*) *africanus*

A recent study of the casts of the wrist bones of this Miocene ape has been reported by the author (Lewis, 1971a, 1971c). The ulnar styloid process has a hooklike appearance and is clearly articular on its distal and peripheral aspects—a hominoid characteristic—and not on the surface facing the interior of the joint as in monkeys. The articular area shows indications of subdivision into a large peripheral area (presumably for a meniscus) and a small facet at the tip (for the triquetral); this arrangement is comparable to that found in *Hylobates* or *Pan*. Thus, the appearance of the ulnar styloid process (Fig. 10), even considered in isolation, strongly suggests a meniscus-containing joint of hominoid type.

The hamate is narrowly wedge-shaped, vertically aligned, and retains only a restricted apical articulation for the lunate. Its facet for the triquetral is markedly spiral. This bone again, in these functionally significant characters, is comparable to that of *Pan*.

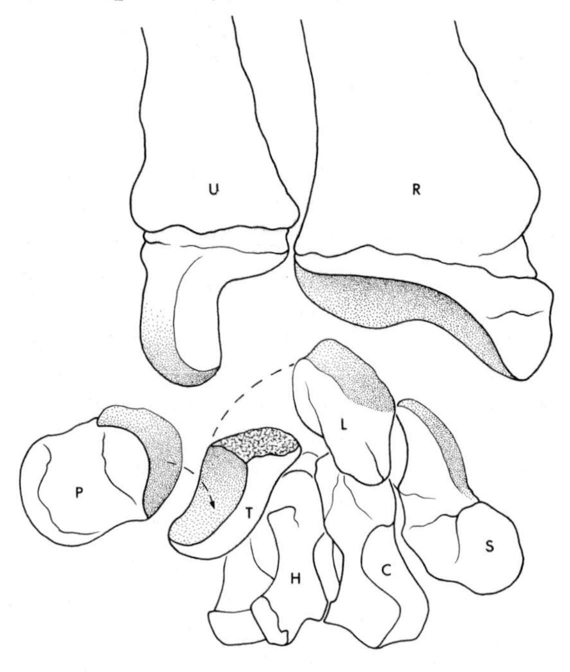

Fig. 10. The left wrist of *Dryopithecus* (*Proconsul*) *africanus* showing the region comparable with that illustrated in Figs. 2 and 6, similarly labeled. The missing portion of the triquetral is indicated by a broken line. [From Lewis (1971a) *Nature* (*London*) **230**, 577–579.]

The triquetral is incomplete and the proximal portion is lacking. The remaining part, however, clearly conveys the triangular pyramidal shape of the bone with the palmar surface presenting a large shallowly concave area for the pisiform, which here articulates by a large facet on its dorsal aspect; this facet is confluent with another articular area (meniscal) on the proximal aspect of the base. The form and relationships of these two bones strikingly recall the disposition in *Pan;* no monkey possesses an arrangement even remotely similar.

The scaphoid retains a slight indication of a dorsal concavity on its proximal articular surface and is broadly grooved at the situation where one could expect the imprint of a bifascicular palmar radiocarpal ligament of hominoid type.

The centrale was not recovered but was clearly a separate entity.

The spiral character of the triquetral facet is, in itself, indicative of a rotatory mechanism during extension, similar to that operative in *Pan.* The form of the capitate reinforces this conclusion. It is obviously "waisted," although less obtrusively so than in the chimpanzee. There can be little doubt that a similar midcarpal mechanism operated in the two forms although the stabilizing effect of this against tensional forces was presumably less effective in the fossil due to the lack of fusion between the centrale and the scaphoid.

Phylogeny of the Hominoidea

The description given above seems to establish beyond reasonable doubt that *D. africanus* possessed a wrist which already had all the basic and unique structural attributes found in living hominoids and must have been similarly adapted to suspensory locomotion. Brachiation therefore appears to have been a feature of the earliest emergent phases of the Hominoidea and may well have been the decisive factor in the establishment of this superfamily in a new ecological niche of the arboreal habitat, thus initiating divergence from the remaining catarrhines. By using the term brachiation it is not the intention to imply gibbonlike ricochetal arm swinging but rather a whole repertoire of arboreal activities dominated by the use of efficient, mobile, grasping forelimbs which play the leading role in climbing and in suspensory locomotor and feeding activities.

Understanding of the wrist morphology of *D. africanus* and its comparison with that of the living Anthropoidea provides new insight into a number of issues relating to the early radiation of the Hominoidea. There can be little doubt that the gibbon line must have diverged before the *D. africanus* grade of structure had been achieved. The form of the gibbon triquetral is clearly derived from a bone similar to that of living monkeys and is quite unlike the reduced triangular pyramidal carpal element characteristic of all other living hominoids (and of *D. africanus*). Further, in the set of the hamate and in the relatively rudimentary nature of the midcarpal locking mechanism gibbons are less characteristically hominoid than even *D. africanus*. The indications then are that the earliest hominoids, ancestral to *D. africanus*, and the presumptive source from which the gibbons were derived, already possessed a meniscus-containing wrist joint. This leads to fresh doubts about the ancestral gibbon status claimed for *Pliopithecus* (*Epipliopithecus*) *vindobonensis* of the late Miocene. The bones of the proximal carpal row of this form are unfortunately unknown but an intact ulna has been recovered. The present author has not had access to this material but the illustrations published by Zapfe (1960) seem to demonstrate convincingly that the distal ulnar extremity had the characteristic form found in monkeys and could not have participated in a meniscus-containing wrist joint. If this is taken in conjunction with the indications that *P. vindobonensis* possessed a long tail (Ankel, 1965) there would seem to be considerable evidence to support the suggestion (based on tooth morphology) made by Von Koenigswald (1969) that "this form probably might be a last survivor of the original group which gave rise to the bilophodont Cercopithecoidea." The very early divergence of the gibbon lineage seems beyond doubt; on the other hand, the osteological evidence of the wrist points to the origin of the orangutan line from a structural grade approximating more closely to, but still inferior to, that of *D. africanus*. These findings are in accord with widely held views on the phylogeny of the Asiatic apes (Simpson, 1964) and substantiate the results obtained from serology and karyology.

Using the criteria of wrist morphology, divergent lines leading to the African great apes, on the one hand, and *Homo*, on the other, could be plausibly derived from the *D. africanus* grade of structure. Implicit in this view, however, would be the notion that fusion between the centrale and the scaphoid has been acquired in parallel in these lines, despite the different selection pressures exerted by their dissimilar locomotor habits. If, in fact, such fusion represents the final perfection of a mechanism apparently adaptive to suspensory locomotion, parallel evolution in the hominid line seems unlikely. The time of fusion of the centrale, or its cartilaginous anlage, with the scaphoid in extant hominoids presents a sequence leading from the Asiatic apes (fusion in old age), through the African apes (late fetal or early juvenile fusion) to man (fusion early in fetal life). It is tempting to interpret this sequence as the outcome of developmental acceleration (de Beer, 1951) indicating derivation of the hominid line from an ape population in which the midcarpal joint was fully remodeled for suspensory locomotion, as in *Pan*. Moreover, the range of variations in the human radiocarpal joint (Lewis *et al.*, 1970) gives the overriding impression of derivation from a structural grade rather more advanced than that of *D. africanus*—more like *Pan,* in fact. Final clarification of this issue must await recovery of relevant postcranial material to fill the long gap between the middle Miocene and the beginning of the Pleistocene. There now seems little doubt, however, that the hominids were derived from brachiating apes already possessing the key modification of a hominoid type of wrist joint, specialized at least to the grade achieved by *D. africanus*. One shred of fossil evidence suggests that the brachiating phase of human evolution persisted into more recent times than the middle Miocene. This is the Sterkfontein capitate (TM1526) discovered and originally described by Broom and Schepers (1946). Certain highly significant features of this bone have been ignored or misinterpreted; indeed, it has been stated that the bone "is similar to that of man in its essential features" (Day, 1965). Yet Le Gros Clark (1947, 1967) remarked upon the strikingly "waisted" appearance of the fossil and its resemblance therein to the chimpanzee; he interpreted the constriction in *Pan* (and, by inference, that in the fossil) as the site of attachment of a strong interosseous ligament binding together capitate and trapezoid, but this was mere speculation and can be shown to have no foundation in fact. As in other primates, a very strong interosseous capitate–trapezoid ligament does occur in *Pan* but it lies more distally, and the indented waist (as noted above) receives the scaphoid during extension and forms a fundamental part of the midcarpal locking mechanism. Somewhat surprisingly, therefore, it seems that the australopithecines had retained virtually unchanged the hallmarks of a significant adaptation to suspensory locomotion; furthermore, secondary changes which somewhat obscure the nature of this mechanism in *Homo sapiens* had not by then appeared (Lewis, 1973). This suggests that the hominid precursors did not abandon suspensory locomotion until a time more recent than that of *D. africanus*.

In the last decade a central tenet of paleoanthropology has been the notion that the Hominidae were derived in the early or middle Miocene (Le Gros Clark, 1959, 1967; Heberer, 1959; Napier, 1964) or perhaps in the Oligocene (Leakey, 1964; Simons, 1965, 1967; Schultz, 1966; Campbell, 1967; Pilbeam, 1970) from an ape population—so-called "dental apes"—combining a hominoid dentition with the essential limb morphology of arboreal quadrupeds and behaving, at best, as mere unspecialized dabblers at suspensory locomotion. The description of the forelimb bones of *D. africanus* by Napier and Davis (1959), indicating that this form preserved an essentially quadrupedal type of wrist articulation, provided the keystone of this hypothesis. Clearer understanding of the morphology of the primate wrist requires abandonment of this idea and a return to the brachiating theory of human origins originally championed by Keith (1923) and Gregory (1928, 1934).

References

Ankel, F. (1965). Der Canalis sacralis als Indikator für die Länge der Caudalregion der Primaten. *Folia Primatol.* 3, 263–276.

Avis, V. (1962). Brachiation: The crucial issue for man's ancestry. Southwestern *J. Anthopol.* **18**, 119–148.

Broom, R., and Schepers. G. W. H. (1946). "The South African Fossil Ape-Men: The Australopithecinae." Transvaal Museum, Pretoria.

Campbell, B. G. (1967). "Human Evolution: An Introduction to Man's Adaptations." Heinemann, London.

Daubenton, L. (1766). "Histoire Naturelle Générale et Particulière avec la Description du Cabinet du Roi." Imprimerie Royale, Paris.

Day, M. H. (1965). "Guide to Fossil Man." Cassell, London.

de Beer, G. R. (1951). "Embryos and Ancestors." Clarendon Press, Oxford.

Gregory, W. K. (1916). Studies on the evolution of the Primates. *Bull. Amer. Mus. Natur. Hist.* **35**, 239–355.

Gregory, W. K. (1928). The upright posture of man: a review of its origin and evolution. *Proc. Amer. Phil. Soc.* **67**, 339–374.

Gregory, W. K. (1934). "Man's Place among the Anthropoids." Clarendon Press, Oxford.

Heberer, G. (1959). The descent of man and the present fossil record. *Cold Spring Harb. Symp. Quant. Biol.* **24**, 235–244.

Jones, R. T. (1967). The anatomical aspects of the baboon's wrist joint. *S. Afr. J. Sci.* **63**, 291–296.

Jouffroy, F. K., and Lessertisseur, J. (1960). Les spécialisations anatomiques de la main chez les singes à progression suspendue. *Mammalia* **24**, 83–151.

Keith, A. (1923). Man's posture: its evolution and disorders. *Brit. Med. J.* **1**, 451–454.

Leakey, L. S. B. (1964). East African fossil Hominoidea and the classification within this super-family. *In* "Classification and Human Evolution" (S. L. Washburn, ed.), pp. 32–49. Methuen, London.

Le Gros Clark, W. E. (1947). Observations on the anatomy of the fossil Australopithecinae. *J. Anat.* **81**, 300–333.

Le Gros Clark, W. E. (1959). "The Antecedents of Man." The University Press, Edinburgh.

Le Gros Clark, W. E. (1967). "Man-Apes or Ape-Men. The Story of Discoveries in Africa." Holt, Rinehart & Winston, New York.

Lewis, O. J. (1964). The homologies of the mammalian tarsal bones. *J. Anat.* **98**, 195–208.

Lewis, O. J. (1965). Evolutionary change in the primate wrist and inferior radio-ulnar joints. *Anat. Rec.* **151**, 275–286.

Lewis, O. J. (1969). The hominoid wrist joint. *Amer. J. Phys. Anthropol.* **30**, 251–268.

Lewis, O. J. (1970). The development of the human wrist joint during the fetal period. *Anat. Rec.* **166**, 499–516.

Lewis, O. J. (1971a). Brachiation and the early evolution of the Hominoidea. *Nature (London)* **230**, 577–579.

Lewis, O. J. (1971b). The contrasting morphology found in the wrist joints of semibrachiating monkeys and brachiating apes. *Folia Primatol* **16**, 248–256.

Lewis, O. J. (1971c). Osteological features characterizing the wrists of monkeys and apes, with a reconsideration of this region in *Dryopithecus (Proconsul) africanus. Amer. J. Phys. Anthropol.* **36**, 45–58.

Lewis, O. J. (1973). The hominoid os capitatum with special reference to the fossil bones from Sterkfontein and Olduvai gorge. *J. Hum. Evol.* **2**, 1–11.

Lewis, O. J., Hamshere, R. J., and Bucknill, T. M. (1970). The anatomy of the wrist joint. *J. Anat.* **106**, 539–552.

MacConaill, M. A. (1950). The movements of bones and joints. The synovial fluid and its assistants. *J. Bone Joint Surg.* **32B**, 244–252.

MacConaill, M. A. (1953). The movements of bones and joints. The significance of shape. *J. Bone Joint Surg.* **35B**, 290–297.

MacConaill, M. A., and Basmajian, J. V. (1969). "Muscles and Movements. A Basis for Kinesiology." Williams & Wilkins, Baltimore, Maryland.

Midlo, C. (1934). Form of hand and foot in Primates. *Amer. J. Phys. Anthropol.* **19**, 337–389.

Napier, J. R. (1964). The locomotor functions of hominids. *In* "Classification and Human Evolution" (S. L. Washburn, ed.), pp. 178–189. Methuen, London.

Napier, J. R., and Davis, P. R. (1959). The Forelimb Skeleton and Associated Remains of *Proconsul africanus. Fossil Mammals of Africa, No.* **16**. Brit. Mus. Natur. Hist., London.

Nayak, U. V. (1933). The articulations of the carpus in *Chiromys madagascarensis* with reference to certain other lemurs. *J. Anat.* **68**, 109–115.

Parsons, F. G. (1899). The joints of mammals compared with those of man. *J. Anat.* **34**, 41–68.

Pilbeam, D. (1970). "The Evolution of Man." Thames and Hudson, London.

Robertson, D. F. (1944). Anatomy of the South American woolly monkey (*Lagothrix*). Part 1. The forelimb. *Zoologica* **29**, 169–192.

Schultz, A. H. (1936). Characters common to higher Primates and characters specific for man. *Quart. Rev. Biol.* **11**, 259–283.

Schultz, A. H. (1944). Age changes and variability in gibbons. *Amer. J. Phys. Anthropol.* **2**, 1–129.

Schultz, A. H. (1966). Changing views on the nature and interrelationships of the higher Primates. *Yerkes Newsletter* **3**, 15–21.

Schwarz, W. (1938). Das Os pisiforme. *Morphol. Jahrb.* **81**, 187–212.

Simons, E. L. (1962). Fossil evidence relating to the early evolution of primate behavior. *Ann. N. Y. Acad. Sci.* **102**, 282–295.

Simons, E. L. (1965). New fossil apes from Egypt and the initial differentiation of the Hominoidea. *Nature (London)* **205**, 135–139.

Simons, E. L. (1967). The earliest apes. *Sci. Amer.* **217**, 28–35.

Simpson, G. G. (1964). The meaning of taxonomic statements. *In* "Classification and Human Evolution" (S. L. Washburn, ed.), pp. 1–31, Methuen, London.

Tuttle, R. H. (1967). Knuckle-walking and the evolution of hominoid hands. *Amer. J. Phys. Anthropol.* **26**, 171–206.

Tuttle, R. H. (1969a). Terrestrial trends in the hands of the Anthropoidea. *Proc. 2nd Inter. Congr. Primatol. Atlanta, Georgia, 1968,* Vol. 2, pp. 192–200. Karger, New York.

Tuttle, R. H. (1969b). Quantitative and functional studies on the hands of the Anthropoidea. I. The Hominoidea. *J. Morphol.* **128**, 309–364.

Tuttle, R. H. (1970). Postural, propulsive, and prehensile capabilities in the cheiridia of chimpanzees and other great apes. *In* "The Chimpanzee" (G. H. Bourne, ed.) Vol. 2, pp. 167–253. Karger, New York.

Von Koenigswald, G. H. R. (1969). Miocene Cercopithecoidea and Oreopithecoidea from the Miocene of East Africa. *In* "Fossil Vertebrates of Africa" (L. S. B. Leakey, ed.), Vol. 1, pp. 39–52. Academic Press, New York.

Zapfe, H. (1960). Die Primatenfunde aus der miozänen Spaltenfüllung von Neudorf an der March, Tschechoslowakei. *Schweiz. Paleontol. Abhandl.* **78**, 1–293.

6

Structure and Function of the Primate Scapula

DAVID ROBERTS

Introduction

This chapter reviews the relationship between the mechanical features of the scapula and shoulder region and the characteristics of locomotor behavior.

Previous studies of the primate scapula are represented by a variety of approaches. Only a few of these have attempted to relate form and function. Some of the more important studies are Mivart (1868) and Schultz (1930) on descriptive morphology; Davis (1949), Müller (1967) and Oxnard (1967, 1968, 1969) on comparative morphology; Miller (1932), Ashton and Oxnard (1963) and Ashton *et al.* (1971) on comparative myology; Gregory (1912), Osborn (1900, 1929), Smith and Savage (1956), Müller (1967), and Gray (1968) on biomechanical analysis; Inman *et al.* (1944), Basmajian (1967), and MacConaill and Basmajian (1969) on electromyography; Howell (1917) and Wolffson (1950) on experimental investigation of development; Ashton and Oxnard (1963) and Ashton *et al.* (1971) on statistical analysis; and Oxnard (1969) on stress analysis.

Experimental investigation of development clearly indicates that scapular structure is determined largely by the stresses generated by its associated musculature. In the absence of muscular stress on the scapula of the dog, only the infraspinous fossa develops in a manner approaching normal, although even in this part the structure is poorly defined (Fig. 1). The supraspinous fossa, the vertebral border, and the spine (including the acromion) develop abnormally or not at all (Howell, 1917).

Multivariate analysis of various parameters of the scapula shows that

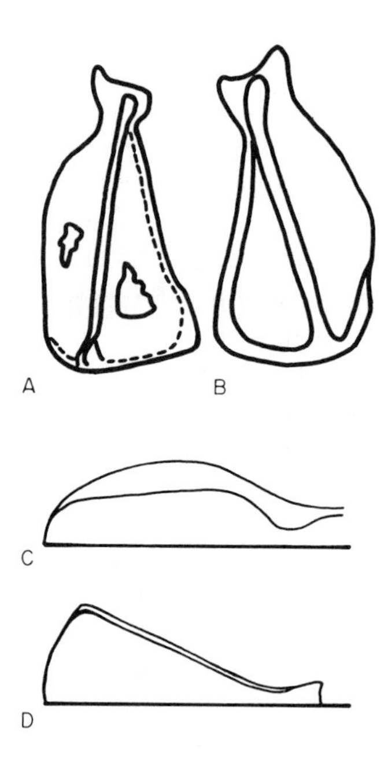

Fig. 1. Abnormal scapular development resulting from the elimination of muscular stresses. (A) Scapula of a 17-week-old dog showing abnormal development as a result of transecting the associated brachial plexus, thus removing muscular stress from the bone during postpartum development. (B) Normal scapula of a 17-week-old dog. (C) Comparison between the supraspinous fossae of the scapulae in (A) and (B). The area of the fossa of the abnormal bone is considerably less than that of the normal bone. (D) Comparison between the infraspinous fossae of the scapulae shown in (A) and (B). The area of the fossa of the abnormal bone is less than that of the normal bone, but comparison between (C) and (D) shows that the supraspinous fossa of the abnormal bone is relatively less well developed than the infraspinous fossa. (Adapted from Howell, 1917.)

"grouping" occurs in the spectrum of primate scapulae (Ashton *et al.* 1971). The groups correspond closely with natural locomotor divisions, indicating quite clearly that a deterministic relationship exists between scapular form and function. Statistical analysis by itself is not, however, a complete method of study because it does not explain the relationship in functional terms.

Structure of the Scapula

The scapula consists of a flat or slightly concavoconvex plate which is divided on its dorsal surface into two fossae by the scapular spine. The spine itself is a flat plate of bone developed in a plane roughly perpendicular to that of the blade. This perpendicular arrangement of the planes of the blade and spine imparts considerable mechanical rigidity to the structure.

Both blade and spine are thickened along their free margins. The plates of bone enclosed within the marginal thickenings are extremely thin and may even be fenestrated. The arrangement of bony plates and marginal thickenings in effect produces a three-dimensional framework, which is clearly a structural arrangement designed to withstand patterns of stress imposed by weight and by forces associated with muscle contraction (Fig. 2). The framelike nature of the bone has been noted before but seems to have been regarded as merely a mechanical system by means of which the body weight could be transmitted to the limb. The bone obviously does undertake this latter function, and must be structured to do so, but consideration of the multiple functions of the shoulder girdle, together with the complex structural morphology of the scapula, suggests that the framework is concerned with more than simple weight transmission.

Determination of Structural Units within the Scapula and Their Function

Structural strengthening of the scapula as a result of localized thickening of the bone was investigated directly by measuring the thickness of the blade over its entire surface. A pair of sliding callipers calibrated to 0.1 mm was used, measurements being made perpendicular to the subscapular surface. The thickness values thus obtained were plotted onto a profile of the blade (true area projection) and contoured at intervals of 1 mm. By this method quantitative results were obtained which not only

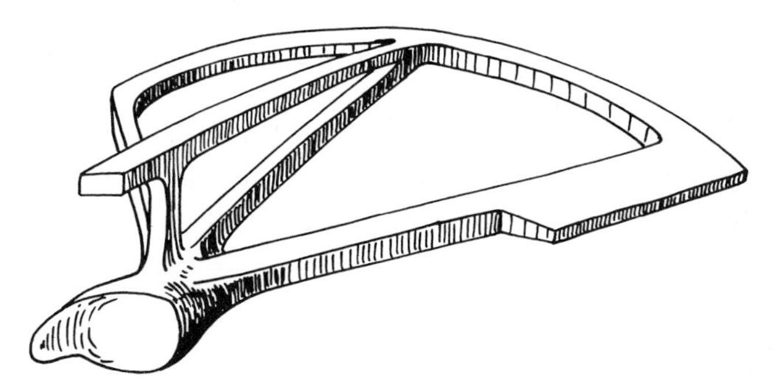

Fig. 2. The structural framework of the scapula. The marginal thickenings of the scapular blade and spine form a framework which resists the stresses generated by the contractions of the associated musculature.

demonstrated the pattern of structural thickenings but also permitted comparison of absolute thickness between various bones. Qualitative results may be obtained by photographing scapular blades using transmitted light (Fig. 3). Thinner areas of bone are indicated by their translucency, in contrast to the more opaque thickened areas. The lateral (axillary) border of the primate scapula is invariably thickened. In many genera, it is further strengthened structurally by being concave along its free edge, producing a fluted appearance. This latter feature also increases the surface area available for muscle attachment. The portion of the vertebral border that encloses the infraspinous fossa is also generally thickened, although not to the same extent as the lateral border. The intersection of the scapular spine with the blade forms a third structural member, so that the infraspinous fossa is surrounded by a subtriangular framework of thickened bone.

Development of a similar structural framework about the supraspinous fossa is more variable, tending to be poorly developed in some genera. Similarity of scapular structure among genera with disparate body weights but with similar locomotor habits (such as, *Ateles* and *Hylobates*) is an indication that structure is related to locomotor function rather than body weight, although it is to be expected that the latter may have some effect.

In *Pongo* and *Gorilla* and to some extent in *Pan* and *Homo,* where the broadly developed infraspinous fossa indicates a massive infraspinatus muscle, the basic structural triangle of the infraspinous fossa is further "triangulated" by the development of subsidiary struts radiating across the fossa from the glenoid region to the vertebral border (Fig. 4, Nos. 4 and 5). These secondary structural thickenings may represent, in part at least, further development of the muscular lineations present on the inferior surface of the scapula in many other genera, and which presumably also act as structural members. In the gorilla, secondary triangulation occurs in both fossae.

In structural mechanics the triangle is regarded as a basic stress unit. An equilaterally triangular structure withstands stresses applied in almost any direction within the plane of the unit. When the direction

---

Fig. 3. The structural frameworks of primate scapulae photographed by means of transmitted light: (A) *Ateles;* (B) *Pan;* (C) *Gorilla;* (D) *Homo;* (E) *Papio;* (F) *Alouatta.* Structures resulting from local thickening of bone are indicated as dark areas. Note the development of secondary thickenings in the fossae of (A), (C), and (D) and the poor development of the structural thickening of the coracoid border in (D) and (E).

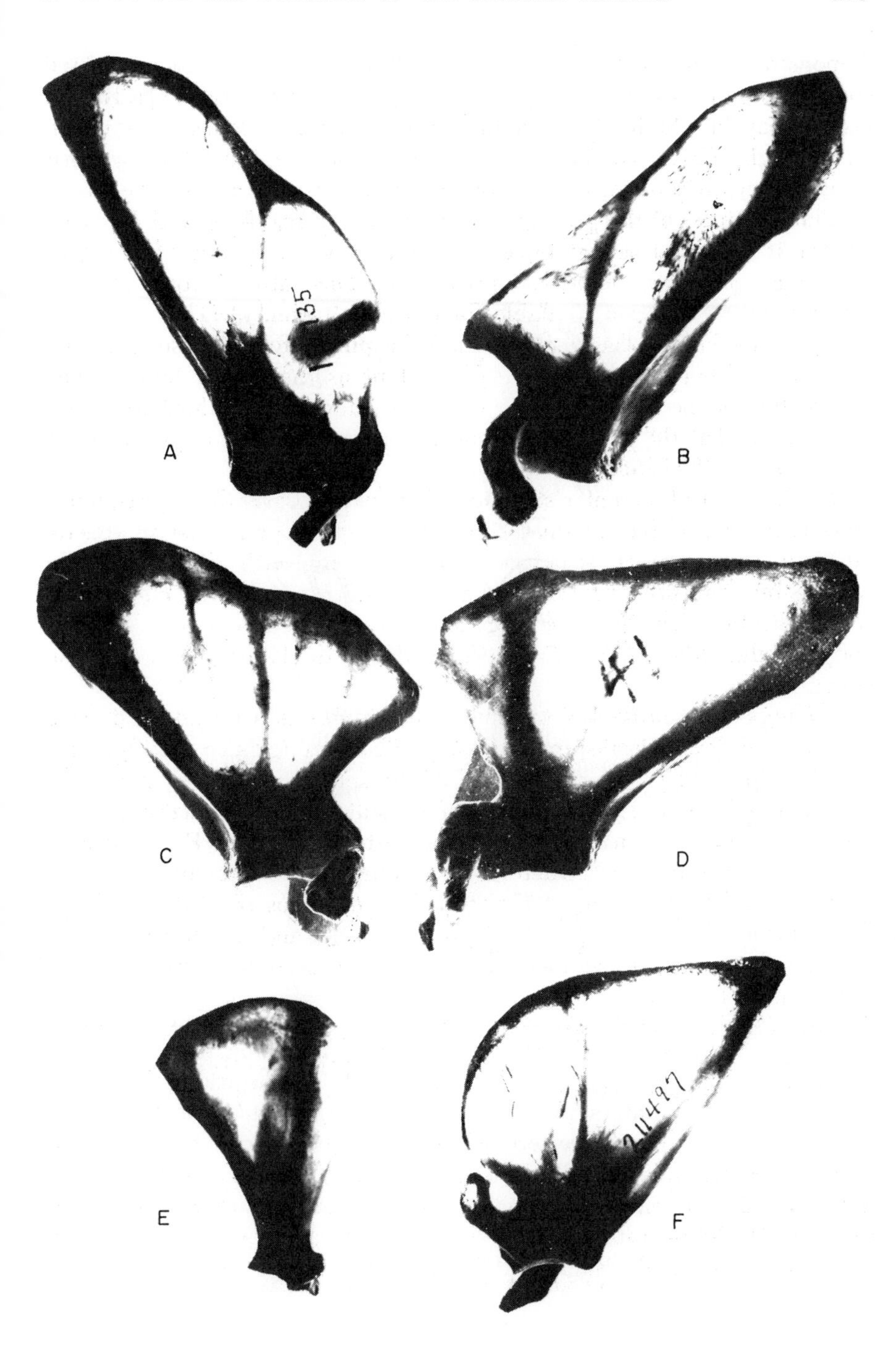
135
A
B
C
41
D
E
21497
F

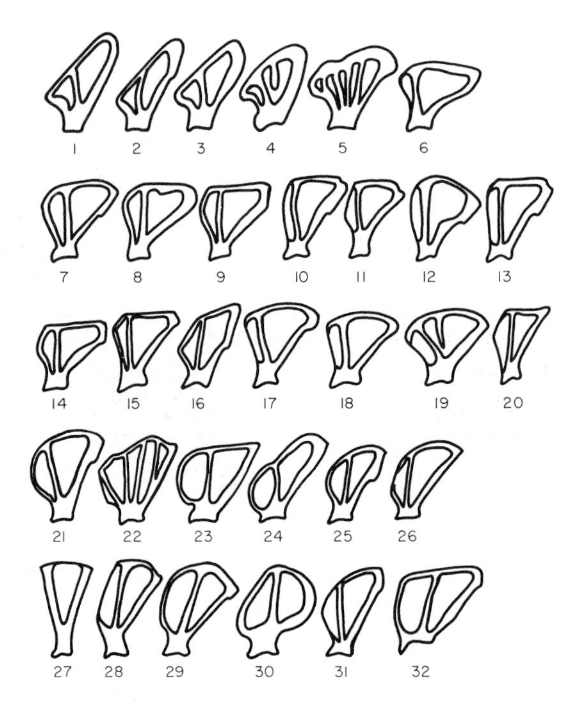

Fig. 4. Framework structures of primate and other mammalian scapulae determined by means of transmitted light, as explained in the text: 1, *Hylobates lar;* 2, *Pan paniscus;* 3, *Pan troglodytes;* 4, *Pongo pygmaeus;* 5, *Gorilla gorilla;* 6, *Homo sapiens;* 7, *Cercopithecus mitis;* 8, *Colobus* sp.; 9, *Presbytis entellus;* 10, *Cercopithecus aethiops;* 11, *Erythrocebus patas;* 12, *Macaca* sp.; 13, *Papio papio;* 14, *Propithecus diadema;* 15, *Indri indri;* 16, *Lemur variegatus;* 17, *Nycticebus coucang;* 18, *Loris tardigradus;* 19, *Perodicticus potto;* 20, *Galago crassicaudatus;* 21, *Cebus albifrons;* 22, *Lagothrix* sp.; 23, *Alouatta belzebul;* 24, *Ateles belzebuth;* 25, *Saimiri sciureus;* 26, *Callithrix chryseleuca;* 27, *Equus equus;* 28, *Canis* sp.; 29, *Felis lynx;* 30, *Ursus* sp.; 31, *Tupaia* sp.; 32, *Myrmecophaga* sp.

of the stress is normally unidirectional, or over a limited range of deviation from the axis bisecting the unit, a structure taking the form of an isosceles triangle is adequate. In primate scapulae, the triangular design varies from subequilateral in the gorilla and orangutan to subisosceles with relatively narrow bases (vertebral border) in the baboon and terrestrial cercopithecines (Fig. 4). The scapula, however, serves a variety of functions, such as providing a framework for the musculature that stabilizes the shoulder joint, establishing leverage for m. teres major, and so on. These functions make slightly different structural demands so that the scapular blade, like most structures in the vertebrate skeleton, is a compromise form.

STRESS ANALYSIS

Oxnard (1967, 1969) showed that the *shape* of the scapula was related to the type of stress normally imposed upon it by compressively or torsionally loading representative scapular silhouettes made of uniformly thick photoelastic material. Interference patterns created in light passing through the stressed plastic were recorded photographically, and indicated the distribution of distortion taking place within the simulated bone. Oxnard concluded that the scapula of quadrupedal primates was best suited to withstanding compressive forces applied along the vertebral border, whereas the form of the hominoid scapula was best suited to withstanding torsional stresses applied across the blade by the muscles which rotate it on the thorax. This study did not consider the distribution of forces with regard to the structural thickenings of the blade noted above, or the force vectors due to simulated muscles.

I have made a photostress study of various scapulae in an attempt to include some of those factors which Oxnard's study neglected. Plastic casts were made of representative scapulae. These were silvered and then coated with a thin photoelastic plastic film on the subscapular surface. Fine copper wires were inserted into the simulated bone along the areas of muscle insertion. Bundles of wires thus simulated the infraspinatus, supraspinatus, serratus magnus, rhomboid, teres major, and levator scapulae muscles. The scapulae were mounted on a board in such a way that they were supported at the glenoid but free to rotate about that point in the plane of the blade. The bundles of simulated muscle fibers were then loaded so as to hold the bone in equilibrium. The force vectors were known, and it could be shown that their components lay within the structural framework of the bone (Fig. 5). The apparatus was then observed by means of a reflection polariscope. Distortion within the simulated bone remained minimal, even at quite high loadings. Conversely, if the simulated muscle forces were arranged so as not to act through the structural framework, considerable distortion took place; even small loads applied in this way were sufficient to cause fracture in some specimens. Attempts to rotate the scapula about the glenoid by increasing the tension on a single muscle also caused distortion unless the force vectors of the other muscles were also adjusted.

Although some objections may be raised to the present technique, the

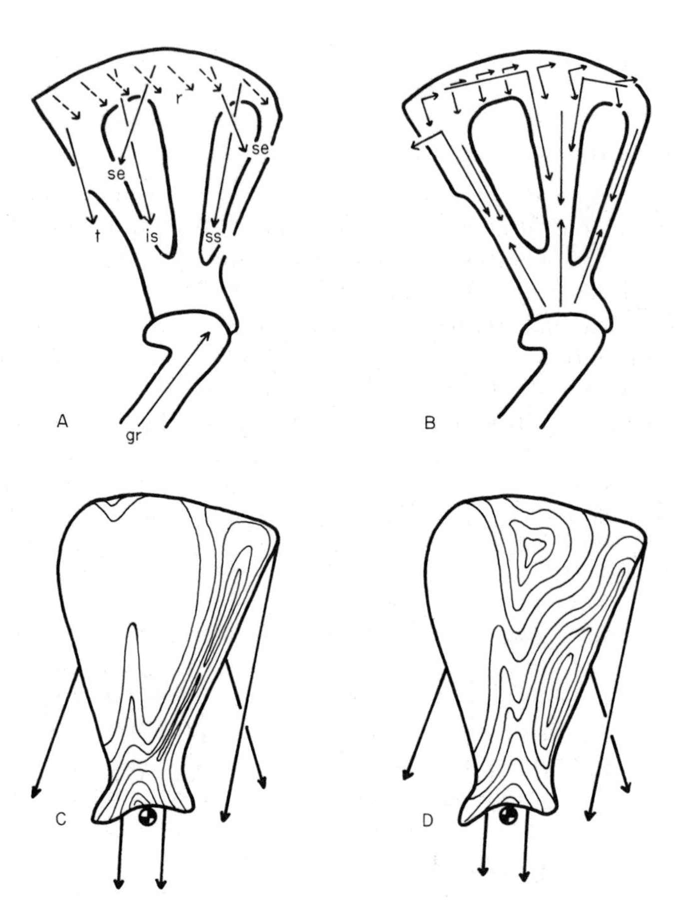

Fig. 5. Stress analysis of the scapula. (A) Distribution of forces within the scapular blade as a result of muscular contractions. gr, Ground reaction; r, rhomboideus; se, serratus; is, infraspinatus; ss, supraspinatus; t, teres major. (B) Components of the forces shown in (A). All components lie within the scapular framework and cancel each other out so that the scapula is in mechanical equilibrium. (C) Birefringence patterns formed by light reflected from a simulated scapula. The bone is in mechanical equilibrium. (D) Birefringence patterns from bone subjected to stresses not in mechanical equilibrium. Strain is indicated in the infraspinous fossa and the bone was at the point of fracture in that region.

experimental simulation does appear to confirm the hypothesized mechanical basis of scapular function. The studies reveal that the structure of the scapula blade reflects a pattern of muscular and gravitational force distribution consistent with joint stabilization and limb function in much the same way as does the areal development of the fossae to be discussed below.

Microstructure of the Scapula

Bone is adapted to withstand patterns of mechanical force not only by the development of gross structures but also by trabecular orientation. The preferred orientation of trabeculae in the cancellous portion of bone is known to correspond closely with the direction of stresses acting on the bone. Scapulae representing major locomotor groups were studied radiographically to determine their trabecular arrangements; these were found to be parallel to the gross structural members already described. At the junction of structural members the trabeculae form an intersecting pattern (Fig. 6). In the vertebral border of some specimens (Fig. 6A and C), some trabeculae are developed perpendicular to the long axis of the border, suggesting that stress acts perpendicularly as well as along the border. In the glenoid region of all the specimens studied, trabecular

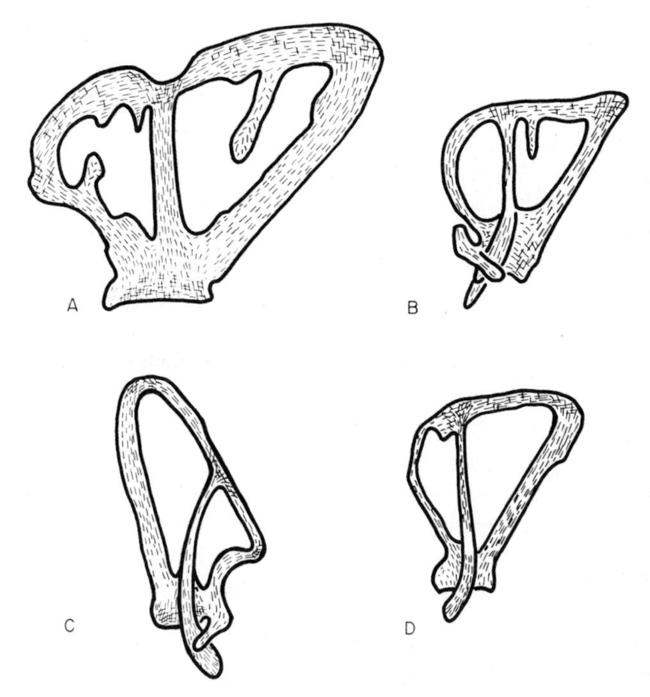

Fig. 6. Trabecular orientation. (A) *Gorilla gorilla;* (B) *Alouatta belzebul;* (C) *Hylobates lar;* (D) *Cercopithecus mitis.*

sheets form concentric layers about the articular facet and intersect with those developed in the borders of the bone. The bone of the infraspinous and supraspinous fossae lacks a cancellous layer and trabeculae are therefore absent.

Scapular Form–Function Relationships

Scapular Fossae

The supraspinatus, infraspinatus, teres minor, and subscapularis muscles comprise the so-called "rotator cuff" of the shoulder. They provide stabilization for the shoulder joint as well as initiating and assisting movement of the forelimb. The supraspinatus and infraspinatus muscles are contained within the supraspinous and infraspinous fossae, respectively (while m. subscapularis lies within the subscapular fossa which embraces the whole of the inferior surface of the bone). Both are multipennate, an indication of a powerful contraction but relatively limited excursion. Insofar as both muscles fill their respective fossae, the areal development of the fossae is therefore an indication of the relative bulk of the muscles.

Analysis of the differential development of the scapular fossae may thus be pertinent to understanding the relationship between scapular form and locomotor habit. By reducing profiles of the scapular blades of different genera to the same scapular spine length, the differential development of the supraspinous and infraspinous fossae may be appreciated. The spine is common to both fossae, as well as providing a measure of the "length" of the blade. Comparison is otherwise meaningless because even in bones which are the same shape (same species) but of different size, the length (and breadth) of the bones increases by a linear function, whereas the corresponding differences in the areas of the fossae increase by a function of the second order.

The "supraspinal" and "infraspinal" indices calculated by Inman *et al.* (1944) are based on scapular length and breadth. The breadth measurement is not easy to define because consistent homologous points cannot be identified on the scapula, and these indices are probably less representative of the true development of the fossae as indicated by measuring their areal extent. Inman *et al.* (1944) have suggested that "the supraspinous portion of the scapula is very conservative and is altered but little in the proportion of the scapula which it makes up." However, Fig. 7 shows that this statement is not true. Across the spectrum of scapulae of primate genera the variation in development of the supraspinous fossa is quite marked. In the proportion of the scapular blade which is made up by the supraspinous fossa, there is considerable difference between, for example, *Pongo* and *Ateles,* or *Perodicticus* and *Galago.* For most of the genera included in Fig. 7 only small samples of scapulae were available, and it was thus necessary to investigate the amount of intraspecific variation which might occur. Supraspinous and infraspinous indices were calculated for a larger sample (sixteen individuals) from *Cercopithecus mitis,* and variation was found to be relatively small.

Figure 7 demonstrates the fact that in primates the area of the infraspinous fossa is invariably greater than that of the supraspinous fossa. Furthermore, there is a trend through certain groups of primate genera for the area of the infraspinous fossa to increase relative to the length of the blade, and also relative to the area of the supraspinous fossa. The New World genera, *Ateles, Alouatta,* and *Lagothrix,* in particular, demonstrate this characteristic. Likewise there is a similar trend for the area of the supraspinous fossa to increase relative to the length of the blade, although not relative to the area of the infraspinous fossa, passing from terrestrial quadrupeds to arboreal quadrupeds and quadrumanous climbers and arm-swingers.

Figure 8 shows an arrangement of primate and nonprimate scapulae based on general structure and morphology. Genera with similar locomotor behavior characteristics tend to be grouped together by this method of arrangement. A relationship between form and function is thus implied.

The locomotor groups may be defined and the characteristics of their scapular fossae described in the following manner:

Group I (*Tarsius* sp., *Galago crassicaudatus*) consists of prosimians that are highly specialized for vertical-clinging and leaping. The scapula is long and narrow, forming a simple strutlike lever arm. The infraspinous fossa may be "hooked," or there may be development of a postscapular fossa to provide leverage for m. teres major. These latter features are often associated with a narrow infraspinous fossa; in other genera where the infraspinous fossa is broadly developed they are usually lacking.

Group II (*Papio papio, Macaca* sp., *Cercopithecus aethiops, Erythrocebus patas*) includes the terrestrial and semiterrestrial monkeys. In these genera the scapula is also long and narrow and tends to lack structural thickening of the coracoid border. In many structural respects the scapulae of this group tend to be intermediate between those of the arboreal quadrupedal monkeys, on the one hand, and those of semicursorial mammals, such as the dog, on the other.

Group III (*Lemur catta, L. macaco, L. variegatus, Indri indri, Propithecus diadema*) comprises those prosimians which are usually described as vertical-clingers and leapers, but which are rather less specialized in this respect than the group I prosimians. The main locomotor divergences between the group I and group III prosimians appear to be the ability of the latter to be terrestrially and arboreally quadrupedal (*Lemur*), and to climb up vertical supports by "shinning" in a bearlike manner (*Indri* and *Propithecus*); it should be noted, however, that on occasion *Galago* walks quadrupedally along large branches. Development of the scapular infraspinous fossa is greater in group III prosimians than in group I, particularly in *Indri* and *Propithecus*. This presumably correlates with the climbing habit mentioned previously. The supraspinous fossa in all group III genera is relatively better developed than in group I, probably reflecting a more consistent habit of reaching the forelimb above the level of the shoulder and in fact greater use of the forelimb in locomotion generally. Group III could probably be further subdivided on the basis of the locomotor differences between the genera mentioned.

Group IV (*Nycticebus coucang, Loris tardigradus, Perodicticus potto*) contains the slow-climbing prosimians, the scapulae of which differ from those of group III by having a slightly less well-developed supraspinous fossa and a broader infraspinous fossa.

Group V (*Cercopithecus talapoin, Cercopithecus mitis, Colobus* sp., *Cebus albifrons, Presbytis entellus*) includes the arboreal quadrupedal monkeys. This group overlaps group III and group IV prosimians in Fig. 7 which may indicate some degree of functional convergence. The development of the infraspinous fossa ranges from about the same as that in the terrestrial monkeys to being about 60% larger. The supraspinous fossa is, however, generally larger than in the terrestrial quadrupeds and shows some evidence of developing structural thickening of the borders.

Group VI (*Saimiri sciureus, Callithrix chrysoleuca*) includes the scapulae of arboreal quadrupedal monkeys which are similar in morphology to those of group V genera. They are separated here partly because *Callithrix* has clawed digits, which must affect the preferred mode of locomotion of that genus in some way, and partly because the scapulae of both genera appear to be somewhat intermediate in form between those of the genera in group V and that of *Tupaia*.

Group VII (*Ateles belzebuth, Lagothrix* sp., *Alouatta belzebul*) contains New World monkeys which have a very wide range of locomotor habits, including quadrumanous climbing and arm-swinging. The area of the infraspinous fossa is not greater than in groups IV and V but the supraspinous fossa is greatly enlarged, being greater than even that of the hominoids. These scapulae commonly have well-developed muscular lineations across the subscapular fossa.

Group VIII (*Hylobates lar, Pongo pygmaeus*) is made up of the predominantly arboreal hominoids. In *Pongo* the infraspinous fossa is very broad, so much so that the supraspinous fossa by comparison appears to be poorly developed. Examination of Fig. 7, however, reveals that the

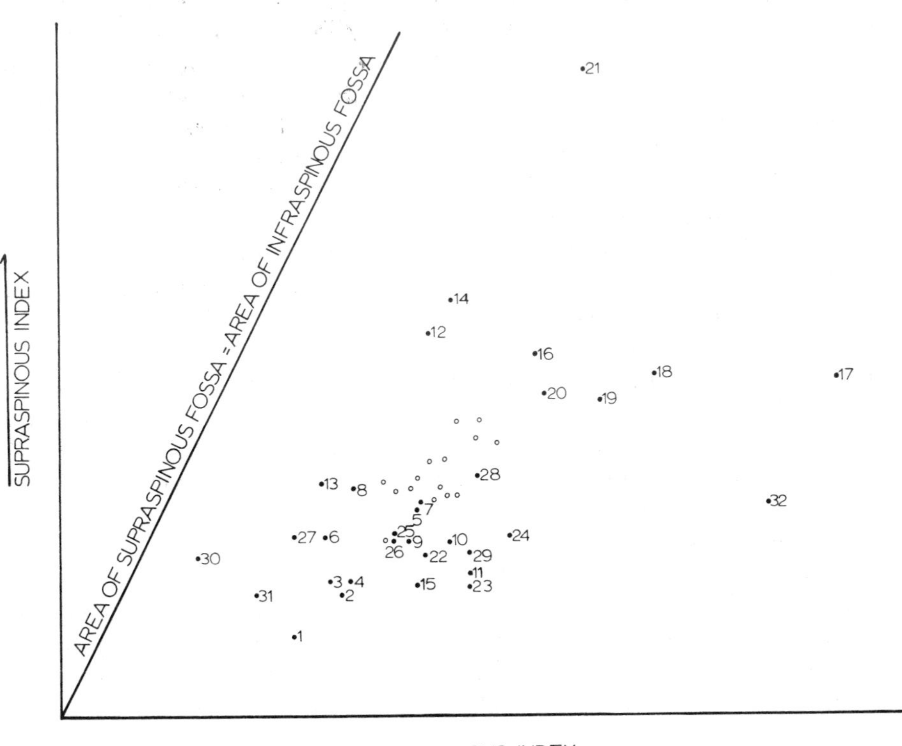

Fig. 7. Bivariate plot of supraspinous and infraspinous indices: 1, *Papio;* 2, *Erythrocebus;* 3, *Cercopithecus aethiops;* 4, *Macaca;* 5, *Callithrix;* 6, *Saimiri;* 7, *Cercopithecus talapoin;* 8, *Cercopithecus mitis;* 9, *Cebus;* 10, *Presbytis;* 11, *Colobus;* 12, *Ateles;* 13, *Aotus;* 14, *Alouatta;* 15, *Cacajao;* 16, *Lagothrix;* 17, *Pongo;* 18, *Pan troglodytes;* 19, *Pan paniscus;* 20, *Hylobates;* 21, *Gorilla;* 22, *Nycticebus;* 23, *Loris;* 24, *Perodicticus;* 25, *Lemur catta;* 26, *Lemur variegatus;* 27, *Lemur macaco;* 28, *Propithecus;* 29, *Indri;* 30, *Galago;* 31, *Tarsius;* 32, *Homo.* The points plotted are means for samples of genera ranging from two to thirty individuals. The circles represent individual indices for fifteen individuals of the species *Cercopithecus mitis.*

supraspinous fossa is, in fact, better developed than in any of the groups described previously with the exception of group VII genera. The very broad infraspinous fossa reflects the quadrumanous climbing habit, while the large, structurally strengthened supraspinous fossa probably reflects the great range of circumduction possible with the forelimb raised above the shoulder level. The general overall increase in the area of the fossae also reflects the much greater muscular force required to secure the shoulder joint in animals which habitually suspend themselves by their forelimbs (Fig. 8). The scapula of *Hylobates* differs from that of *Pongo* in having a relatively narrower infraspinous fossa. This seems to relate to the habit of avoiding lifting the body weight in the manner characteristic of *Pongo* but is also probably involved in the complex rearrangement of the general morphology of the bone in response to a highly specialized form of locomotion. The functional role of the gibbon scapula will be discussed more fully later.

Group IX (*Gorilla gorilla, Pan paniscus, Pan troglodytes*) comprises the knuckle-walking hominoids. Considerable overlap occurs between this group and group VIII demonstrating the primate ability to progress by a variety of different locomotor modes which makes categorization of locomotion so difficult. The knuckle-walkers possess very broad scapulae with large fossae. In *Gorilla,* which is the genus most highly adapted for this mode of quadrupedalism, the supraspinous fossa is relatively larger than in any other primate genus, reflecting the specialized nature of the forelimb—which is very long, presumably to throw the center of gravity of the body back over the hindlimbs (rather than for armswinging) and which is massively built with long proximal elements as in graviportal animals. The feeding habits involve frequent use of the limb in an extended and abducted position which would account in part for the massive development of the supraspinous fossa; but *Gorilla,* as a quadruped, also has special problems in stabilizing the shoulder joint (Fig. 9).

Group X (*Homo sapiens, Australopithecus* sp.) is reserved for the specialized bipedal hominids. *Australopithecus* is not represented in Fig. 7 because the single scapular fragment which has been described lacks both fossae. In *Homo* the infraspinous fossa is very broad and is bordered by thick bony margins. The supraspinous fossa is relatively poorly developed, reflecting the habitual use of the arm with the humerus held below the level of the shoulder. MacConaill and Basmajian (1969) have observed that in man supraspinous muscle activity supplements the action of the capsular ligaments and the coracohumeral ligaments in supporting the weight of the pendant limb.

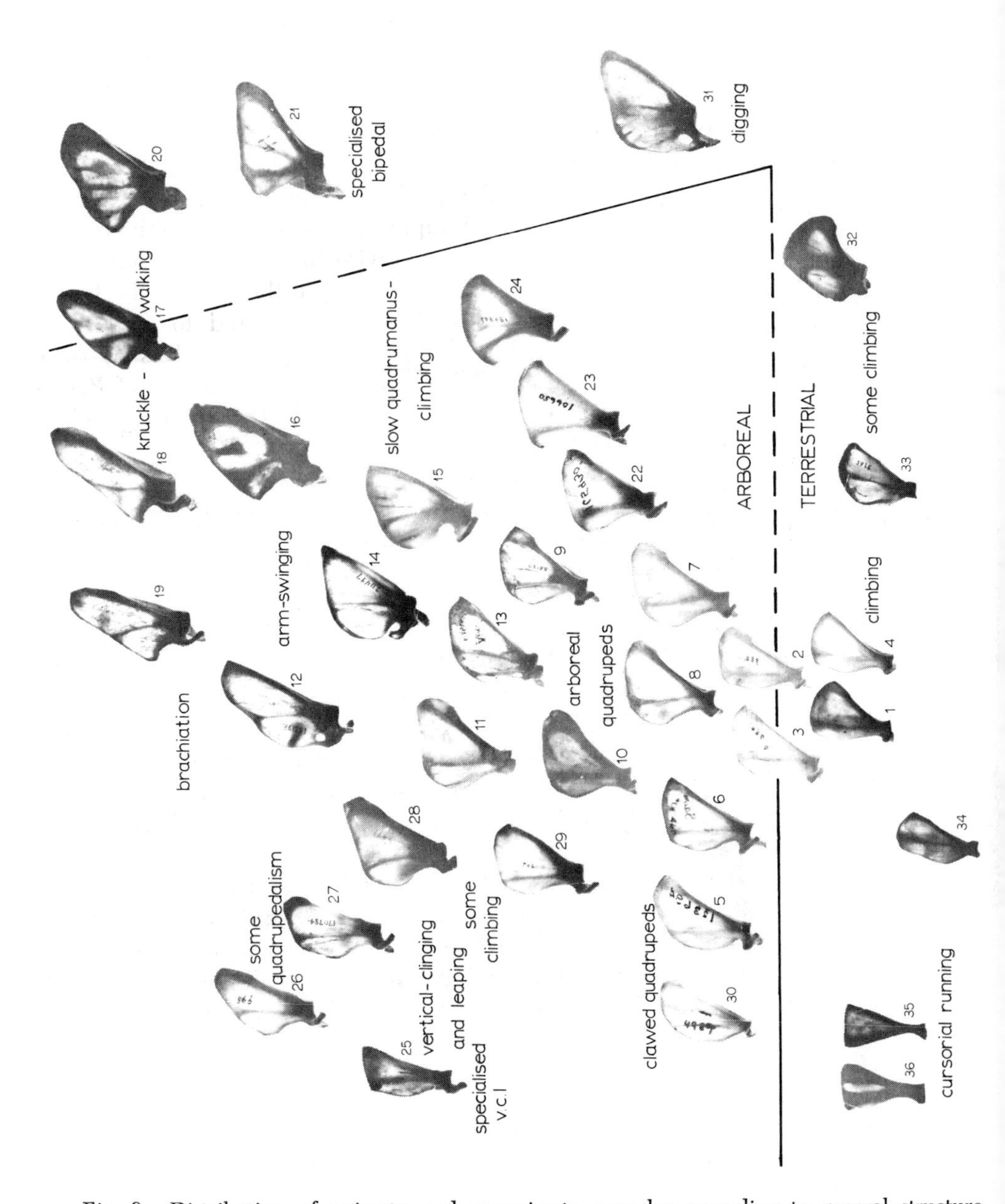

Fig. 8. Distribution of primate and nonprimate scapulae according to general structure and morphology. Scapulae are reduced so that the length of the base of the scapular spine is constant. 1, *Papio;* 2, *Erythrocebus;* 3, *Cercopithecus aethiops;* 4, *Macaca;* 5, *Callithrix;* 6, *Saimiri;* 7, *Cercopithecus talapoin;* 8, *Cercopithecus mitis;* 9, *Cebus;* 10, *Presbytis;* 11, *Colobus;* 12, *Ateles;* 13, *Aotus;* 14, *Alouatta;* 15, *Lagothrix;* 16, *Pongo;* 17, *Pan troglodytes;* 18, *Pan paniscus;* 19, *Hylobates;* 20, *Gorilla;* 21, *Homo;* 22, *Nycticebus;* 23, *Loris;* 24, *Perodicticus;* 25, *Galago;* 26, *Lemur variegatus;* 27, *Lemur macaco;* 28, *Propithecus;* 29, *Indri;* 30, *Tupaia;* 31, *Moropus;* 32, *Ursus;* 33, *Felis;* 34, *Canis;* 35, *Cervis;* 36, *Equus.*

Stabilization of the Shoulder Joint

The shoulder joint is secured against dislocation principally by muscular action, although there are tendinous thickenings in the joint capsule. In quadrupedal animals, it is probable that relatively little muscular effort is required to provide joint stabilization because the reaction between the body weight and ground acts compressively through the joint (Fig. 9). This is so in passive standing and the retraction "power-stroke" of the limb. Potential tensile stress across the joint occurs only when the forelimb is lifted from the ground and will be dependent upon the weight of the limb. Cursorial forms possessing relatively slender, light limbs, usually also have narrow fossae. Noncursorial terrestrial mammals, which have more heavily built forelimbs, typically have larger fossae. Development of the fossae also relates to increased versatility of limb use. For example, in orthograde climbing animals, suspension of the body weight by the arms produces a large and more or less constant potential tension across the shoulder joint (Fig. 9). The rotator cuff muscles are relatively well developed in order to contend with this stress, resulting in broad fossae, typical of the hominoids and certain New

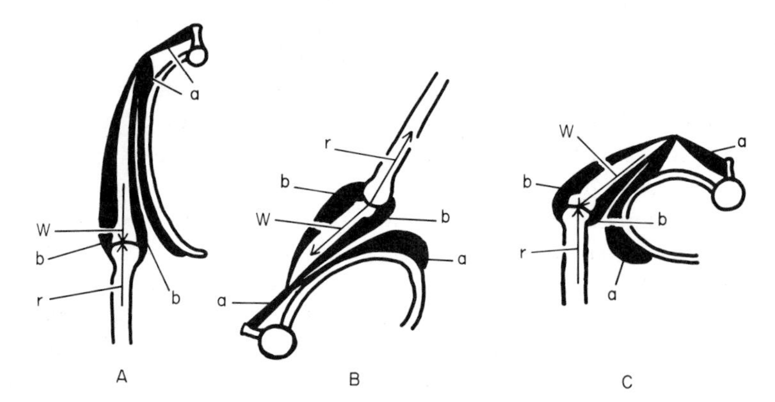

Fig. 9. Potential stresses acting across the shoulder joint in different primates. (A) A quadruped; the potential stress is compressive during the power stroke of the limb. Potential tension is relatively small and occurs only when the limb is lifted from the ground. (B) A quadrumanous climber and arm-swinger; the potential stress across the shoulder joint is mainly tensile. Body weight cannot be supported by m. serratus, this function being performed by m. rhomboideus. (C) A knuckle-walker; reorientation of the thorax prevents the scapula from being oriented as in (A) although knuckle-walking is a form of quadrupedalism. There are thus shearing forces potentially acting across the shoulder joint. a, m. serratus and m. rhomboideus; b, m. subscapularis; r, substrate reaction to body weight; W, body weight.

World genera. The trend, demonstrated in Fig. 7, toward an increase in the relative areas of the fossae in primates, passing from terrestrial quadrupeds through arboreal quadrupeds to quadrumanous climbers, reflects a progression in the amount of potential tension across the joint. The relationship is not exact, however, because other functional requirements are placed upon the shoulder girdle.

LIMB MOVEMENT

The musculotendinous cuff muscles play an important part in limb movement. Inman *et al.* (1944) and MacConaill and Basmajian (1969) have observed that they are active throughout abduction and flexion of the humerus in man, and that "complete paralysis of supraspinatus in man . . . reduces the force of abduction and the power of endurance. In abduction the supraspinatus plays only a quantitative and not a specialized role." However, the size of the infraspinous and supraspinous fossae alone cannot be regarded as being diagnostic of specific forelimb usage because considerable parallelism of mechanical function occurs between animals of dissimilar locomotor habit. Yet it is pertinent to discuss locomotor groups in terms of the characteristic limb posture and movement and relate these to scapular morphology as a whole.

Many nonprimate mammals, such as the horse, cat, and bear possess modes of locomotion which are more stereotyped than is usually the case in primates. It is thus useful to consider these genera briefly before turning to the primates.

The horse, adapted for fast running, possesses scapular fossae that are very narrow, particularly the supraspinous fossa. The structural framework is in effect an isosceles triangle with a very acute apex, and together with the long neck (the region of the scapula between the fossae and the glenoid) forms an elongate, leverlike bone. The glenoid articular facet is nearly hemispherical, in contrast to the pear-shaped facet present in the majority of noncursorial mammals. Associated with this complex of scapular features is a forelimb that has a very restricted range of movement outside a parasagittal plane and that is extremely lightly built considering the body weight of the animal. The limb, which is subject to very rapid acceleration, is elongate in its distal segments so that the center of gravity is situated proximally (Gregory, 1912). The cat, less of a long-distance runner than the horse, possesses a scapula which is relatively much broader, forming two subtriangular units. The supraspinous fossa may account for as much as 50% of the total blade area and appears to be related to the massive forelimbs which not only serve in prey capture, but also absorb the considerable percentage of the kinetic energy generated by the animal in motion which is lost through the shock of the forelimbs contacting the substrate at the end of each leap (Fig. 10).

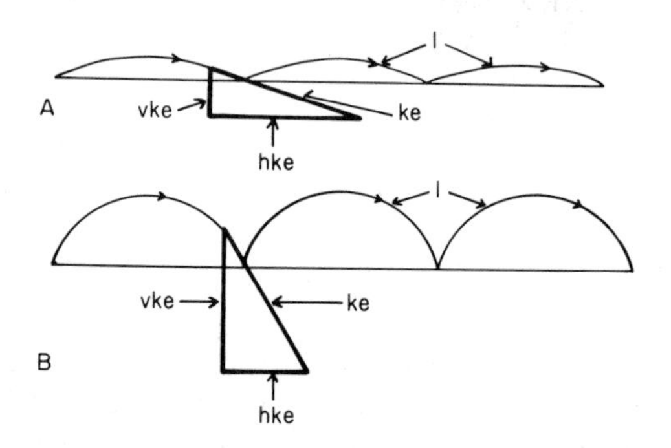

Fig. 10. A comparison of the kinetics of a cursor and a leaping galloper, and their influence upon the length of the vertebral border. (A) A cursor (*Equus*). The amplitude of the vertical displacement of the animal's center of gravity is small. Thus the vertical component of the animal's kinetic energy is also relatively small, the much larger horizontal component being available for continued progression with a minimum of energy expenditure. (B) A leaping-galloper (*Felis*). The amplitude of the vertical displacement of the animal's center of gravity is large. The horizontal component is relatively small and the animal needs a higher output of energy to maintain forward progression than does the horse. ke, kinetic energy of moving body; vke, vertical component of ke; hke, horizontal component of ke; l, locus of the animal's center of gravity during locomotion.

The dog is, in many ways, intermediate in locomotor behavior between the horse and cat and its scapular and forelimb characteristics are likewise intermediate in nature.

Fossorial mammals, such as the anteater and armadillo, possess very broad scapulae and powerful supraspinatus and infraspinatus muscles. This arrangement relates to the use of the limb in a powerful digging stroke and the required stabilization of the shoulder joint. In the anteater, and also in some extinct ground sloths, the scapula is further strengthened by the development of a bridgelike extension of the metacromion over the lateral part of the fossa to fuse with the coracoid border (Fig. 11). Besides imparting structural strength, this bridge also provides leverage for the deltoid muscle, thus enhancing protraction of the limb. A similar mechanical system is present in arm-swinging primates where the

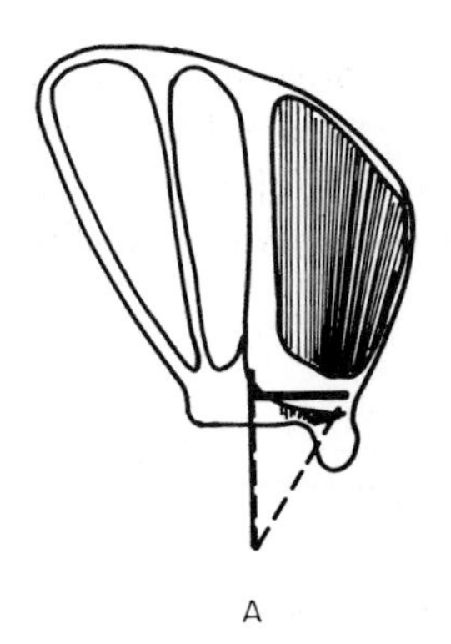

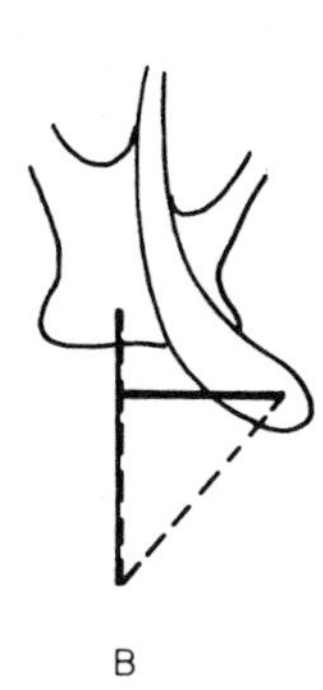

Fig. 11. Comparison of the acromion in *Mylodon* and a hominoid primate. (A) *Mylodon:* the acromion turns cranially to fuse with the coracoid border. This not only strengthens the supraspinous fossa considerably but also moves the origin of m. deltoideus cranially, thus enhancing leverage for extension of the limb. (B) A hominoid primate: a similar mechanical arrangement to that seen in *Mylodon* is seen but the acromion does not fuse with the coracoid border. This permits the fibers of m. deltoideus, which originate on the distal part of the acromion, to assist in abducting the limb as well as extending it. Moving the acromion cranially also permits abduction without the humerus coming into contact with the acromion.

acromion is "swept" cranially providing the necessary leverage for the deltoid muscle.

Davis (1949) noted that in bears the development of a postscapular fossa was related to the climbing habit, being a means of increasing leverage for the teres major muscle. Smith and Savage (1956) discussed generally the development of the inferior angle of the scapula in mammals. They showed that its development was influenced by the need to produce leverage for the teres major, resulting, for example, in the characteristic "hooked" shape of the infraspinous fossa in the seal, *Phocus*. Müller (1967) discussed the shape of the vertebral border in carnivores, concluding that it was determined by the mechanical loading applied by the body weight of the animal. Each of these studies attempted to relate the shape of the scapula to its mechanical functions, either as a means of providing leverage or as a means of absorbing the stresses produced by its associated musculature. The results and conclusions mentioned in this study attempt to expand this concept further, permitting more detailed analysis of scapular morphology in terms of biomechanical function. A very general summary based on the relationship between the size of the scapular fossae, shoulder joint stabilization and limb characteristics may be stated as shown in the following tabulation.

Large fossa	Small fossa
Potentially great shoulder joint stabilization possible	Potentially less shoulder joint stabilization possible
Limb relatively heavy	Limb relatively light
Limb motion relatively powerful	Limb motion relatively less powerful
Limb held or used extensively in an extreme protracted or retracted position	Limb used in a restricted manner

It must be stressed that these are the possible functional characteristics which may be associated with the size of the fossae. The existence of one such characteristic does not imply the existence of any of the others. The gibbon, for instance, requires considerable stabilization of the shoulder joint but has a lightly built limb. Conversely, a quadrupedal, graviportal animal with a heavy forelimb could have a relatively narrow scapula.

The implications of these general conclusions may be applied in a general resume of the morphology of the scapular fossae observed in different primate locomotor groups. In the semiterrestrial primates, especially *Papio* and *Macaca,* the scapula approximates the condition seen in the dog. The supraspinous fossa has little structural development and the blade is generally long and narrow. However, the terrestrial primates still retain considerable powers of climbing ability and as a result their scapulae represent a spectrum of types which range from doglike to forms similar to those of the arboreal quadrupedal monkeys. In the arboreal, quadrupedal monkeys the stresses across the shoulder joint are not always potentially compressive, and thus the muscular requirements for joint stabilization are relatively greater than in terrestrial forms. Furthermore, there is a greater tendency for the forelimb to be used in a protracted position, and this is expressed anatomically in the generally greater development of the supraspinous fossa. It is in the quadrumanous climbers and arm-swingers that the continual use of the forelimb above the shoulder and in lifting the body weight is associated with broad, well structured fossae. Among the hominoids, especially *Pongo* and *Pan,* the broad fossae probably also reflect the relatively heavy limbs and need for joint stabilization. In *Hylobates,* the limbs are lightly built and elongated distally because they must be capable of rapid acceleration in order to reach overhead supports. The scapula is slender and leverlike and thus in some respects the gibbon parallels, mechanically, the forelimbs of a cursorial animal. The major mechanical differences are the much greater range of circumduction of which the gibbon is capable and the degree of joint stabilization apparently required to offset the tensile stress generated in brachiation.

Scapular Motion with Respect to the Humerus

The motion of the scapula with respect to the humerus was observed in *Cercopithecus aethiops* (Fig. 12) using a cineradiographical technique described by Jenkins (1971). During protraction and retraction of the limb, the initial movement of the scapula preceded that of the humerus, i.e., the oscillations of the two bones were out of phase. However, at the end of each stroke, the scapula was observed to pause momentarily, permitting the humerus to catch up, so to speak.

Inman *et al.* (1944) studied glenohumeral movement in man, observing that the early phase of motion is highly irregular as the result of various scapulohumeral relationships which may occur before stabilization is attained. Stabilization here refers to a scapulohumeral orientation which is obtained in the first 30–60° of elevation; thereafter there is a ratio of two degrees of humeral to one of scapular motion. Some doubt has been expressed as to the validity of this observation. However, from a mechanical aspect the attainment of a "stabilized" condition, in which the stresses placed on the scapula form a system which is in equilibrium, seems logical. Inman *et al.* (1944) further stated that "elevation of the extremity . . . is simultaneously accompanied by scapulothoracic movement, an arrangement which critically enhances the power of the attendant muscles." The results obtained from the cineradiographical study of *Cercopithecus aethiops* mentioned above suggest that the advance motion of the scapula over the humerus provides an increase in leverage for the scapulohumeral muscles which are involved in limb movement. It can be shown by simple calculation that

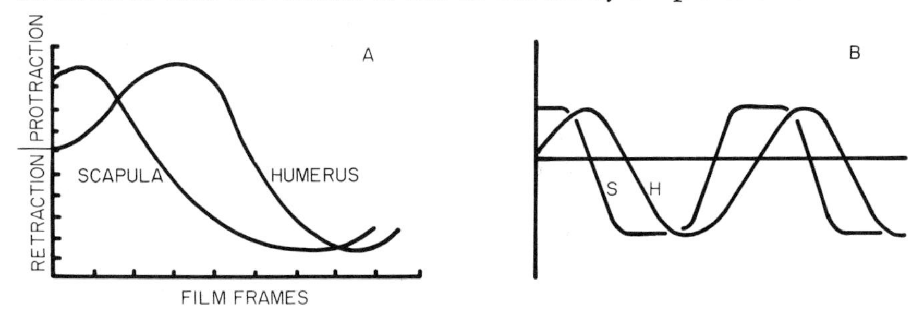

Fig. 12. Scapulohumeral rhythm in *Cercopithecus aethiops.* (A) Scapulohumeral rhythm determined by cineradiographical examination of *Cercopithecus aethiops.* The vertical axis represents anteroposterior movement of the bones. The intervals represent 5° of motion. (B) Schematic diagram to demonstrate interpretation of (A). The anteroposterior motion of the humerus (H) follows a flat-topped curve. The flat part of the scapular (S) curve represents a period of stabilization during which the muscles which are prime movers of the limb contract.

increasing the leverage of the teres major muscle by rotating the scapula caudally about the glenoid articulation is advantageous despite the slight shortening of the muscle (and consequent proportional loss in contracting power) which consequently occurs If the increase in leverage is as much as 100%, the loss of potential contracting power through shortening of the muscle is only about 17% (Fig. 13). The pause observed in the motion of the scapula in *Cercopithecus aethiops* is probably a period during which the stresses imposed on the scapula by the muscle contractions which move the limb are distributed through the structure of the bone in such a way as to produce a minimum distortion (that is, the stress pattern is in equilibrium with the structure). This "pause period" probably corresponds closely to the "stabilization" described by Inman *et al.* It should also be noted that Inman *et al.* (1944) also showed that the contractions of teres major occur predominantly during periods when the scapula is at rest on the thorax. This further supports the contention that the scapula must be in equilibrium with respect to the force components distributed through its structure before any major muscular effort to move the limb can take place.

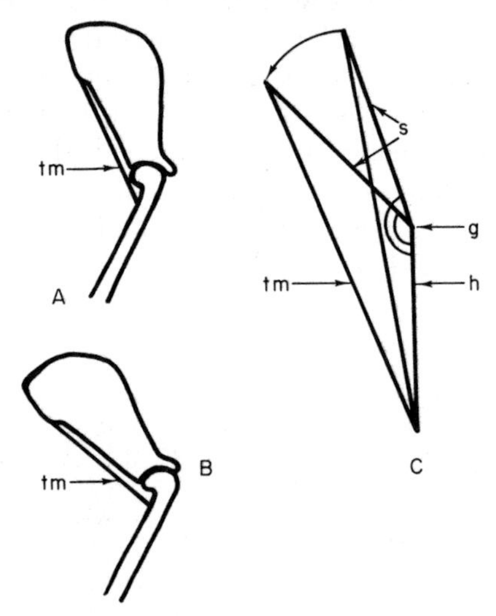

Fig. 13. Mechanical advantage of m. teres major. (A) Relative positions of the scapula and humerus with forelimb protracted but before the scapula has begun to move caudally. The m. teres major (tm) is acting on the humerus through a low angle which reduces its mechanical efficiency. (B) The scapula has "rocked" caudally in advance of the humerus thus widening the angle through which it acts on that bone. (C) Schematic representation of the scapula, humerus, and teres major relationships. The mechanical efficiency of m. teres major is a function of the cosine of the angle at g. As the angle decreases to 90° the efficiency of the lever system increases to a maximum. As the angle becomes less than 90° efficiency decreases.

Functional Implications of Scapular Shape

Inasmuch as scapular motion is so intimately related to forelimb movement, it seems likely that the shape and anatomical relations of the scapula are an important determinant in scapular movement. Although numerous studies have been conceived with the scapula as a structural unit, no studies have attempted to relate the shape of the scapula to its function as a dynamic structure.

It was shown above that the motion of the scapula is related to limb motion and probably varies with different modes of locomotion. In the horse and other cursorial animals the scapula is a slender leverlike bone which has a short, straight vertebral border and acute superior and inferior angles. Although the vertebral border is surmounted by a broad suprascapular cartilage, the scapula lies well beneath the line of the neck as the result of the long "cantilever" neural spines of the thoracic vertebrae and the well-developed neck musculature. The scapula is thus free to oscillate across a broad, relatively flat area. The detailed shape of the bone is therefore probably affected little by the nature of its dynamics. In the cat, however, the situation is rather different. The scapula is very much broader by virtue of the well-developed fossae. It also projects above the level of the neck as defined by the neural spines. Effectively, the scapula is a large structure contained in a small space, in contrast to the condition seen in the horse. In the cat, therefore, rotation of the scapula is to some extent restricted and would be even more so if the superior angle of the blade were not "rounded off" in a smooth curve (Fig. 14A). In the quadrupedal primates the situation is similar to that in the cat.

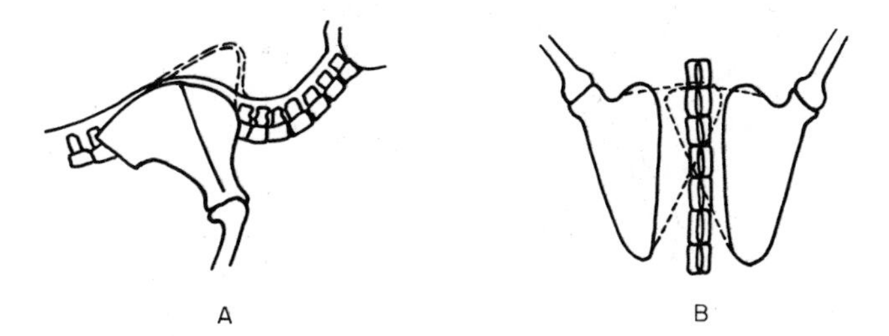

Fig. 14. Ergonomics of the scapula in a quadruped and a hominoid. (A) In quadruped, the superior angle of the scapula is "rounded-off," thus permitting the bone to slide forward beneath the neck without disrupting the neckline. (B) In hominoid, the superior angle is "removed" effectively by shortening the coracoid border, thus producing an "oblique" bone. When the scapula is drawn toward the midline the superior angle thus does not disrupt the structures over the backbone, and the two scapulae do not interfere with each other.

In the Hominoidea, a premium is placed on scapular mobility, but a long vertebral border and relatively large fossae are also required. Motion is thus enhanced, not by "rounding off" the superior angle, but by decreasing the length of the cranial border and increasing the length of the vertebral border. The resulting asymmetrical bone is capable of caudal rotation (depression) without the superior angle overriding the vertebral column or the angles of the blades interfering with each other (Fig. 14B).

The Vertebral Border

Müller (1967) concluded that the curved vertebral border of the scapula in carnivores corresponds with a mathematical model of an ideal structural form for withstanding the stresses imposed by body weight loading. However, in many noncarnivorous mammals, particularly primates, the vertebral border tends to be straight, especially along the infraspinous region; thus, weight loading may not be the only determinant of the form of the border or may be expressed in terms which have less effect on the morphology of the border than is indicated by Müller's model.

Aside from any consideration of the effect of stress distribution or bone dynamics, arching the vertebral border increases its length. Curvature of the border could presumably be related to body weight in terms of the amount of muscular insertion required, especially as the serratus and rhomboid muscles insert along its entire length. However, there is an apparent paradox in the scapula of the horse which has a short vertebral border but which supports a relatively large body weight. This problem is perhaps explicable in terms of gait characteristics. A fast moving horse displaces its center of gravity relatively little in the vertical direction. In a mammal such as the cat, which uses a bounding or leaping gallop, the center of gravity describes a series of arcs with a large amplitude. As a result, the kinetic energy transmitted to the ground at each point of forefoot contact is relatively much greater in the cat than in the horse (Fig. 10). The shoulder girdle of the horse is thus called upon to absorb relatively less shock than that of the cat, which in turn could account for the relatively short vertebral border of the horse scapula.

In arm-swinging primates the kinetic energy absorbed by the shoulder musculature must be relatively large, as in the case of the cat, and is associated with a long vertebral border. Since the vertebral border in many primates (such as, *Ateles, Alouatta, Hylobates,* and *Pongo*) is straight, it appears that the border need not necessarily be arched in order to transmit or absorb large stresses. Furthermore, in arboreal primates, especially quadrumanous and arm-swinging forms, the relatively great development of the fossae permits a long straight vertebral border. It may be tentatively suggested that the arched border in carnivores and other mammal groups is a compromise between the need for a long vertebral border in conjunction with relatively narrow fossae.

Morphology of the Glenoid

The form of the glenoid articular surface in primates varies from a nearly hemispherical to a pear-shaped outline. The pear-shaped outline is produced by a ventral projection of the cranial margin of the facet into a more or less elongate lip. The hemispherical type of articular surface occurs in brachiating primates (sensu stricto) and in cursorial mammals, such as the horse and deer. The similarity of the articular facets points to some form of functional parallelism. Pear-shaped articular surfaces are typical of most other quadrupedal animals, both terrestrial and arboreal. In certain genera, such as *Tupaia* and *Callithrix,* the cranial lip appears to be elongated and narrow. The functional implication of this morphology is not apparent.

In some of the Hominoidea and in the arm-swinging New World monkeys there appears to be an intermediate condition between the two extremes of facet morphology. The functionally important parameter of the glenoid articular facet is almost certainly the craniocaudal curvature. In those genera with hemispherical glenoids the curvature is a smooth, circular arc (Fig. 15B). It therefore permits concentric rotation of the humeral head within the glenoid socket. In those genera in which the facet is pear-shaped, the craniocaudal axis is an arc with a decreasing radius anteriorly. Motion of the articular head of the humerus almost certainly involves sliding as well as rotation in order to maintain contact with the glenoid. The presence of the lip permits a greater amount of retraction of the humerus than would otherwise be possible (Fig. 15A). The lip is thus a bony stop which prevents dislocation of the joint when a high angle is obtained between the axes of the scapula and humerus. This feature apparently occurs in mammals that have a limited amount of scapular rotation. As the lip and pear-shaped outline is present in the vast majority of mammals, including the primitive insectivores and fossil forms such as *Notharctus,* it is probably an ancestral condition. The modification of the glenoid to an ovate or nearly hemispherical form is seen as an adaptation for rapid limb motion (probably with a high acceleration increment), together with a complex of other characteristics

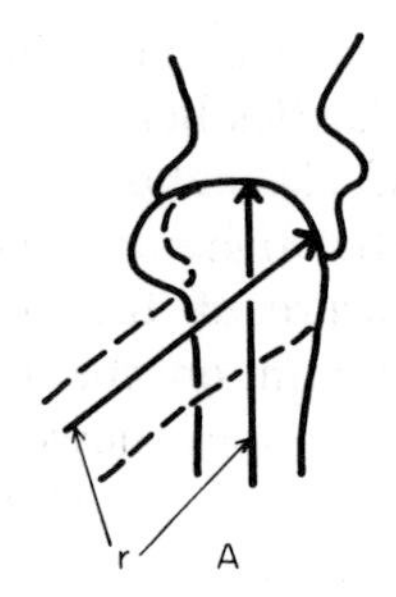

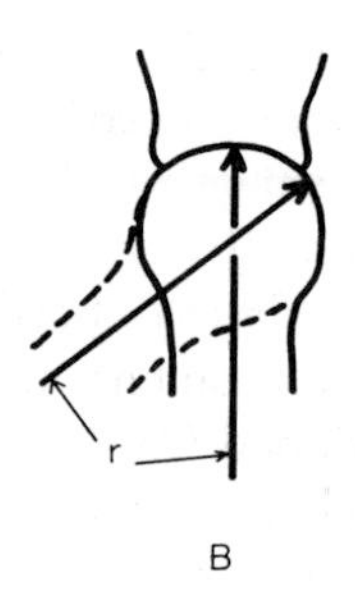

Fig. 15. The functional implications of different glenoid shapes. (A) The glenoid articulation terminates cranially in an elongate lip. This permits the humerus to be retracted through a large angle, as shown, without the ground-reaction forces (r) acting through the limb tending to cause dislocation of the joint. The motion of the scapula is thus reduced to a minimum. (B) Where the glenoid articulation lacks a cranial lip, the ground reaction forces come to act beyond the arc similar to that in (A). To prevent possible dislocation of the joint, the scapula must be rotated (forward) on the thorax.

such as a narrow scapula, light, distally elongated limbs, and proximal concentration of the musculature. It seems probable that in the hominoids the glenoid may have become ovate initially because of the freedom of movement which a ball and socket joint imparts to the forelimb. In the gorilla, orangutan, chimpanzee, and man it is probably retained for this reason but in the gibbon has become part of the complex of features developed for rapid limb motion.

The Scapula of *Australopithecus*

The only described* scapular fragment of man's late Pliocene and early Pleistocene ancestors, the genus *Australopithecus*, is the lateral portion of a bone discovered at Sterkfontein in South Africa and described and figured by Broom *et al.* (1950). The fragment preserves the neck, glenoid articular facet, and coracoid process. The proximal part of the acromion is present but crushed. Broom *et al.* (1950) observed that there were a number of similarities between the Sterkfontein fragment and homologous features in the scapula of *Pongo*. These include a recurved coracoid with a large area of attachment for the biceps muscle, lack of a distinct scapular notch, an ill-defined postglenoid neck, and also the shape of the lateral (axillary) border. The narrowness of

* Further finds have been made at Lake Rudolph in East Africa by Richard Leakey. These have not yet been described.

the coracoid and the appearance of the scapular spine are, however, manlike. Campbell (1966) observed similarities between the Sterkfontein scapula and those of "semibrachiators," i.e., the angle between the glenoid facet and the lateral border, the angle between the scapular spine and the lateral border, the curvature of the coracoid, and the degree of development of the biceps origin on the coracoid. Oxnard (1968) questioned Campbell's interpretation and showed that the angles which he had described were in fact closer to the statistical mean for *Hylobates* than for any other primate genus. It should be noted, however, that the angles are also very close to the means for the orangutan. In the figure published by Broom *et al.* (1950), it is difficult to determine to what degree the acromion embraced the glenoid cavity. There does seem to be some indication that it did so rather more than it does in the orangutan. If this is so it suggests that stabilization of the shoulder joint was an important feature, perhaps to the extent that circumduction of the limb was restricted slightly to ensure it, as seems to be the case in the knuckle-walking apes. Together with the other characteristics of the fragment, this implies an animal capable of slow-quadrumanous climbing, such as the orangutan, but with some adaptation for ground living, where it may have progressed on occasion by a form of "fist-walking" as does the orangutan when forced to become terrestrial (Tuttle, 1969). This interpretation is not incompatible with the position of *Australopithecus* in human evolution. Recent estimates put the age of the South African deposits at under three million years, that is, about one million years older than the Olduvai deposits in East Africa in which foot bones of a clearly bipedal animal (*Homo habilis*) have been discovered.

Conclusions

The analysis in the chapter has been centered on the function of the scapula. This bone, although an important skeletal component in terms of the locomotor behavior, is still only a single component in a highly complex system. Despite the contribution that an understanding of scapular biomechanics may make toward an understanding of the shoulder as a whole, our knowledge of shoulder biomechanics is still far from complete. Furthermore, when it is considered that the shoulder itself is only part of a locomotor system that includes more or less the entire postcranial region of the animal, it will be realized that general conclusions on scapular function must be drawn with great caution.

The conclusions that may be drawn from available data are of two types; those which are concerned with the intrinsic functioning of the shoulder girdle, and those which are concerned with the relationship between the shoulder girdle and locomotion. The latter conclusions may be used either as an aid in categorizing locomotion or as a means of explaining the rationale behind locomotor categorizations.

1. The morphology of the scapula (including the development of macro- and microstructures) is not entirely predetermined on a genetic basis, but is dependent to a large degree on secondary influences during ontogenetic development, including the stresses imposed by its associated musculature.

2. The area of the scapular fossae appears to be related to the stabilization of the shoulder joint and to the requirements of the forelimbs, such as the amount of circumduction and the power or complexity of use in any given position. In quadrumanous and arm-swinging genera, such as the Asian apes and the New World monkeys (together with marginal groups of Old World monkeys), the fossae tend to be broad, thus providing a sufficient area of attachment for the large rotator cuff muscles required for shoulder joint stabilization and at the same time permitting power and range in forelimb circumduction. There appears to be a relationship existing between the development of a broad infraspinous fossa and the habit of lifting the weight of the body. This is demonstrated by the orangutan and slow-climbing prosimians. In genera which habitually hold or use the forelimb above the level of the shoulder, for example, *Ateles, Alouatta, Hylobates, Pan,* and *Gorilla,* the supraspinous fossa also tends to be well developed. In sharp contrast is the poor development of the supraspinous fossa in quadrupeds which do not normally raise the limb above the level of the shoulder, such as, *Papio, Macaca,* and *Erythrocebus.*

3. The relationship between scapular form and locomotor and feeding behavior is expressed also in the thickening of the spine and borders of the scapula. In *Macaca,* where the stresses on the scapula resulting from locomotion and feeding are presumed to be relatively small, the structures are poorly developed when compared to the same structures of the scapula in, for example, *Gorilla,* which are assumed to be subjected to relatively greater stress.

4. The profile of the scapular blade is determined by functional demands, being broad in genera which move their limbs slowly over a wide range of circumduction, and relatively narrow in animals which move their limbs rapidly but with limited ranges of circumduction, such as within a parasagittal plane. Examples of such animals would be the orangutan and the baboons, respectively. In brachiators, both speed and range of circumduction are required. The scapula of the gibbon is thus a

compromise which is functionally narrow but retains large fossae through a reorganization of the morphology of the bone. Other genera and species approach the condition seen in the gibbon (e.g., *Ateles* and *Pan paniscus*).

5. As noted by Smith and Savage (1956), the development of the inferior angle relates to the use of the teres major muscle.

6. The form of the superior angle relates to the motion of the blade in relation to other structures of the shoulder and neck regions.

7. The length of the vertebral border may relate not only to the weight of the animal, but also to the amount of kinetic energy which is periodically absorbed through the shoulder region during locomotion.

8. The scapular blade appears to require stabilizing on the thorax before certain motions of the limb occur. This suggests that the bone must be in a state of mechanical equilibrium during periods of great stress.

Acknowledgments

This work was supported by Grant No. 2735 made by the Wenner-Gren Foundation for Anthropological Research Incorporated. I am indebted to Dr. S. Anderson of the American Museum of Natural History and Dr. E. G. Erikson of Brown University for permission to examine material in their care.

References

Alexander, R. M. (1968). "Animal Mechanics." Sidkwick and Jackson, London.

Ashton, E. H. and Oxnard, C. E. (1963). The musculature of the primate shoulder. *Trans. Zool. Soc. London* **29**, 553–650.

Ashton, E. H., Flinn, R. M., Oxnard, C. E., and Spence, T. F. (1971). The functional and combined classificatory significance of combined metrical features of the primate shoulder girdle. *J. Zool.* **163**, 319–350.

Basmajian, J. V. (1967). "Muscles Alive." Williams & Wilkins, Baltimore, Maryland.

Broom, R., Robinson, J. T., and Schepers, G. W. H. (1950). Sterkfontein apeman *Plesianthropus. Mem. Transv. Mus.* **4**, 11–83.

Campbell, B. G. (1966). "Human Evolution." Aldine, Chicago, Illinois.

Davis, D. D. (1949). The shoulder architecture of bears and other carnivores, *Fieldiana Zool.* **34**, 285–305.

Gray, J. (1968). "Animal Locomotion." William Clowes and Sons, London.

Gregory, W. K. (1912). Notes on the principles of quadrupedal locomotion and on the mechanism of the limbs in hoofed animals. *Ann N. Y. Acad. Sci.* **22**, 267–294.

Howell, J. A. (1917). An experimental study of the effect of stress and strain on bone development. *Anat. Rec.* **3**, 233–252.

Innam, V. T., Saunders, M., and Abbott, L. C. (1944). Observations on the function of the shoulder joint. *J. Bone Joint Surg.* **26**, 1–30.

Jenkins, F. A., Jr. (1971). Limb posture and locomotion in the Virginia opposum (*Didelphis marsupialis*) and in other non-cursorial mammals. *J. Zool.* **165**, 303–315.

MacConaill, M. A., and Basmajian, J. V. (1969). "Muscles and Movements," Williams & Wilkins, Baltimore, Maryland.

Miller, R. A. (1932). Evolution of the pectoral girdle and forelimb in primates. *Amer. J. Phys. Anthropol.* **18**, 1–56.

Mivart, St. George. (1868). On the appendicular skeleton of primates. *Phil. Trans. Roy. Soc. London* **157**, 299–429.

Müller, H. J. (1967). Form und Funktion der Scapula, Vergleichend analytische Studien bei Carnivoren und Ungulaten. *Z. Anat. Entwicklungsgesch.* **126**, 205–263.

Osborn, H. F. (1900). *Patriofelis* and *Oxyaena restudied as terrestrial creodonts. Bull. Amer. Mus. Natur. Hist.* **13**, 269–279.

Osborn, H. F. (1929). The Titanotheres of ancient Wyoming, Dakota, and Nebraska. *U. S. Geol. Surv. Monogr.* **55**, 1, 2:1–953.

Oxnard, C. E. (1967). The functional morphology of the primate shoulder as revealed by comparative anatomical, osteologic and discriminant function techniques. *Amer. J. Phys. Anthropol.* **26**, 219–240.

Oxnard, C. E. (1968). A note on the fragmentary Sterkfontein scapula. *Amer. J. Phys. Anthropol.* **128**, 213–218.

Oxnard, C. E. (1969). The descriptive use of neighborhood limited classification in functional morphology: An analysis of the shoulder in primates. *J. Morphol.* **129**, 127–148.

Schultz, A. H. (1930). The skeleton of the trunk and limbs of higher primates. *Human. Biol.* **2**, 303–438.

Smith, M. J., and Savage, R. J. (1956). Some locomotory adaptations in mammals. *J. Linn. Soc. London Zool.* **42**, 603–622.

Tuttle, R. H. (1969). Knuckle-walking and the problem of human origins. *Science* **166**, 953–961.

Wolffson, D. M. (1950). Scapula shape and muscle function, with special reference to the vertebral border. *Amer. J. Phys. Anthropol.* **8**, 331–341.

7

Postural Adaptations in New and Old World Monkeys

M. D. ROSE

Introduction

Prost (1965) coined the term "positional" to refer to activities performed by an animal in which there is a gross displacement of the animal relative to its surroundings ("locomotion") together with those in which there is no such displacement ("posture"). The fact that this term has not become current in the primatological literature is related, at least in part, to the emphasis placed on locomotion in primate studies. Thus, in investigations of the functional morphology of living or fossil primates, and of behavioral activities, emphasis has often been placed on frequently occurring locomotor activities or on locomotor activities which it is assumed have high survival value ("frequent locomotor habits" and "critical locomotor habits," respectively; Prost, 1965).

The importance of postural activities for primates can be appreciated by considering overall patterns of activity. A number of features can, consequently, be seen. Thus, if one lists the distinct types of positional acts an animal performs (i.e., lists the "positional activities" making up the "positional repertoire"), it is evident that postural activities are equal or greater in number than locomotor activities. [I prefer to use the terms "positional activity" and "positional repertoire" to Prost's (1965) similar terms "bounded positional pattern" and "positional totipotentiality," mainly on the grounds that they are less cumbersome.] Similarly, primates of all types spend more of their time in postural than in locomotor activities. A further point is that on many occasions postural and locomotor activities grade into one another; this is particularly the case in suspended positional activities. None of these features are necessarily sufficient reasons for paying much attention to postural activities. However, when one considers the part played by "positional behavior" (positional activities in their physical context; Prost, 1965) in adaptations for life in a particular habitat it becomes clear that postural behavior forms an indispensable part of a number of these adaptations.

This chapter is mainly concerned with the adaptive nature of postural behavior in the group of quadrupedal primates represented by the New and Old World monkeys. The discussion is based on observations of positional behavior in 46 species (20 of the 30 genera) of New and Old World monkeys. Observations have been made on captive animals, on cine film of wild and captive animals, and on six species of three genera of African monkeys in the wild. About 200 hours have been spent observing wild animals, while about 170 hours were spent observing captive animals. These observations have been supplemented by data gathered from a comprehensive review of the frequently inadequate literature on the positional behavior of monkeys.

Activity Patterns

Haddow (1952) was the first to note that the overall activity patterns of arboreal Old World monkeys tend to be similar. Periods of intensive feeding occur during the morning and afternoon, with a rest period of variable duration during the middle part of the day. Periods of movement precede the morning feeding period and follow the afternoon feeding period as animals move between sleeping and feeding sites. The day range associated with this type of activity pattern may vary from a few hundred meters to about 3 km (Aldrich-Blake, personal communication, *Cercopithecus mitis*; Jay, 1965, *Presbytis entellus*; Rowell, 1966, forest living *Papio anubis*). A similar activity pattern is seen in at least *Ateles geoffroyi* (Carpenter, 1935; Eisenberg and Kuehn, 1966; Richard, 1970) and *Alouatta villosa* (Carpenter, 1934; Chivers, 1969; Richard, 1970) among New World monkeys.

Percentage activity figures given in the literature are difficult to interpret in that activity patterns, such as feeding and resting, are not usually distinguished from particular positional activities such as sitting or the various types of quadrupedal locomotion. However, it is evident from the figures given by the authors referred to above that about 75% of the day is spent either feeding or resting; movement and "other" activities account for the rest of the time.

The activity patterns of terrestrial monkeys are more variable than those of arboreal monkeys. In general, animals feed at more or less the same intensity throughout the day, although peaks of feeding intensity may occur (Hall, 1965, *Erythrocebus patas*). A midday resting period may be present (Kummer, 1968, *Papio hamadryas*) or may be lacking (Hall, 1962, *Papio ursinus*). About 50% of the time may be spent feeding and resting, but this figure is not strictly comparable with that for arboreal monkeys since with terrestrial monkeys time is spent feeding on the move. Day range figures are also difficult to interpret as animals may feed in trees for part of the day, or may spend the whole day on the ground. However, animals may travel up to 20 km in a day, although the average day range is about 7 km (Hall, 1965, *Erythrocebus patas*; Hall and DeVore, 1965, *Papio anubis;* Altmann and Altmann, 1970, *Papio cynocephalus;* Kummer, 1968, *Papio hamadryas*).

Daily activity patterns and day ranges may be influenced by a variety of factors such as seasonal variation in climate (Aldrich-Blake, personal communication, *Cercopithecus mitis*), food availability (Gautier-Hion, 1966, *Miopithecus talapoin*), type of food eaten (Thorington, 1968, *Saimiri sciureus*), and diurnal temperature and the availability of shaded sites (Rahaman and Parthasarathy, 1969, *Macaca radiata;* Chivers, 1969, *Alouatta villosa*). However, the concensus of opinion is that food availability constitutes the prime determinant. Altmann and Altmann (1970) note further that although other factors may account for day-to-day variability within an area, food availability accounts for long-term differences between areas.

Feeding and resting therefore predominate during diurnal activities. Resting is obviously associated with postural activities. The degree to which postural activities contribute to feeding in a number of African monkeys is shown in Table I.

It is clear from the table that in an arboreal setting postural activities (mainly sitting) account for about 95% of the time spent feeding. For animals on the ground, sitting and standing postures account for about 60% of the time spent feeding. Using data from Table I, Aldrich-Blake (personal communication), Chalmers (1968), and Crook and Aldrich-Blake (1968), one can roughly estimate that at least some arboreal monkeys might spend about 75% of daylight hours in various types of postural activity, while the equivalent figure for terrestrial monkeys would be about 40%. If the fact that the nighttime sleeping period is spent in some form of postural activity is taken into account, then postural activities might occupy 90% of the time of arboreal monkeys and about 70% of the time of terrestrial monkeys. These figures are, at best, a very

TABLE I
PERCENTAGE TIME SPENT IN VARIOUS ACTIVITIES DURING FEEDING[a]

Situation	Species	Sitting	Standing	Walking	Running
Arboreal	*Colobus guereza*	95 (475)	1 (5)	3 (15)	1 (5)
	Cercopithecus mitis	94 (470)	1 (5)	4 (20)	1 (5)
	Cercopithecus ascanius	95 (475)	1 (5)	3 (15)	1 (5)
	Cercopithecus aethiops	94 (188)	1 (2)	4 (8)	1 (2)
	Papio anubis	90 (450)	4 (20)	6 (30)	0
Low bushes	*Cercopithecus aethiops*	72 (144)	10 (20)	16 (32)	2 (4)
Terrestrial	*Cercopithecus aethiops*	38 (76)	25 (50)	32 (64)	5 (10)
	Papio anubis	30 (60)	35 (70)	30 (60)	5 (10)

[a] Each species was observed for a total of either 500 or 200 minutes. Figures in parentheses refer to time in minutes that each activity was observed. For any one species the figures represent pooled data from different individuals, different times of day, and from individuals feeding from sources of varying abundance. Arboreal feeding was observed in sympatric species in Budongo Forest, Uganda. Feeding in other situations was observed in Murchison Falls National Park, Uganda.

rough approximation, but they do highlight the importance, on a time basis, of postural activities in the lives of monkeys.

Postural Adaptations in Arboreal Monkeys

SIZE AND THE SMALL-BRANCH SETTING

As a result of her studies on caged catarrhine primates, Avis (1962) developed the concept of the "small-branch setting," which is that part of the arboreal environment situated at the periphery of tree crowns and made up of small-diameter flexible branches. Avis came to the conclusion (the "small-branch hypothesis") that only the apes are capable of operating successfully in the small branch setting and that their success stems from the use of various suspended locomotor activities. Avis believes that monkeys are adapted for movement on and between firmer supports.

The characteristics of a particular branch as a substrate or superstrate for positional activities vary for animals of different sizes, a point well made by Washburn (1957). Some of the consequences of an increase in body size in arboreal primates have been discussed, in evolutionary terms, by Napier (1963, 1967). Thus for small animals, where the branch to body size ratio is high, problems of balance are minimal, especially if the animal can lower its center of gravity toward the supporting branch. For larger animals, problems of balance become critical. Napier believes that in the course of primate evolution these problems have been solved by the development of prehensile extremities by which an animal can suspend itself below branches.

I believe that an examination of the positional behavior of New and Old World monkeys leads to slightly different conclusions to those of Avis and Napier. I agree with Grand (1972) that nearly all primates, not just apes, operate successfully within the small-branch setting. It is this zone which contains a number of concentrated food sources for most noninsectivorous primates (food sources also utilized at least in part by terrestrial primates). As Ripley (1967) has indicated, there are a number of zones within the arboreal environment, all of which place certain demands on an animal's positional capabilities. However, the demands of the small-branch setting are in many ways more pressing than those of other zones. Adaptations for life in the small-branch setting are related to body size and in the monkeys, at least, involve both above and below branch activities. Among the larger forms especially, postural activities form an essential part of these adaptations. Prehensile extremities are of as much use in solving problems of balance during above as they are during below branch activities.

Small-Sized Arboreal Monkeys

Increase in body size leads to problems associated with the relative strength of the muscular system as well as to problems associated with the maintenance of balance. The power of muscle is related to its cross-sectional area, which increases as the square of the linear dimensions of an animal. Volume, and in effect weight, increases as the cube of the linear dimensions. This means that larger animals must possess morphological and/or behavioral adaptations to perform as efficiently as smaller ones. Most of the unique features of the postural behavior of small-sized monkeys can be related in one way or another to the possession of a relatively powerful muscular system. The callitrichid New World monkey species of the genera *Callimico, Callithrix, Cebuella, Saguinus,* and *Leontideus* are examples of this type of animal.

These animals all possess relatively short limbs, a feature which is probably related to the possession of a relatively powerful muscular system. The limbs are maintained in a fairly flexed position so that the already low center of gravity is lowered even more toward the supporting branch. This, in combination with a high branch to body size ratio, reduces to a minimum the problems of balance. Due to the shortness of the forelimbs, the trunk is held horizontally during quadrupedal sitting postures. Upright sitting postures are adopted, usually during feeding. The forelimbs are, of necessity, removed from the supporting surface and the trunk is characteristically held in a near vertical position.

Clinging postures are adopted on and under branches of all orientations. The prehensile power of the claws is usually sufficient to stabilize postures of this type. The hallux may grip during clinging postures on near vertical branches and during the bipedal suspended postures which are adopted during feeding.

Medium-Sized Arboreal Monkeys

Animals of this size are included in three of the subfamilies of the Cebidae: the aotines, *Aotus* and *Callicebus*; the pitheciines, *Cacajao*, *Chiropotes*, and *Pithecia*; and the cebines, *Cebus* and *Saimiri*. There are, nevertheless, variations of size within the group and some of the variability of positional behavior within the group can be attributed to this fact. However, all the members of the group are of a size where claws are inadequate as prehensile organs. The digits bear nails and a variety of hand and foot grips may be used for stabilization (Bishop, 1964).

Postural behavior in the aotines is fairly similar to that of the callitrichids, although a variety of grips may be taken with the hands and feet. However, hook grips are not usually taken during suspended activities. In the pitheciines the relative length of the forelimbs is such that semierect sitting postures are possible in which the hands retain contact with the supporting branch. Body size is still compatible with the adoption of clinging postures on vertical branches, although this is not a common activity. As with the previously described groups, sleeping postures include hunched sitting and prone lying postures. Suspended postures are adopted relatively frequently, at least by *Cacajao*. Both quadrupedal and bimanual suspended postures may be adopted, both the hands and the feet taking hook grips. These postural activities are associated with quadrupedal suspended walking and suspended bimanual locomotion (arm swinging), respectively.

Saimiri and *Cebus* differ in body size and in the relative length of their limbs, although their intermembral indices are similar. The trunk is held fairly horizontally during sitting in *Saimiri*, although semierect postures may be adopted. Semierect postures are commonly adopted by *Cebus*.

Pollex and hallux grips are taken during above branch activities, although the pattern of hand grips is more variable in *Saimiri*. Prehensile tail use is common during postural activities. Bipedal standing postures and bipedal locomotion are fairly common activities, especially when animals come to the ground.

When feeding, *Saimiri* may adopt bipedal suspended postures in association with a tail grip. *Cebus* frequently adopts quadrupedal, bipedal, or bimanual suspended postures, especially during feeding. Hook grips are taken by the hands and/or feet and a tail grip is usually taken to provide additional stability.

In general, the body size of both small- and medium-sized monkeys is such that a variety of activities, both postural and locomotor, both above and below branches are possible in the small-branch setting. The use of specialized hand and foot grips and the prehensile use of the tail in some species represent the only specific adaptations necessary for successful operation within this zone.

Large-Sized Arboreal Monkeys

With large-sized animals there is a clear dichotomy between forms adapted in terms of their above branch activities and those adapted in terms of their below branch activities. There are further specializations of particular activities within the groups.

Above Branch Adaptations

All Old World monkeys can, and almost all do enter the small-branch setting. Postural adaptations to this end are perhaps best seen in arboreal cercopithecines, such as *Cercopithecus*. During the day, animals enter the small-branch setting to feed and occasionally to rest, or during passage from one tree to the next as part of normal progression, or during escape. Most resting, daytime sleeping, and overtly social interactions take place on larger, more central branches. Activity patterns during feeding tend to be similar for all Old World monkeys in trees. The animal walks peripherally along a branch, stands briefly to inspect the vegetation for food objects, then sits and feeds on all food objects within the range of its forelimbs. Having exhausted the food at a particular site, it walks centrally along the branch until it comes to the next branch leading to the periphery and repeats the previous sequence of actions. Having repeated this pattern at various sites within a tree the animal will move into another tree and start the sequence again. In heavily laden trees, an animal may spend a whole morning or afternoon working its way systematically through a tree. On less well-laden trees, an animal will sample a few sites, while spending less time at each one. Ripley (1970) has described feeding of this sort in *Presbytis entellus*. Ripley terms feeding at one site a "feeding bout"; a series of bouts at adjacent sites is a "feeding interval" and a series of feeding intervals, with major moves between each interval, is a "feeding period."

The sitting postures adopted during feeding are characteristic. An animal may sit along the length of a branch, facing centrally or peripherally, or across a branch, at right angles to its length. Wherever possible, contact is made between the ischial callosities and the branch. The hindlimbs are abducted at the hip and contact the supporting branch or nearby branches. The feet make plantigrade contact with the branches, or at more unstable sites the halluces may grip. Thus the animal sits on a stable triangular base the apices of which are formed by the two feet and the ischial calosities. When a posture is adopted along a branch it is frequently only possible for an animal to gain contact with one of the the ischial callosites. In all these postures, the trunk is inclined forward slightly to bring the center of gravity well within the triangular base. However, trunk movements over quite a wide range are still compatible with a balanced posture. The forelimbs are also free to move throughout their range and gather food.

It is frequently impossible for an animal to gain ischial callosity support when it is sitting among small branches. In this case it sits across a branch with the feet resting on the branch. The lateral toes flex round the branch from above and the halluces grip from below. The hindlimbs are highly flexed, so that the rump is at or below the level of the feet, lowering the animal's center of gravity toward the branch. This again is a stable posture, leaving the forelimbs free for food gathering. However, the trunk is less free to move than in postures in which there is ischial callosity support, as it must be flexed forward to bring the center of gravity over the supporting branch. Postures of this type are not squatting postures, and there is usually a tensile force acting in the foot rather than the more usual compressive force experienced by the foot during positional activities.

The tail plays both an active and a passive part in stabilizing sitting (and other) postures. The use of the tail in postural activities is discussed in detail in a later section.

The sitting postures adopted on larger branches always involve ischial callosity support. Foot or hand grips are rarely taken. Prone or lateral lying postures may be adopted by resting animals. Quadrupedal standing postures form a minor part of the postural repertoire. They are adopted as transitional postures between other activities or as alert postures. The hands and feet are placed in palmigrade and plantigrade positions, respectively. Bipedal standing postures are very rarely adopted. They occur when an animal is gathering food from an overhead position or as alert postures. Suspended postures are similarly rare and are usually adopted by an animal prior to dropping to a branch at a slightly lower level or after short upward leaps.

The postural behavior of colobine monkeys among small branches is in all respects similar to that of arboreal cercopithecine monkeys. On larger branches, the hindlimbs may be stretched out along a branch or allowed to hang free during sitting postures. Resting animals frequently adopt lying postures in situations where a cercopithecine monkey would adopt a sitting posture.

Papio, Mandrillus, and *Erythrocebus* enter the small branch setting during the day to feed. Sitting postures are usually restricted to those in which there is ischial callosity support. On all but the smallest branches there is a tendency to adopt tripedal standing postures similar to those adopted during ground feeding. The hands are placed in digitigrade positions during tripedal and quadrupedal standing postures adopted on branches. *Macaca* and *Cynopithecus* favor sitting postures in which there is ischial callosity support, but are capable of adopting postures stabilized by foot grips alone. Tripedal standing postures are not adopted, but the hand may be placed in a digitigrade position during quadrupedal standing postures in some species.

Ischial Callosity Function and Adaptations for Sleeping in the Small-Branch Setting

Washburn (1957) suggested that ischial callosities are specific adaptations related to the adoption of sitting postures for sleeping. The data given above suggests that ischial callosity usage also forms an essential part of daytime postural activity, both among small and on larger branches. Washburn (1957) identified a characteristic sitting posture in *Papio anubis* resting in trees during the day which he also observed in a range of caged Old World monkeys and gibbons resting during the day and sleeping at night. I have identified this posture in five species of wild *Cercopithecus* monkeys and in wild *Colobus guereza* and *Papio anubis* resting and sleeping during the day, and in a wide range of caged Old World monkeys resting and sleeping during the day and sleeping at night. In this posture an animal sits with the ischial callosities resting on a branch, the feet propped up on either vertical branches or on a horizontal branch at about the level of the animal's head or shoulders. The trunk is inclined forward and the hands may rest on the branch on which the ischial callosities rest or may take grips on other branches. This posture has been termed the "foot-prop posture." As with some of the postures described above, the foot-prop posture is frequently adopted in the small-branch setting.

Although there are a number of postures in which the animal sits on a stable triangular base it is only the foot-prop posture which seems to be specifically related to sleeping. I believe there is enough evidence available to formulate a hypothesis which explains the function of this posture. Thus Lumsden (1951) and Bert *et al.* (1967) have shown that at least some Old World monkeys sleep at night in the small-branch setting, presumably as a means to avoid predators. Bert *et al.* have recorded electroencephalograms from a number of sleeping primates. Whereas man (and also *Galago, Pan,* and probably *Gorilla*—all nest sleepers) have four stages of normal, slow-wave sleep, passing from light to profound, *Papio papio* and at least some *Macaca* species (Bert *et al.,* 1970) only show stages one and two (light sleep) and occasionally stage three. However, all these species have periods of fast-wave, or rapid eye movement sleep, an important concomitant of which is the relaxation of the postural musculature. Thus for an animal sleeping in the small-branch setting the pattern of slow-wave sleep allows for rapid arousal so that the animal can change or adjust its sitting posture. Any sitting posture with a stable base would presumably satisfy the demands of balance during sleep of this type. However, during periods of fast-wave sleep the animal must be in a posture which can be passively maintained during periods of muscular relaxation. If muscular relaxation takes place while an animal is in the foot-prop posture two things tend to happen. The already inclined trunk tends to rotate forward and the feet tend to either slide down a vertical support or slip off a horizontal support. This is initially associated with flexion at the knee. However, the trunk is prevented from rotating any further forward by coming into contact with the fronts of the thighs. The hindlimbs are prevented from folding at the knee by the weight of the trunk, so that a new, passive state of equilibrium is produced.

There are no doubt many ways in which an animal can wedge itself among a tangle of small branches so that it does not overbalance during periods of fast-wave sleep. However, the ubiquity of occurrence and the mechanical characteristics of the foot-prop posture strongly suggest that it is functionally significant.

If the explanation of ischial callosity function given above is correct, one may reasonably ask why New World monkeys have not, like the hylobatids, evolved ischial callosities in parallel with the Old World monkeys. There are several factors which may be involved. Thus apart from the ateline and alouattine species, New World monkeys have a smaller body size than almost all Old World monkeys. Problems associated with balance and the effects of body weight on the supporting tissues are thus minimal. Alouattine and ateline species use suspended postures among small branches rather than the sitting postures adopted by Old World monkeys. New World monkeys also tend to favor lying, rather than sitting postures when they sleep, and the larger forms tend to sleep on larger branches than those used by most Old World monkeys.

Below Branch Adaptations

The members of two of the subfamilies of the Cebidae, *Ateles, Brachyteles* and *Lagothrix* (Atelinae), and *Alouatta* (Alouattinae) show specializations for life in the small-branch setting mainly in terms of their below branch activities. Postural activities are important, but are more closely associated with locomotor activities than in arboreal Old World species.

Animals enter the small-branch setting to feed, but a variable amount of daytime resting also takes place in this zone. Animals also rest during the day, and probably sleep at night on larger, more central branches. A variety of suspended activities are performed in the small-branch setting, using all combinations of the four limbs in association with the prehensile tail. Food is gathered by means of one or both hands or by the tail. The variety of suspended postures is such that an animal may be orientated in virtually any position. In general, hook grips are taken with the hands. Hook grips are taken with the feet on larger branches, but hallux grips are usually taken on smaller branches. During quadrupedal suspension the head and body face upward. This is a fairly common posture in *Alouatta.* Bipedal suspension is in general a characteristic activity of smaller animals and is not a common activity in the large sized New World monkeys. However, such postures, in association with a tail grip, may be adopted on vertical branches so that the animal may extend its body horizontally to gather food with the hands. A common posture involves suspension by one hand and the tail, so that the head and body face downward. This posture is adopted by both feeding and by resting animals and has the advantage that the animal has a wide field of view. Suspension by the tail alone is also common, especially by resting animals, and may be maintained for long periods. In the atelines at least, the tail acts in all ways like a fifth limb.

These postures are associated with complex locomotor sequences among small branches which involve both below and above branch quadrupedal walking and arm swinging. They may also be associated with the dropping, bridging, and arm-swinging usually used in preference to leaping across gaps.

Sitting is the main postural activity on larger branches. It is characterized by the fact that the gluteal region almost always contacts the branch, the trunk is held in a nearly upright position, and the back is highly flexed. The hindlimbs may be flexed or folded in a number of ways which seem to have more to do with the demands of comfort than with the demands of stability. Animals usually sleep in hunched sitting postures or lateral lying postures on larger branches.

As with arboreal Old World monkeys, quadrupedal standing postures are not frequently adopted, and when they are the limbs are held fairly flexed at the elbows and knees. Hand and foot contact is palmigrade and plantigrade and, on larger branches, a modified palmigrade hand position may be adopted (Tuttle, 1967) in which the second to fifth fingers are flexed so that the dorsal surfaces of the terminal phalanges rest on the supporting branch.

In the ateline species at least, bipedal standing and walking are fairly common activities, both on branches and on the ground. The tail may grip when animals become bipedal on branches and acts as a counterweight during bipedal activities on the ground. The bipedal activities of the ateline monkeys are more skillfully and economically performed than those of any other group of monkeys. This may at least in part be related to the great mobility of the hindlimbs which is primarily associated with the use of suspended activities.

Tail Use and Postural Activities

The tail may be used both actively and passively and in both prehensile and nonprehensile ways during postural activities. When an animal sits on a branch the tail usually hangs straight below the animal and must to a certain extent lower the animal's center of gravity toward the branch. Active movements of the tail may be made to counteract small changes in the animal's equilibrium. On occasion, the tail may be draped over a particular branch in order to correct a tendency for the animal to overbalance in a certain direction. In *Cercopithecus aethiops* and *Saimiri sciureus* the tail may be braced against a vertical branch as an aid to stability.

The use of the tail in the above types is also a feature of many lying postures and tail movements help to stabilize quadrupedal standing postures. In *Cercopithecus aethiops, Cercopithecus campbelli, Erythrocebus patas, Callithrix* species, *Saimiri sciureus,* and ateline and alouattine species the tail may act as the third leg of a tripod during bipedal standing postures.

Ankel (1962) divides monkeys into three groups on the basis of tail use: those with a prehensile tail with a sensitive surface terminally (Greifschwanzaffen mit Tastflache am Schwanzende), those with a hair-covered prehensile tail (Greifschwanzaffen, Schwanz allseitig behaart), and those with a relaxed, nonprehensile tail (Schlaffschwanzaffen). The first group consists of ateline and alouattine species, the second of *Cebus* and *Saimiri,* and the third of all other monkeys. Although it lacks the power and refinement of that of the first two groups, prehensile tail use is in fact quite common in Schlaffschwanzaffen other than colobine monkeys. The tail may be used in a prehensile way by adult monkeys during postural and locomotor activities and by infants, who wrap their tails round their mother's body or tail base while being carried in a ventral position. Adult animals sitting, standing, or walking side by side may twine their tails together, an activity which presumably serves a social rather than a positional function. The species for which some type of prehensile tail use has been noted are listed in Table II.

It is evident from Table II that prehensile tail use in Schlaffschwanzaffen is associated mainly with postural activities. It will almost certainly be possible to extend this list further when observations have been made of other species. The extent to which the tail is used in a prehensile way varies considerably. Variations in the incidence of prehensile and other types of tail use in a number of Schlaffschwanzaffen are listed in Table III.

Although no figures are available, the incidence of prehensile tail use among alouattine and ateline species (and probably cebine species as well) is almost certainly greater than those listed in Table III. Prehensile tail use in these New World forms is, of course, an essential part of below branch activities and also plays a stabilizing role during above branch activities as in Old World species (e.g., *Cercopithecus lhoesti*).

Postural Adaptations in Terrestrial Monkeys

The positional behavior of monkeys on the ground is related to the physical characteristics of the environment in a much less specific way than is the case with arboreal monkeys. As Table I indicates, a terrestrial monkey feeding on the ground may spend about equal periods of time sitting, standing, and walking. As far as postural activities are concerned, it is the relatively large contribution of standing postures to the positional

TABLE II

Occurrence of Different Types of Prehensile Tail Use in Monkeys

Species	Postural	Locomotor	Infant	Tail twining
Cercopithecus pogonias[a]	X			
Cercopithecus albogularis[m]				X
Cercopithecus nictitans[a]	X			
Cercopithecus mona[a]	X			
Cercopithecus campbelli[b]	X	X		
Cercopithecus ascanius[m]	X		X	
Cercopithecus mitis[m]	X			X
Cercopithecus lhoesti[m]	X			
Cercopithecus hamlyni[c]	X			
Cercopithecus aethiops[d]	X			X
Cercopithecus talapoin[e]				X
Macaca mulatta[f]	X		X	
Macaca fascicularis[e]	X			
Macaca radiata[e]			X	
Erythrocebus patas[a,g,h]	X		X	
Cercocebus galeritus[i]	X			
Cercocebus albigena[i]	X	X		
Cercocebus aterrimus[i]	X			
Theropithecus gelada[m]				X
Colobus verus[j]			X	
Papio anubis[m]			X	
Saguinus spp.[m]	X			
Callithrix spp.[m]	X		X	
Callimico goeldi[e]	X			
Callicebus moloch[k]				X
Aotus trivirgatus[m]				X
Saimiri sciureus[m]	X		X	
Cebus spp.[m]	X	X	X	
Alouatta spp.[m]	X	X	X	
Ateles spp.[m]	X	X	X	
Brachyteles arachnoides[l]	X	X		
Lagothrix spp.[m]	X	X	X	

[a] Struhsaker and Gartlan (1970).
[b] Bourlière *et al.* (1969).
[c] Dandelot (1956).
[d] Struhsaker (1967).
[e] Karrer (1970).
[f] Rahaman and Parthasarathy (1968).
[g] Goswell and Gartlan (1965).
[h] Hall (1965).
[i] Allen (1925).
[j] Booth (1956).
[k] Moynihan (1967).
[l] Erickson (1963).
[m] Observed by the author.

TABLE III

PERCENTAGE INCIDENCE OF VARIOUS TYPES OF TAIL USE DURING THE SITTING POSTURES OF A NUMBER OF OLD WORLD MONKEY SPECIES[a]

Species	Hanging	Draping	Prehensile use
Cercopithecus lhoesti	6.5 (3)	69.5 (38)	24.0 (11)
Cercopithecus ascanius	71.8 (121)	26.5 (40)	1.7 (3)
Cercopithecus aethiops	70.2 (52)	29.8 (22)	0
Colobus guereza	90.0 (144)	10.0 (16)	0
Papio anubis (in trees)	94.0 (47)	6.0 (3)	0

[a] Number in parentheses refer to actual number of observations. Animals were observed in Budongo Forest, Uganda, with the exception of *Cercopithecus lhoesti,* which was observed in the Impenetrable Forest, Uganda.

repertoire which is a characteristic of this group of monkeys. *Theropithecus gelada* tends to spend more time feeding in sitting postures than *Papio* species (Crook and Aldrich-Blake, 1968). This is related to the fact that *T. gelada* tends to feed on objects which require two-handed gathering. *Papio* incorporates more sitting into its feeding pattern when seasonal changes in food availability make its diet similar to that of *T. gelada* (DeVore and Hall, 1965).

When quadrupedal or tripedal standing postures are adopted on the ground, the hands are placed in a digitigrade position in *Papio, Theropithecus, Erythrocebus, Mandrillus, Cynopithecus,* most *Macaca* species, and some *Cercopithecus* species (including *C. aethiops*). A palmigrade hand position is maintained by *Presbytis entellus* on the ground. Animals adopt a tripedal posture during feeding, the free hand being used to gather food. The supporting limbs characteristically become as strutlike as possible during these postures.

Bipedal standing postures are common in terrestrial species, but are more common in semiterrestrial species such as *Cercopithecus aethiops, Presbytis entellus,* and some *Macaca* species. The incidence of bipedal locomotion is low compared to that of bipedal postures. Bipedalism occurs in a number of situations. By far the most common is when an animal needs to increase its visual horizon, for orientation, food location, or as an alert posture. *Cercopithecus aethiops, Presbytis entellus,* and *Erythrocebus patas* may incorporate bipedal hops into locomotor sequences, usually for the purpose of orientation. Bipedalism is also associated with the transport of objects, usually food, but sometimes an infant. Bipedalism of this type is locomotor, and when associated with food carriage frequently serves to remove an animal from possible aggressive encounters. In fact, Hewes (1961) elicited this type of bipedalism in caged animals by threatening the individuals concerned. This type of behavior is most marked in semiterrestrial species and is possibly related to the breakdown of the mechanisms which operate to reduce the incidence of aggression when animals feed in trees (Ripley, 1970). When feeding at food sources which are close to the ground, such as bushes or low branching trees, animals may stand bipedally to feed rather than climb into the bush or tree and sit. Although this type of bipedalism does not occur as frequently as bipedalism to increase the visual horizon, the postures are maintained for relatively long periods of time when it does occur. If the food sources are fairly close together, an animal may walk bipedally between them. Brief periods of bipedal standing and occasionally of bipedal locomotion occur in a variety of other situations such as greeting, sexual activity, aggression, and play.

The sitting postures adopted during feeding and resting on the ground are always with a triangular base. The hindlimbs are quite widely abducted during feeding, to allow the animal to search for and gather food from the area immediately in front of it. In *Papio, Theropithecus,* and to a lesser extent in some *Macaca* species, there are areas of thickened hairless skin associated with the ischial callosities. In *Papio* these areas extend dorsolaterally around the anus and tail base, while in *Theropithecus* they extend ventrally on either side of, and in the male across the perineum. In sitting postures adopted on the ground, a larger area of skin contacts the substrate than is the case during arboreal sitting. The areas of thickened skin therefore act as secondary callosities and it is probably their wear-resistant properties rather than their frictional properties which are the more important. *Papio* tends to spend time sitting at rest as well as sitting feeding and the trunk is held in a fairly erect position so that the body weight is transferred to the ground through the region of the ischial callosities and the skin areas dorsal to them. *Theropithecus* spends more time feeding in sitting postures in which the trunk is inclined forward, so that the body weight is transferred to the ground through the ischial callosities and the skin areas ventral to them. Animals also shuffle forward while maintaining a sitting posture, dragging these skin areas over the ground. The behavioral and morphological adaptations for feeding in *Theropithecus* have been discussed in detail by Jolly (1970).

Posture, Functional Morphology, and the Evolution of Positional Activities

In this section, I shall suggest a few of the ways in which a consideration of postural behavior may be of help in solving some of the problems which have generally been considered mainly in terms of locomotor behavior. Many features of the positional apparatus of an animal may be analyzed either in terms of activities which place maximal stress on the apparatus or in terms of activities which place a more or less continuous stress of a certain type on the apparatus. The limbs are not generally exposed to maximal stresses during postural activities, nor are they exposed to stresses of a different type to those occurring during locomotor activities. A possible exception to this is provided by the foot, which may be exposed to tensile forces during certain sitting postures. However, the axial skeleton, including to a certain extent the pelvic girdle, is exposed to forces during some postural activities which are different from those acting during locomotor activities. Moreover, it is possible that these forces may occasionally approximate the maximum forces which the axial skeleton must withstand.

The ubiquity of trunkal erectness in primates is a well-established fact. The antiquity of trunkal erectness and its association with postural behavior has been emphasized by Napier and Walker (1967). Slijper (1946) has shown that the dimensions of the vertebral bodies of all terrestrial mammals must be such as to withstand the compressive forces generated during the assumption of postures in which the trunk is held in an erect position. Preliminary studies of my own indicate that the area of the vertebral bodies, especially in the lumbar segment of the vertebral column, is relatively greater, even in small primates, than in most other mammals. It is possible that the ischial component of the pelvis, especially in primates possessing ischial callosities is similarly adapted to act as a weight-bearing column.

The external form and the histological structure of the ischial callosities have been studied by Pocock (1925) and Miller (1945), respectively. Their structure is basically that of fibro-fatty cushions bound down to the underlying ischial tuberosities and covered with a tough, thick, highly frictional surface layer. I am at present studying the ischium and ischial callosities of Old World monkeys in an attempt to relate their structure more closely to their function in the sitting postures of this group.

Locomotor activities of living primates have been used in various ways as models for the understanding of the locomotor behavior of fossil forms. I have suggested above that postural behavior is important in understanding the positional adaptations of at least some living primates. A more realistic reconstruction of the positional behavior of fossil primates could result from a consideration of the degree to which postural behavior might have contributed to the total pattern. Similarly, the change of locomotor behavior with time might be better understood if consideration is given to the possible postural as well as the possible locomotor activities involved in that change.

A number of authors (e.g., Hewes, 1961; Snyder, 1967) have emphasized the locomotor aspects of bipedalism either in the context of the evolution of human bipedalism or as a general phenomenon. In the monkeys at least, bipedalism is as much a postural phenomenon as it is a locomotor phenomenon. It is possible to use monkey bipedalism of this type as a model for stages in the evolution of human bipedalism. Stated very simply, the evolution of human bipedalism can be considered in three stages: the evolution in arboreal animals of features, both behavioral and morphological, preadaptive to ground bipedalism; the evolution of habitual bipedalism in a preadapted form living a terrestrial or semiterrestrial life; the perfection of bipedalism in a habitually bipedal form. These stages differ slightly from those suggested by Sigmon (1971). The behavior of living monkeys suggests models for the first two of these stages. The predominantly arboreal ateline monkeys use both postural and locomotor bipedalism in trees and on the ground. It is of interest that the proficient bipedalism in this group owes much to the mobility of the hindlimbs which is an essential component of the postural and locomotor suspended activities which contribute extensively to the positional repertoire.

Factors which have been suggested as being important in the development of habitual ground bipedalism include food transport (Hewes, 1961), other object carrying, the need to increase the visual horizon, and the need to free the hands for offense and defense (Napier, 1964). Wescott (1967) emphasizes the occurrence of bipedalism during displays. Sigmon (1971) uses the bipedalism of *Pan* as a model. She suggests that bipedalism in early hominids, as in *Pan*, must have occurred in a variety of situations and that no one factor is more important than others since they all have survival value. Napier (1964) also makes this point. The point is a valid one; however, it is still possible that one factor, or group of factors may have had a greater potential for the further development of bipedalism.

As mentioned above, although bipedalism associated with object carrying and the increase of the field of view is common in terrestrial and especially semiterrestrial monkeys, bipedalism during feeding from low food sources is of relatively long duration when it does occur. This type of bipedalism is mainly postural but may be associated with periods of bipedal locomotion. Napier (1967), Sigmon (1971), and others have suggested that woodland savannah is the type of environment in which hominid bipedalism could have become established. It is in just this type of habitat that a large number of low food sources exist. Thus, given the amount of preadaptation to bipedalism which is found in some living monkeys, together with the right habitat, economical bipedalism could occupy large periods of an animal's time without compromising its ability to escape or travel longer distances using quadrupedal locomotion.

Carpenter and Durham (1969) have emphasized the fact that there are many types of suspended activities used by primates, of which brachiation is but one type. As with bipedalism, authors speculating on the evolution of brachiation (e.g., Avis, 1962) have thought mainly in terms of locomotor activities and also in terms of bimanual suspended locomotion. Again, it is evident from an examination of the positional behavior of some living monkeys that bimanual suspension is associated with other types of suspension and that the postural element is also important. An animal such as *Ateles*, whose bimanual suspended locomotion on occasion approximates that of brachiating apes (Erikson, 1963), uses a wide range of suspended locomotor activities which involve the use of all combinations of the four limbs. It also uses suspended postural activities to an extent which is equal or greater than the use of suspended locomotor activities. In species less adapted for below branch life, suspended postural activities predominate over suspended locomotor activities and quadrupedal or tripedal suspended postures predominate over bimanual suspended postures. In all these animals, suspended positional activities are associated mainly with feeding. It might, therefore, be useful to consider a change from postural to locomotor suspension and from quadrupedal to bimanual suspension as at least one pathway for the development of advanced bimanual suspension.

Acknowledgments

Field work was financed by a Grant-in-Aid from the Royal Society. Other work was financed in part by a Pre-Doctoral Fellowship from the Smithsonian Institution and by the Boise Fund, Oxford.

References

Allen, J. A. (1925). Primates collected by the American Museum Congo expedition. *Bull. Amer. Mus. Natur. Hist.* **47**, 283–499.

Altmann, S. A., and Altmann, J. (1970). Baboon ecology. *Biblio. Primatol. No.* 12.

Ankel, F. (1962). Vergleichende Untersuchungen über die Skelettmorphologie des Greifschwanzes südamerikanischer Affen (Platyrrhina). *Z. Morphol. Oekol. Tiere* **52**, 131–170.

Avis, V. (1962). Brachiation: the crucial issue for man's ancestry. *Southwest. J. Anthropol.* **18**, 119–148.

Bert, J., Ayats, H., Martino, A., and Collomb, H. (1967). Le sommeil nocturne chez le babuin *Papio papio.* Observations en milieu naturel et donées électrophysiologiques. *Folia Primatol.* **6**, 28–43.

Bert, J., Pegram, V., Rhodes, J. M., Balzano, E., and Naquet, R. (1970). A comparative sleep study of two Cercopithecinae. *Electroencephalogr. Clin. Neurophysiol.* **28**, 32–40.

Bishop, A. (1964). Use of the hand in lower primates. *In* "Evolutionary and Genetic Biology of Primates" (J. Buettner-Janusch, ed.). Academic Press, New York.

Booth, A. H. (1956). Observations on the natural history of the olive colobus monkey, *Procolobus verus* (van Beneden). *Proc. Zool. Soc. London* **129**, 421–430.

Bourlière, F., Bertrand, M., and Hunkeler, C. (1969). L'écologie de la mone de Lowe (*Cercopithecus campbelli lowei*) en Cote D'Ivoire. *Terre Vie* **23**, 135–163.

Carpenter, C. R. (1934). A field study of the behaviour and social relations of howling monkeys. *Comp. Psychol. Monogr.* **10**, 1–168.

Carpenter, C. R. (1935). Behaviour of red spider monkeys in Panama. *J. Mammal.* **16**, 171–180.

Carpenter, C. R., and Durham, N. M. (1969). A preliminary description of suspensory behaviour in nonhuman primates. *Proc. 2nd Intern. Congr. Primatol.*, Vol. 2. S. Karger, Basel.

Chalmers, N. R. (1968). Group composition, ecology and daily activities of free living mangabeys in Uganda. *Folia Primatol.* **8**, 247–262.

Chivers, D. J. (1969). On the daily behaviour and spacing of howling monkey groups. *Folia Primatol.* **10**, 48–102.

Crook, J. H., and Aldrich-Blake, P. (1968). Ecological and behavioural contrasts between sympatric ground dwelling primates in Ethiopia. *Folia Primatol.* **8**, 192–227.

Dandelot, P. (1956). Le comportement de deux cercopithèques de l'Hoest en captivité. *Mammalia* **20**, 330–331.

DeVore, I., and Hall, K. R. L. (1965). Baboon ecology. *In* "Primate Behavior" (I. De Vore, ed.). Holt, Rinehart and Winston, New York.

Eisenberg, J. F., and Kuehn, R. E. (1966). The behaviour of *Ateles geoffroyi* and related species. *Smithson. Misc. Collect.* **151**, 1–63.

Erikson, G. E. (1963). Brachiation in New World monkeys and in anthropoid apes. *Symp. Zool. Soc. London* **10**, 135–164.

Gautier-Hion, A. (1966). L'écologie et l'éthologie du talapoin, *Miopithecus talapoin talapoin. Biol. Gabon.* **2**, 311–329.

Goswell, M. J., and Gartlan, J. S. (1965). Pregnancy, birth and early infant behaviour in the captive patas monkey, *Erythrocebus patas. Folia Primatol.* **3**, 189–200.

Grand, T. I. (1972). A mechanical interpretation of small branch feeding. *J. Mammal.* **53**, 198–201.

Haddow, A. J. (1952). Field and laboratory studies on an African monkey, *Cercopithecus ascanius schmidti* Matschie. *Proc. Zool. Soc. London* **122**, 297–394.

Hall, K. R. L. (1962). Numerical data, maintenance activities and locomotion of the wild chacma baboon, *Papio ursinus. Proc. Zool. Soc. London* **139**, 181–220.

Hall, K. R. L. (1965). Behaviour and ecology of the wild patas monkey, *Erythrocebus patas,* in Uganda. *J. Zool.* **148**, 15–87.

Hall, K. R. L., and DeVore, I. (1965). Baboon social behaviour. *In* "Primate Behavior" (I. DeVore, ed.). Holt, Rinehart and Winston, New York.

Hewes, G. W. (1961). Food transport and the origin of hominid bipedalism. *Amer. Anthropol.* **63**, 687–710.

Jay, P. (1965). The common langur of North India. *In* "Primate Behavior" (I. DeVore, ed.). Holt, Rinehart and Winston, New York.

Jolly, C. J. (1970). The seed-eaters: a new model of hominid differentiation based on a baboon analogy. *Man* **5**, 1–26.

Karrer, R. (1970). The use of the tail by an Old World monkey. *Primates* **11**, 171–175.

Kummer, H. (1968). Social organization of Hamadryas Baboons. *Biblio. Primatol.* No. 6.

Lumsden, W. H. R. (1951). The night-resting habits of monkeys in a small area on the edge of the Semliki forest, Uganda. A study in relation to the epidemiology of sylvan yellow fever. *J. Animal Ecol.* **20**, 11–30.

Miller, R. A. (1945). The ischial callosities of primates. *Amer. J. Anat.* **76**, 67–91.

Moynihan, M. (1967). Comparative aspects of communication in New World primates. *In* "Primate Ethology" (D. Morris, ed.). Wiedenfelt and Nicolson, London.

Napier, J. R. (1963). Brachiation and brachiators. *Symp. Zool. Soc. London* **10**, 183–195.

Napier, J. R. (1964). The evolution of bipedal walking in the hominids. *Arch. Biol.* **75**, 673–708.

Napier, J. R. (1967). Evolutionary aspects of primate locomotion. *Amer. J. Phys. Anthropol.* **27**, 333–342.

Napier, J. R., and Walker, A. C. (1967). Vertical clinging and leaping—a newly recognised category of locomotor behaviour of primates. *Folia Primatol.* **6**, 204–219.

Pocock, R. I. (1925). The external characters of catarrhine monkeys and apes. *Proc. Zool. Soc. London,* pp. 1479–1579.

Prost, J. H. (1965). A definitional system for the classification of primate locomotion. *Amer. Anthropol.* **67**, 1198–1214.

Rahaman, H., and Parthasarathy, M. D. (1968). The expressive movements of the bonnet macaque. *Primates* **9**, 259–272.

Rahaman, H., and Parthasarathy, M. D. (1969). The home range, roosting places and day range of the bonnet macaque (*Macaca radiata*). *J. Zool.* **157**, 267–276.

Richard, A. (1970). A comparative study of the activity patterns and behaviour of *Alouatta villosa* and *Ateles geoffroyi*. *Folia Primatol.* **12**, 241–263.

Ripley, S. (1967). The leaping of langurs: a problem in the study of locomotor adaptation. *Amer. J. Phys. Anthropol.* **26**, 149–170.

Ripley, S. (1970). Leaves and leaf monkeys. *In* "Old World Monkeys" (J. R. Napier and P. H. Napier, eds.). Academic Press, New York.

Rowell, T. E. (1966). Forest living baboons in Uganda. *J. Zool.* **149**, 344–364.

Sigmon, B. A. (1971). Bipedal behaviour and the emergence of erect posture in man. *Amer. J. Phys. Anthropol.* **34**, 55–60.

Slijper, E. J. (1946). Comparative biologic-anatomical investigations on the vertebral column and spinal musculature of mammals. *Kon. Ned. Akad. Wetensch. Verh.* **52**, 1–128.

Snyder, R. C. (1967). Adaptive values of bipedalism. *Amer. J. Phys. Anthropol.* **26**, 131–134.

Struhsaker, T. T. (1967). Behaviour of vervet monkeys (*Cercopithecus aethiops*). *Univ. Calif. Publ. Zool.* **82**, 1–74.

Struhsaker, T. T., and Gartlan, J. S. (1970). Observations on the behaviour and ecology of the patas monkey (*Erythrocebus patas*) in the Waza Reserve, Cameroun. *J. Zool.* **161**, 49–63.

Thorington, R. W., Jr. (1968). Observations of squirrel monkeys in a Colombian forest. *In* "The Squirrel Monkey" (L. A. Rosenblum, and R. W. Cooper, eds.). Academic Press, London.

Tuttle, R. H. (1967). Knuckle walking and the evolution of hominoid hands. *Amer. J. Phys. Anthropol.* **26**, 171–206.

Washburn, S. L. (1957). Ischial callosities as sleeping adaptations. *Amer. J. Phys. Anthropol.* **15**, 269–276.

Wescott, R. W. (1967). Hominid uprightness and primate display. *Amer. Anthropol.* **69**, 738.

8

Origins, Evolution, and Function of the Tarsus in Late Cretaceous Eutheria and Paleocene Primates

FREDERICK S. SZALAY
and
RICHARD LEE DECKER

Introduction

Many aspects of the origins of primates as well as other ordinal groups of the Mammalia are still shrouded in mystery. To solve some of the problems, as an ever increasing number of fossils of new and known taxa revealing new anatomical parts are recovered, it becomes necessary, in addition to systematic description, to order and interpret this wealth of material in terms of evolutionary morphology and phylogeny. Thus the study of known Paleocene primate postcranials, along with the earliest known eutherian ones, can be approached by asking a series of interrelated, virtually inseparable questions. To which Cretaceous and Tertiary groups are they most closely related? What morphological features do they share either as primitive retentions (symplesiomorphies), shared advanced traits (synapomorphies) or independently acquired traits, and what characters differentiate them from their closest relatives? What particular mechanical differences exist between the bones and joints of these related taxa? Do the respective parts of the skeleton allow inferences concerning their habitual use which in turn can be correlated with a way of life either as terrestrial or arboreal? Finally, what particular locomotor, feeding, or possibly other behavioral functions might have been performed by the animals possessing the known fossil parts? The answers to these questions must result in new syntheses of the phylogeny, mode of life, and of the probable nature of selection pressures that molded the once living fossils.

Biological characters can be weighted to construct phylogenies without knowledge of function and biological role. A glance into the evolutionary history of past lineages beyond a knowledge of phylogeny, however, necessitates functional studies. Our purpose is to analyze representative specimens of late Cretaceous eutherians and Paleocene paromomyiform primate tarsals, and present an account of the morphology and the inferred mechanics (function) of the astragalocalcaneal complex. The study is restricted to the astragalus and calcaneum because no other tarsal elements of these mammals are as yet known.

Major research efforts on late Cretaceous deposits in Montana by Robert E. Sloan and associates, begun in 1962, have resulted in the collection of an unprecedented wealth of eutherian specimens. Some of the studies on these forms have been published while others are in preparation. A comprehensive report by Szalay, Nelson, and Sloan is in preparation on all the therian postcranials from the Bug Creek Cretaceous. Some of the evidence on the astragalus and calcaneum of two of the several species, *Procerberus formicarum* and *Protungulatum donnae* (see Sloan and Van Valen, 1965), is used here as representative of the ancestral levels of organization achieved by two lineages of early eutherians.

Not until 1935 was the most complete North American Paleocene primate skeleton, the entire sample of *Plesiadapis gidleyi* along with the dentition and fragmentary skull, described in detail by Simpson (1935). He compared the astragalus and calcaneum with tupaiids in general, and in detail with *Lemur, Pelycodus,* and *Notharctus.* The comparison to lemuriforms was significant, and Simpson astutely stressed the similarity of *Plesiadapis* with the Eocene lemuriforms not only in dentition but also in postcranial anatomy. Although Simpson has clearly and unequivocally classified *Plesiadapis* as a primate, some recent students of primates have outrightly rejected this (Cartmill, 1972; Charles-Dominique and Martin, 1970; Martin, 1972) or ignored the remains of *Plesiadapis* and other paromomyiforms, mainly, we suspect, because of unfamiliarity with the early Tertiary record.

The paromomyiform tarsal remains now known in the Paleocene range from astragali and calcanea of paromomyids, picrodontids, and carpolestids, some less than 4 mm long, to those of some species of *Plesiadapis* 12 mm long. The astragalar and calcaneal morphology of all known North American and European paromomyiforms is similar, and it is reasonable to assume that in the latest Cretaceous and early Paleocene the morphology shown by the paromomyiforms represented the earliest level of primate tarsal organization.

Our analysis, developed below, strongly suggests that the Cretaceous eutherians were primarily adapted to a terrestrial adaptive zone. Thus, divergence of the earliest known primates appears to have been from a terrestrial stock toward an arboreal way of life, suggesting that the acquisition of the ordinal characters of primate postcranial anatomy correspond with the invasion of an arboreal milieu. Arguments that the paromomyiforms are not really primates—either adaptively (other than their dentition) or phylogenetically—now more than ever must somehow refute the fossil evidence that reveals that the petrosal bulla, the tarsal morphology (indicative of a habitually inverted pes), and the molar morphology of the paromomyiforms are phylogenetically related to those of the Eocene and younger "real" primates.

Methods

Our approach to problems surrounding the tarsus has involved extensive morphological comparisons between living Insectivora, Carnivora, Primates, and the relevant known fossils of Cretaceous and early Tertiary Eutheria.

Estimation of the axes of rotation at certain joints have aided our own visualization of movements as well as the descriptive aspects of our discussions. There are a number of excellent contributions in the literature that deal with various aspects of the joints and the axes of rotation of the joints in mammalian feet and which may be most profitably consulted on this subject (Elftman and Manter, 1935; Manter, 1941).

In order to estimate the axes, we have manipulated the moving parts, where possible, and used curvatures of the articular facets. One useful method, when both tibia and astragalus are available, as in the extant taxa examined, brought out some significant differences in the upper ankle joints examined. In principle it resembles Manter's (1941) method for the lower ankle joint. We attached a pointed instrument to the tibia so that together they formed a single rigid body. The astragalus is passively moved on the tibia throughout the entire permissible range, being careful that it is the morphology which dictates the direction of movement. The instrument point is adjusted until it coincides with a single point on the astragalus throughout the range of movement. This is done for both sides of the astragalus for two points. These points define a unique line, the axis of joint rotations. By definition, an axis is a line of points showing zero velocity during movement. If one is unable to obtain a single coincident point, the changing position of the axis through the movement is indicated, as in the upper ankle joint of man (Barnett and Napier, 1952). Needless to say, this method could only be employed in those fossil taxa for which the crus is known.

Once determined, the axes also permit proper assessment of the structure of bones for other mechanical properties. For example, we have assumed that differential dimensions of articular surfaces influence the ability of a joint to distribute loads and its stability in different positions as well as its mobility. The axes make sensible some differences in muscular relationships, a few of which leave their imprints on the bone.

In particular we addressed ourselves to the question of whether a tarsus, given its morphology, was capable of adjusting to and withstanding the stresses of various orientations or was rather suited to more nearly level orientations, correlated with arboreality and terrestriality, respectively.

The conceptual framework provided by Bock and Wahlert (1965) has strongly influenced our analysis. These authors clearly distinguish between mechanical function of a morphological complex and the one or several biological roles performed by that function.

Substrate

As a medium for locomotion, each major habitat, the terrestrial, subterranean, arboreal, aerial, or the aquatic, has its distinctive characters. As this has been often noted the branches of an arboreal environment can be distinguished from the ground below in that the branches "are (1) discontinuous, (2) mobile, (3) variable (and always restricted) in width, and (4) oriented at all angles to the pull of gravity" (Cartmill, 1972, p. 107). Also, the geometry of a single branch, curved rather than plane, is itself important. In contrast, terrestrial substrates may vary more in composition, but whether level or irregular, a horizontal plane of the substrate predominates and offers a potentially broad base for support. Barriers are usually easily bypassed for preferred substrates by a terrestrial animal, whereas fewer possible pathways of progression from any given point are available to an arboreal vertebrate. For these reasons, an arboreal mammal must often adapt to a greater relative variety of substrate or physical situations involving the substrate in their environment.

Differences in substrate should receive emphasis in discussions of postcranial morphology, function, and adaptive significance. The adaptive relationship to arboreal substrates of such characters as tail prehen-

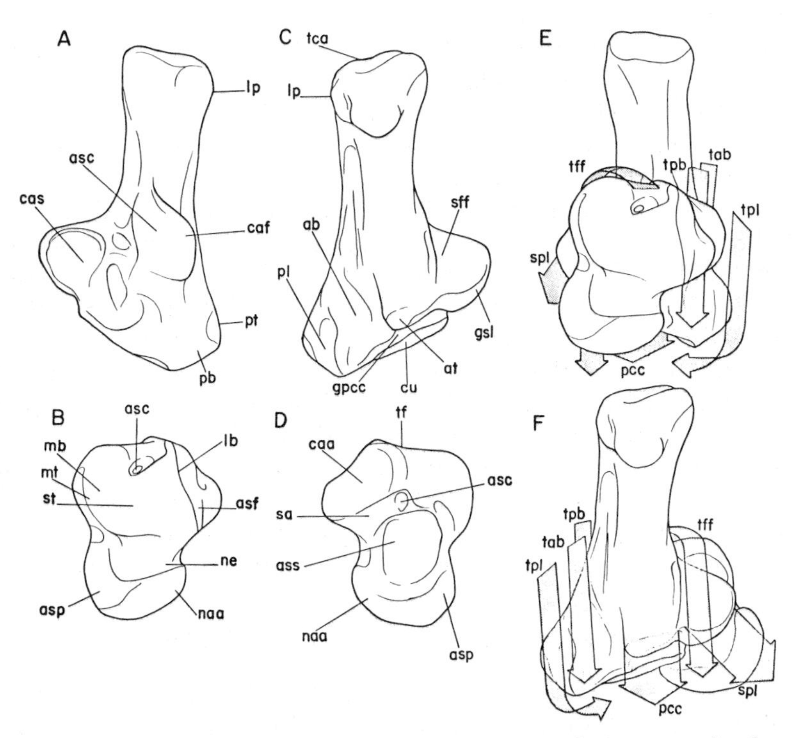

Fig. 1. The calcaneum (A, C) astragalus (B, D), and the astragalocalcaneal complex (E, F) of *Protungulatum* to demonstrate the terminology used in this paper. Dorsal (A, B, E) and plantar (C, D, F) views.

Abbreviations on the calcaneum: ab, groove for m. abductor digiti quinti?; asc, astragalocalcaneal facet; at, anterior plantar tubercle; caf, calcaneal fibular facet; cas, calcaneal sustentacular facet; cu, cuboid facet; gpcc, groove for plantar calcaneocuboid ligament; gsl, sustentacular groove for attachment of "spring" ligament; lp, lateral process of tuber calcanei; pb, groove for tendon of m. peroneus brevis; pl, groove for tendon of m. peroneus longus; pt, peroneal tubercle; sff. sustentacular groove for tendon of m. flexor digitorum fibularis; tca, tuber of the calcaneum.

Abbreviations on the astragalus: asc, astragalar canal; asf, astragalar fibular facet; asp, astragalar "spring" ligament facet; ass, astragalar sustentacular facet; caa, calcaneoastragalar facet; lb, lateral border (crest) of trochlea; mb, medial border (crest) of trochlea; mt, medial tibial facet of trochlea; naa, naviculoastragalar facet; ne, neck of astragalus; sa, sulcus astragali; st, superior tibial facet of trochlea; tf, trochlear groove for tendon of m. flexor digitorum fibularis.

Abbreviations on the astragalocalcaneal complex: pcc, (long) plantar calcaneocuboid ligament; spl, "spring" ligament (calcaneonavicular ligament); tab, tendon or fleshy fibers of abductor digiti quinti or other muscles; tff, tendon m. flexor (digitorum) fibularis; tpb, tendon m. peroneus brevis; tpl, tendon m. peroneus longus.

sility, and opposability of the hallux and pollex among marsupials and primates is clear. Differences in evenness of the substrate have been employed to explain functions of the lower and upper ankle joints and of the crus (Barnett and Napier, 1953; Barnett, 1955, 1970).

The primary functions of the pes are to maintain a stable relationship of the organism to its substrate and to propel, whereupon the former role becomes even more important. Successful adaptations in arboreal mammals in the pes include an ability to adjust to unevenness and variation in the substrate and orient to a number of physical situations. Orientations are accomplished in part by inversion and eversion of the foot effected by rotations of joints at odd angles to the animal's direction of progression or that of the pes. Joints of the tarsus must permit the desired movements while remaining stabilized and distributing loads in various positions, especially during stance phases of the limb. Inverted as well as everted orientations of the foot may be required of arboreal mammals to adjust to the substrate while resistance remains directed through the hip. For this reason near level orientations of the pes are better adapted on terrestrial substrates, and different sets of morphological and functional requirements prevail.

Relationships of Cretaceous and Early Tertiary Eutheria

The past decade has seen an unprecedented surge of interest in the study of Late Cretaceous and Paleocene Mammalia (for some recent contributions and reviews see Clemens, 1963, 1966, 1970; Lillegraven, 1969; McKenna, 1969; Russell, 1964; Sloan, 1969; Szalay, 1969; Szalay and McKenna, 1971; Van Valen, 1966; Van Valen and Sloan, 1966). Important discoveries in the field have multiplied the known fossil specimens, and equally productive efforts in laboratories have resulted in a large number of publications, too numerous to list. However, almost all of the published studies dealing with phylogenetic, ecological, and systematic aspects of the Late Cretaceous to Early Tertiary mammalian radiation concentrated on the morphological data derived from the cranium and the dentition.

It is of importance to present some of the phylogenetic conclusions (Fig. 2) derived from a preliminary survey of and a detailed study of selected early eutherian feet. It has obvious implications for evolutionary studies of early Tertiary mammals, including primates. In many ways the most surprising preliminary result of this study is a picture of phylogenetic relationships not revealed by the dental evidence, and one which sometimes contradicts previous inferences derived from the latter.

The conclusion that the astragalocalcaneal complex of the latest

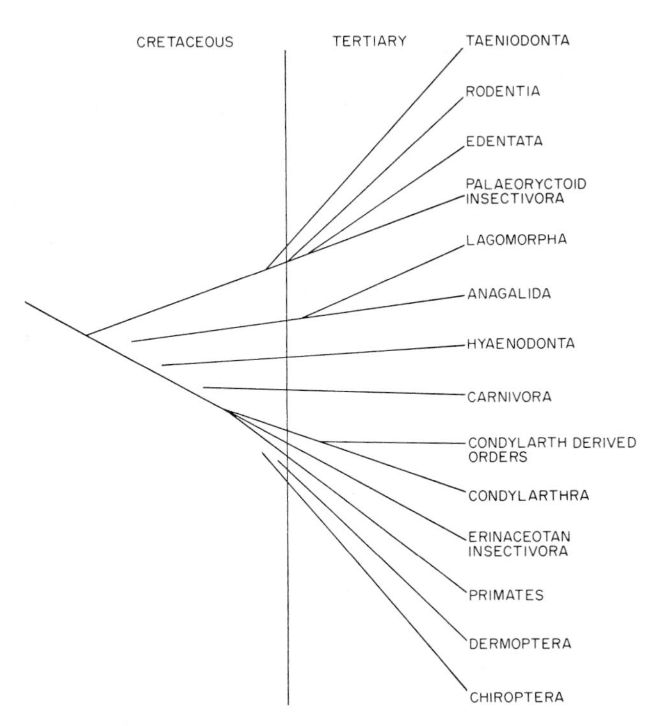

Fig. 2. Schematic phylogeny of most of the Cretaceous-Tertiary orders of eutherians based on the total available skeletal and dental evidence. All, approximately fifteen, of the condylarth-derived orders are pooled together. Uncertainty exists as to the derivation of primates from erinaceotan Insectivora or Condylarthra. An early eutherian diversification, possibly in pre-Campanian times, is inferred from the tarsus.

Cretaceous palaeoryctoids (a superfamily which should include the Palaeoryctidae and Leptictidae) is relatively advanced, and that it probably did not give rise to those of the Condylarthra, Carnivora, Hyaenodonta, Primates, and Anagalida and Lagomorpha, is, in general, contradictory to most recent phylogenetic considerations (Clemens, 1970; Lillegraven, 1969; McKenna, 1969; Szalay, 1969; Van Valen, 1966, 1967, 1970). Considering the tarsal structure, as well as the dental evidence, it appears that the known latest Cretaceous palaeoryctids are ancestors or near the ancestry of the following families: Leptictidae, Pantolestidae, and probably the Tupaiidae. Origins of the Rodentia and the Microsyopoidea from either the Palaeoryctidae or the Leptictidae, the various pantolestoids from the pantolestids appears probable. The palaeoryctid-derived pantolestids might have been ancestral to both the Metacheiromyidae and Epoicotheriidae (neither of the Pholidota, contra Emry, 1970, although these are not necessarily edentates). Origins of the Mixodectidate are obscure although, as suggested by Van Valen (1966), special ties between the former and the Dermoptera are possible. The evidence from the tarsus supports the contention of Van Valen (1966) that the Taeniodonta evolved from palaeoryctids, whereas the Tillodontia was condylarth-derived.

McKenna (1969) and Van Valen (1970) have recently repeated, following the original suggestion by Wood (1962) and McKenna (1961), that the Rodentia was derived from early primates. The paramyid rodent tarsus is so similar to the primitive palaeoryctoid one that derivation from either palaeoryctid or leptictid insectivorans seems very probable. This inference is supported by all the available basicranial evidence.

Because the vague and ill-defined concept Insectivora is central in any schema of eutherian evolution, the suggested phylogenetic relationships within this nebulous category are summarized below. The combined available morphological evidence supports an ancient phyletic distinctness and thus either the subordinal, or eventual ordinal, separation of the probably palaeoryctoid-derived Insectivora (which are probably, in addition to the Leptictidae and Palaeoryctidae, the Microsyopidae, Pantolestidae, Tupaiidae, possibly the Tenrecoidea and the Chrysochloridae) from those derived from an erinaceotan, nyctitheriid *Batodon–Leptacodon*-like ancestry. The status of the Apheliscidae, Pentacodontidae, and Ptolemaiidae, although possibly pantolestid derived may be left "incertae sedis." The status of the Deltatheridiidae, in spite of its recent allocation to the Palaeoryctoidea by Szalay and McKenna (1971), is best considered, along with *Kennalestes* and the Mixodectidae (Szalay, 1969), undetermined at present. The recently named and diagnosed order Anagalida (Szalay and McKenna, 1971) might have evolved from an ancestry with a dentition not unlike that of *Kennalestes,* but from forms with an unreduced astragalar canal (see the pes of *Anagale* in Simpson (1931) or that of *Pseudictops* in Sulimsky (1968).

A preliminary comparative study of the tarsal morphology of living tupaiids indicates that neither calcanea nor astragali have significant primitive eutherian characters nor do they appear to share homologous advanced features with the paromomyiforms. In fact, the tarsus of living tupaiids is a very advanced one, suggesting a palaeoryctoid derivation of this family.

A recent derivation by McKenna (1969), Sloan (1969), and Szalay (1969), of the Primates and erinaceotan Insectivora from the leptictids or palaeoryctids is not considered probable, nor is the latest Cretaceous differentiation of *Protungulatum* from a *Cimolestes*-like form as suggested by Lillegraven (1969). Common ancestors of these and of palaeoryctoid Insectivora, Carnivora, Hyaenodonta, and Anagalida must be sought in Campanian or equivalent, or older deposits of North America or Asia.

The Tarsus of Late Cretaceous Eutherians and Paleocene Primates

Morphological Aspects

To describe all the relevant early eutherian evidence is beyond the scope of this paper; it is pursued in detail elsewhere. Instead, the astragalus and calcaneum of the late Cretaceous *Procerberus* and *Protungulatum* are compared with the paromomyiform astragalocalcaneal complex. The Bug Creek tarsals of *Protungulatum, Procerberus* and *Cimolestes,* the latter two differing only in size from one another, represent some of the earliest known astragali and calcanea of the Condylarthra and Insectivora.

From the total available evidence the following might be adduced as primitive characters of the eutherian astragalocalcaneal complex: a distal peroneal tubercle, a cuboid facet oblique to the long axis of the calcaneum, a posterior astragalocalcaneal articulation forming a relatively large angle (35–40°) with the long axis of the calcaneum, an oblique axis of rotation of the astragalocalcaneal articulation, a short calcaneal body anterior to the astragalocalcaneal facet, a plantar anterior tubercle on the calcaneum, a low, slightly grooved tibial trochlea on the astragalus, a relatively short tibial trochlea, a large astragalar canal piercing the body, possibly cuboid contact with the astragalus (alternating type of tarsus), an astragalar sustentacular facet not continuous with other facets, and a wide astragalar head laterally thicker and rotated slightly dorsolaterally. Of the known Cretaceous tarsal remains these features are present primarily in *Protungulatum* and some are retained in *Procerberus* and paromomyiform primates. Judged by the sum of the suggested primitive eutherian features, *Protungulatum* possessed an astragalocalcaneal complex more primitive than that of any of the known palaeoryctoids or other known Eutheria.

Figures 3–13 of this paper show the morphology as well as the axes of rotation of the joints examined in *Protungulatum, Procerberus* (tarsally virtually identical to *Cimolestes*), and *Plesiadapis.* The fitted astragalus and calcaneum of *Procerberus formicarum* unfortunately display size disparities commonly encountered in individuals of different size within a species. Because of lack of associated bones from one animal this cannot be helped, yet the reader should take this into consideration when viewing the figures.

Morphology as well as joint movements of a relatively primitive eutherian foot, that of the Paleocene arctocyonid *Arctocyon* (incl. *Claenodon*), was treated by Schaeffer (1947), and therefore some of the discussion by him is relevant.

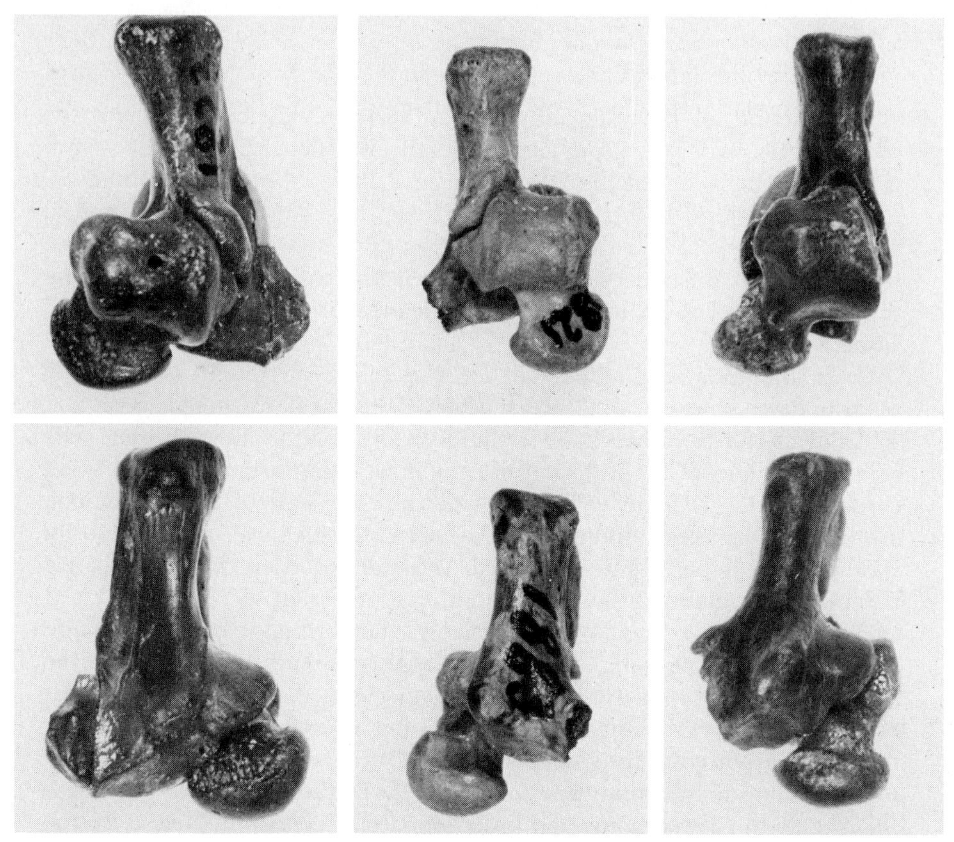

Fig. 3. Dorsal (top) and plantar (bottom) views of astragali and calcanea in an everted position. Left: *Protungulatum donnae,* latest Cretaceous, Bug Creek Anthills locality, U.M.V.P. Nos. 1914 (astragalus) and 1823 (calcaneum); Middle: *Procerberus formicarum,* latest Cretaceous, Bug Creek Anthill locality, U.M.V.P. Nos. 1821 (astragalus) and 1883 (calcaneum); Right: *Plesiadapis,* c.f., *P. gidleyi,* early late Paleocene Saddle locality, A.M.N.H. (H.C.) Nos. 89533 (astragalus) and 89534 (calcaneum). The peroneal tubercles of the calcanea of *Procerberus* and *Plesiadapis* are broken off.

Astragalus

The proportions of the astragalus are different in the genera compared. The trochlea is relatively narrower and longer in paromomyiforms, the ratio of trochlear length to astragalar length in Plesiadapis being 0.66, whereas in *Protungulatum* and *Procerberus* these ratios are 0.54 and 0.49, respectively. The tibial trochlea is relatively the longest and least broad in *Plesiadapis* compared with *Protungulatum* and *Procerberus;* it is not only relatively longer in paromomyiforms, but it is laterally very high compared to the latest Cretaceous genera. The body of paromomyiform astragali is relatively much more robust than that of the known Cretaceous forms due to the great relative thickness of the body laterally. Unlike in *Procerberus,* but as in *Protungulatum,* the tibial side of the paromomyiform trochlea is not keeled but rounded, whereas the high fibular side is sharply crested, as in the former genera.

The astragalar canal is significant not only for phylogenetic inference but it also places certain limitations and conditions on function. Although the astragalar canal persists in some paromomyiforms (and in some lemuriforms, such as the subfossil *Archaeolemur* or *Megaladapis*) it tends to be relatively smaller than in *Protungulatum.* Nevertheless, when it is present as in the large *Plesiadapis tricuspidens,* or even in forms in which it is clearly vestigial, the trochlear articular surface extends posteriorly only medially, as in *Protungulatum.* The astragalus of *Procerberus,* and other known astragali of palaeoryctoids, have no astragalar canals and the articular surface extends more posteriorly and laterally than in *Protungulatum* or in paromomyiforms.

Schaeffer (1947, p. 4) summarized the following concerning this canal (called generally a foramen): "It has been stated that this foramen, or whatever passed through it, was responsible for greatly restricted movement in the upper ankle joint. The trochlear articular surface, however, extends posteriorly on both sides of the foramen, as in *Orycteropus,* which, incidentally, has a primitive ungulate-like postcranial skeleton. . . . The tibia covers the foramen during complete plantar flexion in the aardvark and must have done so in the primitive ferungulates." In the Cretaceous *Protungulatum,* the articular surface of the trochlea extends posteriorly only on the medial side and a groove running posterolaterally gave sufficient protection for the vessel(s) or nerve(s) traversing the astragalus. The contents of this canal probably did restrict plantar flexion; blood vessels or nerves traversing it were probably present only in forms which did not require the extremes of plantar flexion. This, then, was probably characteristic of the primitive eutherian condition which probably required a range of movement not more than 50–70°.

One of the most critical differences of paromomyiforms from the late Cretaceous eutherians lies in the construction of the head of the astragalus. Both in *Protungulatum* and *Procerberus* the lateral part of the articular surface on the head is much broader than the medial one. The laterally extensive articular facet indicates that, as in terrestrial plantigrade quadrupeds, level orientation and eversion are the most frequently performed functions as the plantigrade pes contact the more or less flat substrate. It is during eversion that the lateral side of the articular facet of the astragalar head is subjected to the greatest and most frequent stresses. In contrast to the latest Cretaceous eutherians, in paromomyiforms the medial side of the articular facet of the astragalar head is relatively broader than these surfaces in *Protungulatum* and *Procerberus*. The facet for the spring ligament is strongly developed in paromomyiforms, moderately in *Procerberus,* and poorly in *Protungulatum*. The medial broadening, therefore, indicates that this side in the known early primates was subject to relatively greater sum total of forces and therefore stressed more frequently than the same side in *Protungulatum*. A great deal of variation occurs in this respect among paromomyiform astragali, but the usual condition is away from that represented by *Protungulatum* and *Procerberus*. Given a horizontal orientation of the astragalar tibial trochlea the head of the astragalus is oriented in a dorsolateral direction in the Cretaceous eutherians, whereas in paromomyiforms the orientation is opposite, in a dorsomedial direction.

Calcaneum

The large peroneal process or tubercle in *Protungulatum* (Fig. 1) and *Procerberus* (Fig. 3) is located on the distal end of the calcaneum, off the lateral edge of the cuboid facet. In paromomyiforms it is somewhat more proximal (posterior) to the cuboid facet. As in the Cretaceous eutherians the groove on the lateral border of the peroneal tubercle is for the tendon of the peroneus longus running in a slightly posterodorsal-anteroplantar direction. On the dorsal surface the less distinct groove is for the tendon of the peroneus brevis both in paromomyiforms and the Cretaceous eutherians. The plantar surface of the peroneal tubercle in *Protungulatum* is bordered by the curved lateral edge of the peroneal tubercle and the anterior plantar tubercle in such a way that a depression is formed, possibly for the abductor digiti quinti or other muscles. In paromomyiforms this groove is neither as prominent nor as distinct. Unlike *Protungulatum* which has a narrow crest as the posterior base of the peroneal tubercle, the Paleocene primates have a robust base for the tubercle. These differ-

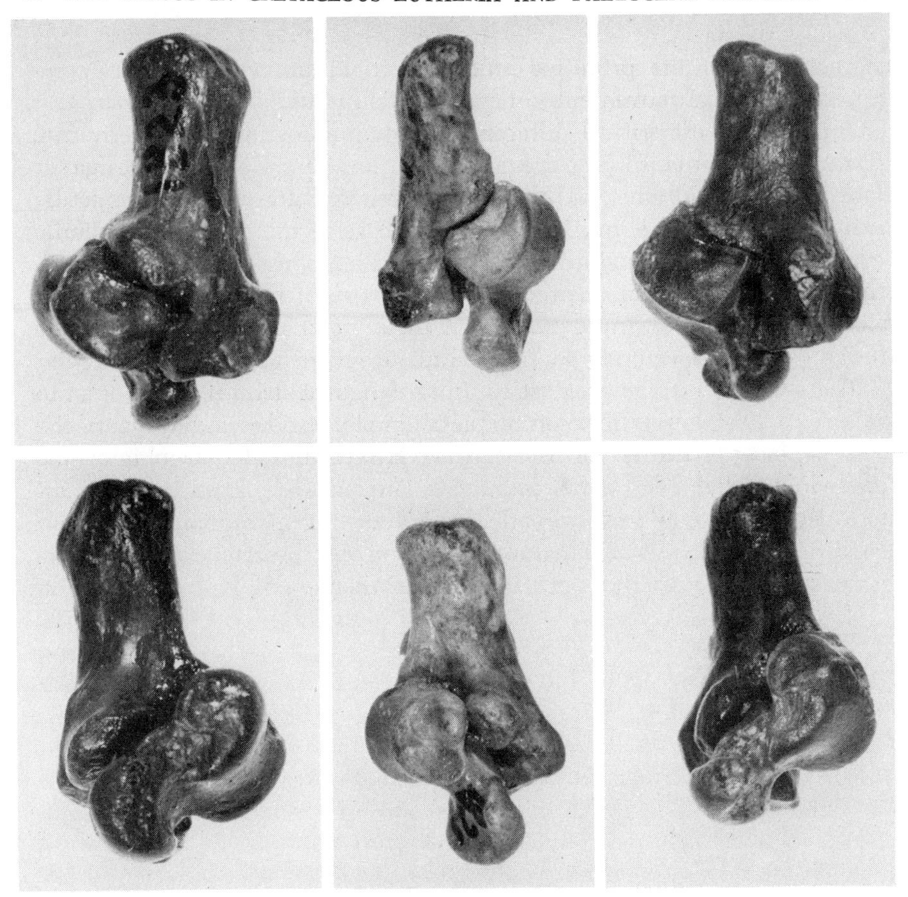

Fig. 4. Lateral (top) and medial (bottom) views of astragali and calcanea in an everted position of *Protungulatum donnae* (left), *Procerberus formicarum* (middle), and *Plesiadapis*, c.f., *P. gidleyi* (right). For data on the provenance of the specimens see Fig. 3.

ences possibly reflect differences in the stresses on the tubercle caused by the peroneus muscles. In spite of the robustness of the base of the peroneal tubercle in Paleocene primates, there is an apparent decrease in the relative size, which appears to be simultaneous with the proximal migration of the tubercle.

On the posterior third of the lateral side of the tuber of the calcaneum there is a robust dorsally located tuberosity in paromomyiforms, the lateral process as in other primates, for the origins of the abductor digiti quinti. Although this tuberosity is present in *Protungulatum,* it is smaller and significantly more toward the plantar side of the tuber.

The cuboid facets of the genera compared, all parts of imaginary cones of differing heights and base diameters in *Protungulatum, Procerberus,* and paromomyiforms, are all concave to differing degrees, permitting gliding movements. In *Protungulatum,* but less so in *Procerberus,* the surface of the cuboid facet and the anterior tubercle are oblique in a medial and plantar direction, probably representative of the primitive eutherian condition. In paromomyiforms the surface is more nearly perpendicular to the long axis of the calcaneum. The cuboid facet is more extensive medially than laterally in *Protungulatum,* whereas in the archaic primates it is nearly equally developed on both sides.

There is a deep transverse groove in *Protungulatum* on the medioplantar side of the distal end, between the cuboid facet and the anterior tubercle. The groove, probably for the partial attachment of the plantar calcaneocuboid ligament, shifts proximally and medially, and it is reduced to a small groove and pit on both *Procerberus* and paromomyiforms. The condition in *Protungulatum,* also shared by the earliest miacoid Carnivora, is probably the primitive eutherian arrangement.

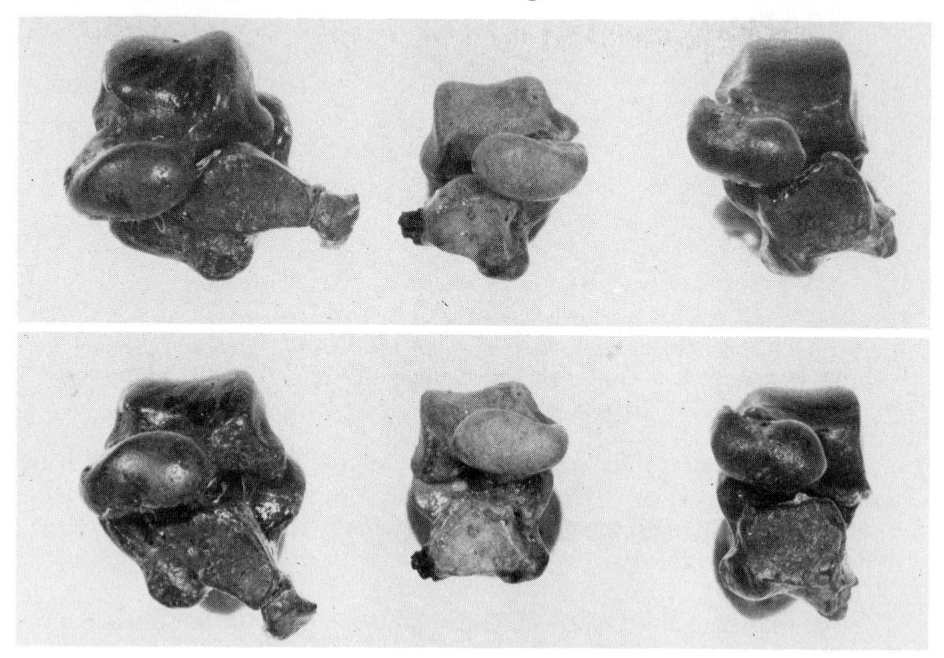

Fig. 5. Distal views of the everted (top) and inverted (bottom) positions of the astragali and calcanea of *Protungulatum donnae* (left), *Procerberus formicarum* (middle), and *Plesiadapis,* cf. *P. gidleyi* (right). For data on the specimens see Fig. 3.

There is a very prominent anterior plantar tubercle on paromomyiform calcanea for the attachment of plantar calcaneocuboid ligament(s); it is relatively more prominent on known taxa of primate calcanea than on those of known Cretaceous eutherians. In the latter it is the prominent distal termination of the plantar ridge, whereas in paromomyiforms it is offset medially from the ridge running down on the plantar side of the tuber.

Related Features of the Astragalus and Calcaneum

A number of morphological characters of the astragalocalcaneal complex are functionally inseparable and therefore can be discussed together.

The path of the tendon of the flexor digitorum fibularis can be seen on the distal border of the astragalus and plantar surface of the sustentaculum of the calcaneum. The paths for the tendon of this digital flexor is moderately marked on *Protungulatum* and *Procerberus,* but it is deeply excavated in paromomyiforms. One cannot but infer that this flexor of the digits and plantar flexor of the foot might have been relatively more important in the earliest primates than in the known Cretaceous Eutheria, and/or there were movements performed tending to upset its alignment maintained by the groove.

The articulation of the fibula with the tarsus in *Protungulatum* is with both the astragalus and the calcaneum, and this is undoubtedly primitive for the Eutheria and other Mammalia. The calcaneum of *Procerberus* appears to lack the fibular facet and hence contact with the fibula. It appears that in paromomyiforms, as in Eocene primates, the fibula was not in contact with the lateral side of the astragalocalcaneal facet. There is, however, a small band immediately lateral to the superior edge of the astragalocalcaneal facet which might or might not be considered a remnant of the primitive mammalian fibular facet. Yet, the restrictions imposed on the upper ankle joint by the primitive calcaneofibular contact, well displayed by *Protungulatum,* were probably not present in paromomyiforms. In both *Protungulatum* and *Procerberus* the fibular facet on the lateral side of the astragalar body is a crescent which is slightly deeper anteriorly than posteriorly. In paromomyiform astragali, however, this facet is best described as a half circle, occupying the anterior half of the lateral side of the body, the condition seen in most other primates.

Unlike *Protungulatum* and *Procerberus,* in which the astragalocalcaneal facet forms a large angle with the long axis of the calcaneum, this facet of paromomyiforms forms a very small angle with the latter. *Protungulatum* has a gently arched facet, part of an imaginary quasi-cone with a large radius, whereas the representative *Plesiadapis* has a facet that can be described as part of a quasi-cone with a relatively smaller radius. *Procerberus* is similar to paromomyiforms in this important feature. Much of the astragalocalcaneal facet is distinctly more anterior to the posterior margin of the calcaneal sustentacular facet in *Protungulatum* than in *Plesiadapis*. Correspondingly, the calcaneoastragalar facet of the astragalus is more anterior to the astragalar sustentacular facet in *Protungulatum* than in *Plesiadapis; Procerberus* is intermediate in these repects.

The astragalar sustentacular facet is separate from the naviculoastragalar facet in the known Cretaceous eutherians, but in paromomyiforms the two facets tend to merge. The two facets become continuous in the large *Plesiadapis tricuspidens* of the Cernay late Paleocene of France, as well as in later primates and in some small paromomyids of the North American late Paleocene. The plane of the calcaneal sustentacular facet of paromomyiforms forms a relatively small angle with the transverse plane of the calcaneum, unlike that of *Protungulatum* which forms a relatively larger one.

Virtually all the mentioned differences of *Plesiadapis* from the Cretaceous eutherians are characters shared with Eocene and later primates. Although the presence of a large peroneal tubercle and the presence of a simple, gliding calcaneocuboid articulation in paromomyiform primates are differences from later representatives of the order, these features show intermediate characters between the primitive eutherians, on the one hand, and early representatives of other primate higher categories, on the other.

The Joints and Their Axes of Rotation

Schaeffer (1947), more than anyone else, has used information from the joint axes for analyzing function in fossil mammals, and he has successfully incorporated his findings into an explanation of the origins of the order Artiodactyla. As noted under Methods, observing the changes in the orientation of the axes of the tarsal joints, or other parts of the skeleton, allows a great appreciation of the changes in function that might have occurred from ancestral to the various derived levels of organization. Changes in joint dispositions and mechanics allow comparison of function with living species, and these in turn supply invaluable insight into the nature of the substrate on which the animals lived, and sometimes into the mode of locomotion itself.

It is even more important for making the correct inferences, that one realizes the compromise of any joint surface between differential load bearing, stability, and mobility in various directions. There is a greater enlargement of parts of the articular facets where the resultants of

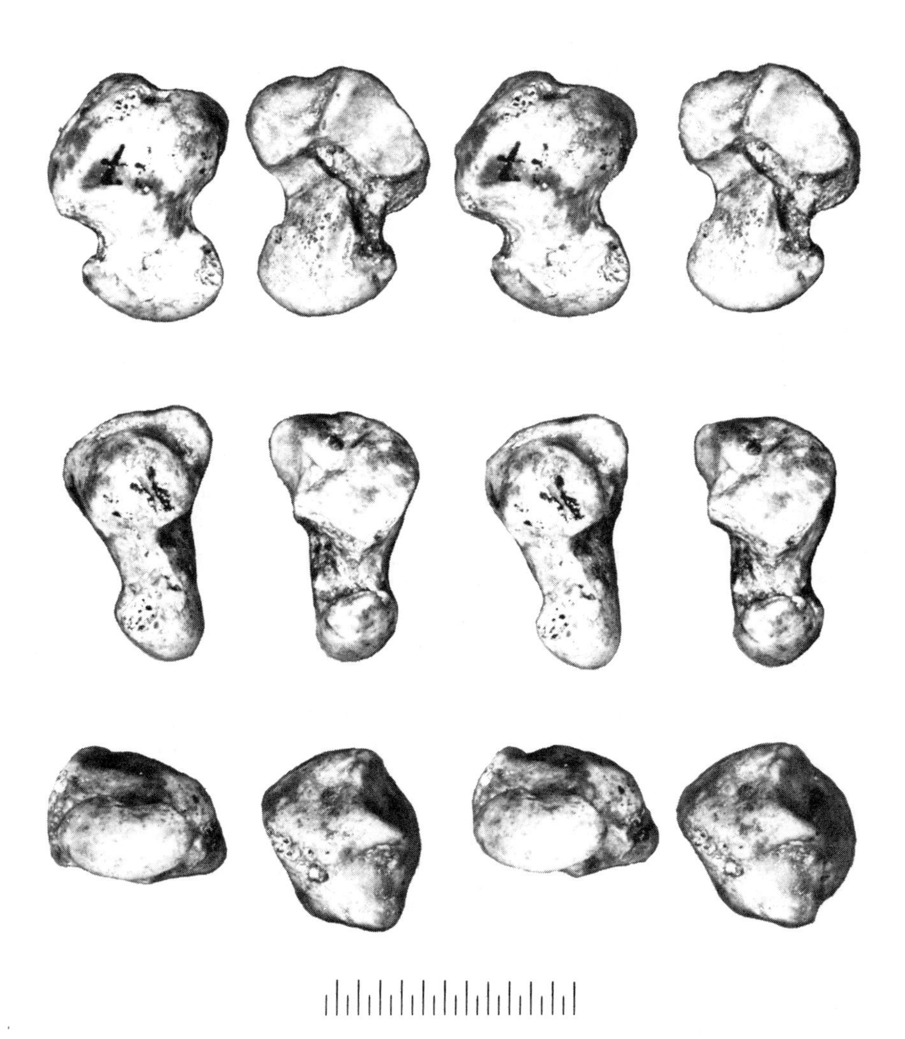

Fig. 6. Stereoscopic pairs of the right astragalus of *Plesiadapis tricuspidens* from the late Paleocene Cernay locality, France. Top row, dorsal view on the left and plantar on the right; middle row, medial view on the left and lateral one on the right; bottom row, distal view on the left and proximal view on the right. Subdivisions on the scale represent 0.5 mm.

forces during a particular type of motion is greater or where stability must be maintained in directions where no mobility occurs. Thus, the proportion of the articular surfaces of the upper ankle, lower ankle, calcaneocuboid, and astragalonavicular joints seem to reflect capabilities for

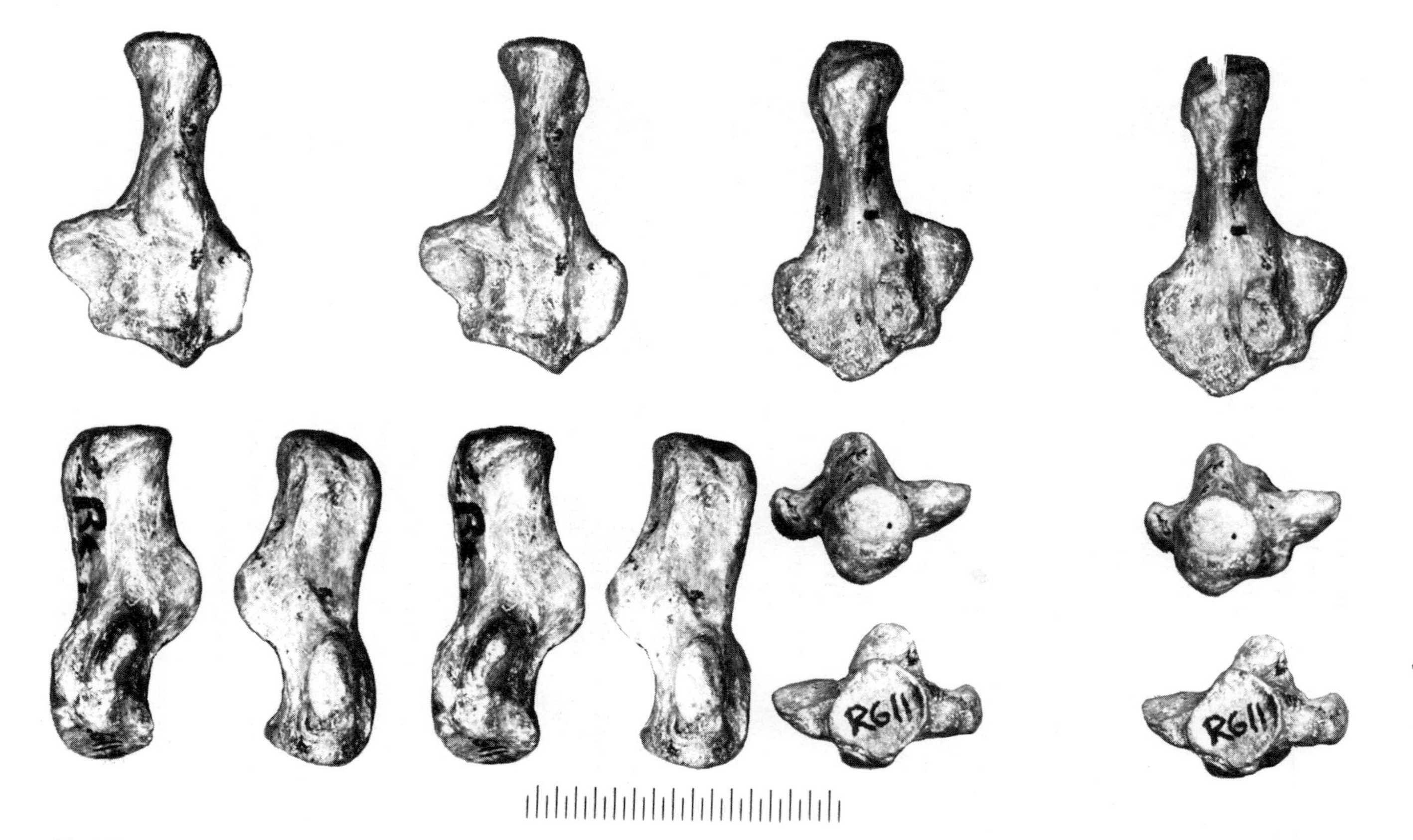

Fig. 7. Stereoscopic pairs of the left calcaneum of *Plesiadapis tricuspidens* from the late Paleocene Cernay locality, France. Top row, dorsal view on the left and plantar one on the right; bottom row, from left to right, medial, lateral, proximal (above), and distal (below) views. Subdivisions on the scale represent 0.5 mm.

different orientations of the foot. This line of analysis applied to other parts of postcranial remains can prove to be similarly revealing.

We will examine the upper ankle, lower ankle, calcaneocuboid, and astragalonavicular joints of *Protungulatum, Procerberus,* and *Plesiadapis.* The availability of fossils restricts an inquiry into other joints of the pes, yet mechanical function of the joints between the crus, astragalus, calcaneum, and between the latter two and the cuboid and navicular are reflected on the astragalocalcaneal complex. Figures 5 through 10 show the axes of these joints, except that of the astragalonavicular, in *Protungulatum, Procerberus,* and *Plesiadapis.*

Upper Ankle (Crurotarsal or Cruropedal) Joint

Movement at this joint involves the distal ends of the tibia and fibula and the body of the astragalus. When movement occurs here the astragalus is functionally part of the foot.

In *Protungulatum,* the axis of the upper ankle joint passes from the middle of the medial wall of the astragalar body to the dorsolateral border of the astragalocalcaneal facet of the calcaneum. This is different from the upper ankle joint axis established by Schaeffer (1947) for the Paleocene *Arctocyon* (including *Claenodon*), a more advanced, larger, and probably more cursorially adapted, arctocyonid condylarth. In *Procerberus* the upper ankle joint axis exits higher laterally than in *Protungulatum,* although it remains more similar in this respect to *Protungulatum* than to paromomyiforms. In the latter, the upper ankle joint axis passes from a point on the anterior half near the ventral edge of the medial wall of the body obliquely through the astragalocalcaneal articulation to a point higher on the lateral wall of the calcaneum.

The articular surface of the tibial trochlea of paromomyiforms goes past the astragalar canal medially and extends far posteriorly, but there is no sign of trochlear articulation posterolateral to the astragalar canal.

Measuring from the axis of the upper ankle joint, the trochlea covers an arc of approximately 120, 140–150, and 135° in *Protungulatum, Procerberus,* and *Plesiadapis,* respectively. This, of course, is not the arc covered by the crus from a dorsiflexed to a plantarflexed position (Fig. 11). In *Protungulatum* and *Plesiadapis,* the estimated tibial axis covered an arc of about 70 and 80° respectively, whereas in *Procerberus* it moved across about 95°, all generously estimated. The palaeoryctoid astragalar body with its relatively deep, extensive trochlea, nearly equally sharp tibial and fibular crests, and with its absence of the astragalar canal, represents an advanced condition compared to the inferred primitive

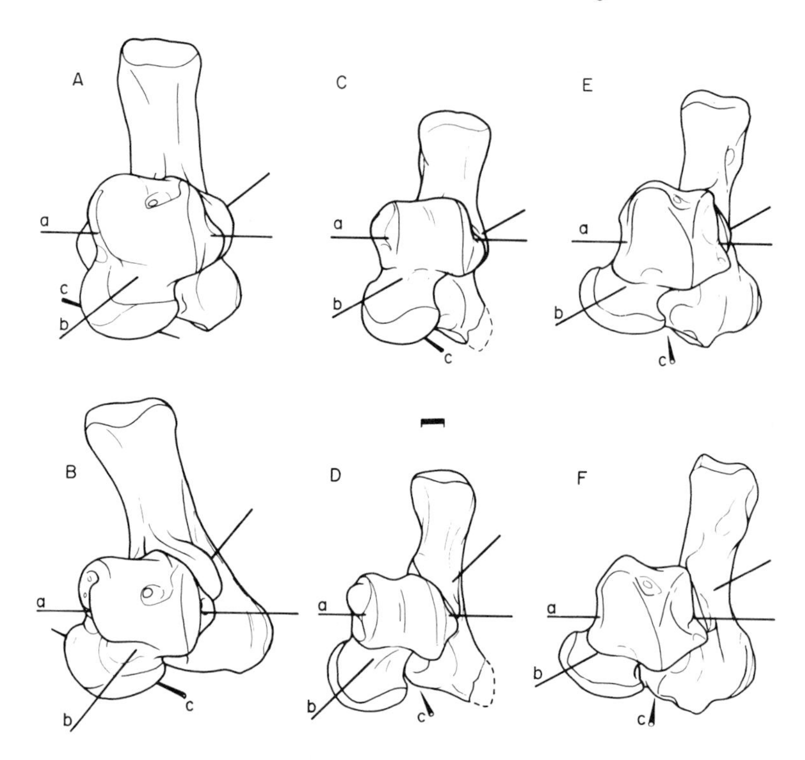

Fig. 8. Dorsal view of the left astragalocalcaneal complex and the axis of rotation of the upper ankle (a), lower ankle (b), and calcaneocuboid (c) joints in *Protungulatum* (A, B), *Procerberus* (C, D), and *Plesiadapis* (E, F) in inverted (top row) and everted (bottom row) positions. The scale represents 1 mm.

condition best represented by *Protungulatum*. The large measured arc from the posterior to the anterior limits of the medial side of the tibial trochlear facet indicates a probable selective advantage for some form of locomotor activity where extreme plantarflexion of the foot was required.

Lower Ankle (Astragalocalcaneal or Subtalar) Joint

This joint involves articulations and movements between the calcaneoastragalar facet, the astragalar sustentacular facet, and the distal calcaneoastragalar facet of the astragalus and the corresponding facets of the calcaneum. The axis of this joint in *Protungulatum*, a compromise axis of three pairs of articulations, passes from a point distal to the tibial trochlea and slightly medial on the dorsal surface of the neck of the astragalus to

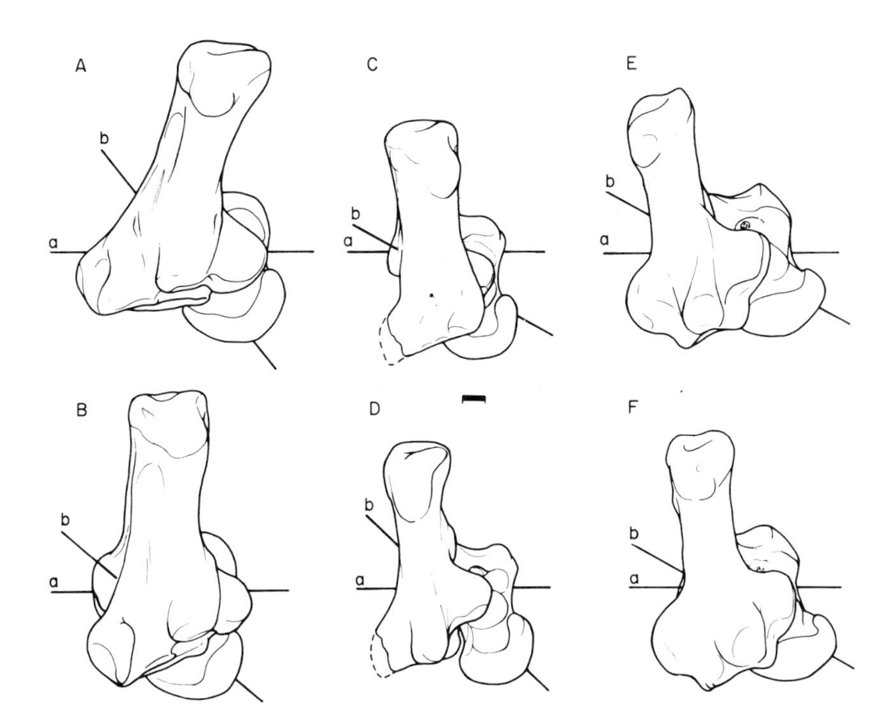

Fig. 9. Plantar view of the left astragalocalcaneal complex and the axis of rotation of the upper ankle (a) and lower ankle (b) joints in *Protungulatum* (A, B). *Procerberus* (C, D), and *Plesiadapis* (E, F) in inverted (top row) and everted (bottom row) positions. The scale represents 1 mm.

a point on the lateral side of the calcaneum beneath the astragalocalcaneal facet.

The lower ankle joint axis in paromomyiforms passes from a point distal to the tibial trochlea and on the medial side of the neck, similar to that in *Protungulatum* and *Procerberus,* and exits on the calcaneum at a point not unlike in *Protungulatum.* This axis, although relatively similar in the three genera compared, displays differences in the degree of anteroposterior orientation. In *Protungulatum* it is the most dorsoplantarly and least anteroposteriorly oriented, whereas the opposite is true for *Plesiadapis. Procerberus* displays an intermediate condition.

One of the differences in the lower ankle joint in living terrestrial versus arboreal forms is the relatively flat, as opposed to curved, articular surfaces. This may manifest itself in the orientation of the axis of the subtalar joint. In the terrestrial forms this axis may be slightly more dorso-

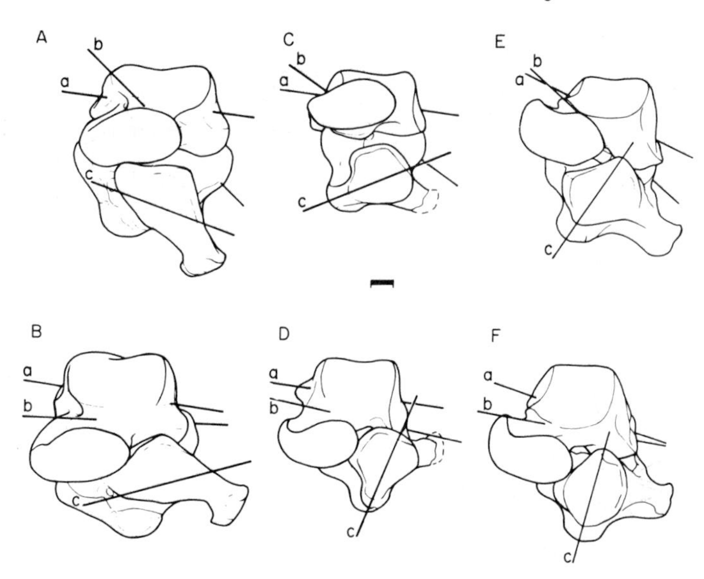

Fig. 10. Distal view of the left astragalocalcaneal complex, and the axis of rotation of the upper ankle (a), lower ankle (b), and calcaneocuboid (C) joints in *Protungulatum* (A, B), *Procerberus* (C, D), and *Plesiadapis* (E,F) in an inverted (top row) and everted (bottom row) positions. The scale represents 1 mm.

plantar, forming a larger angle with the long axis of the calcaneum. This occurs in such predominantly terrestrial genera as *Ursus, Archaeolemur, Theropithecus,* or *Homo.* As seen, *Protungulatum, Procerberus,* and *Plesiadapis* form a functional series in regard to the lower ankle joint axis. In *Protungulatum* the movements occur around an axis that is quite dorsoplantarly oriented, whereas in paromomyiforms, as in later primates, the axis of rotation is closer to the orientation of the long axis of the calcaneum.

Calcaneocuboid Joint

Primitively the axis of rotation of this joint passes through the cuboid, without traversing the calcaneum.

The axis of this joint in *Protungulatum* may be described as that of a very long and very large cone, the base of which was anterior and lateral to the calcaneum and cuboid contact. It appears from the relative position of the axis that flexion and extension were particularly significant. In *Procerberus* the calcaneocuboid joint axis is the axis of a cone relatively shorter than that of *Protungulatum* and the axis is slightly more anteriorly and more dorsally oriented. This axis allowed a greater degree of mediolateral rotation than that seen in *Protungulatum.*

In paromomyiforms it appears that the calcaneocuboid joint axis was

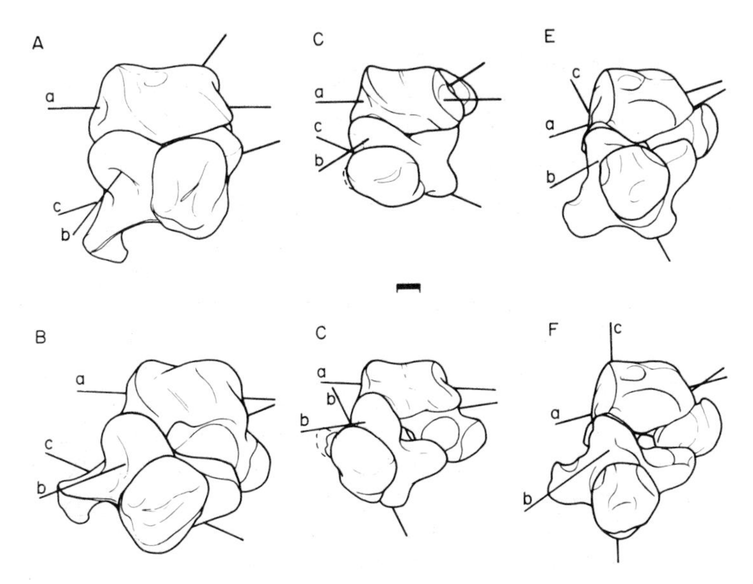

Fig. 11. Proximal view of the left astragalocalcaneal complex and the axis of rotation of the upper ankle (a), lower ankle (b), and calcaneocuboid (c) joints in *Protungulatum* (A, B), *Procerberus* (C, D), and *Plesiadapis* (E, F) in an inverted (top row) and everted (bottom row) positions. The scale represent 1 mm.

the long axis of a rather short, broad based cone. The axis of this cone was oriented slightly anteriorly, dorsally, and laterally with respect to the upper ankle joint axis. It is in many ways intermediate between the *Protungulatum*-like axis orientation, the latter probably primitive for eutherians, and the cuboid pivot of other primates. In most primates, except in paromomyiforms and *Tarsius*, for example, the calcaneocuboid joint axis is more nearly parallel to the long axis of the calcaneum. This condition, as that of the paromomyiforms, facilitates mediolateral rotations more nearly along the long axis of the foot.

Astragalonavicular Joint

There are no eutherian naviculars known from the Cretaceous and none have been described for any paromomyiforms. The astragalar heads, however, can be informative concerning this joint.

The articular facet for the spring ligament, running from the sustentaculum of the calcaneum to the navicular, is very small in *Protungulatum*, exceptionally well developed in paromomyiforms, and moderate in size in *Procerberus*. As described above, the medial half of the astragalar head is relatively thick, unlike that of *Protungulatum* and *Procerberus*. These differences in morphology suggest that paromomyiform feet were capable of greater degrees of inversion than those of the known Cretaceous Eutheria.

Mechanical Considerations

A ratio of measurements taken approximately along the long axis of the calcaneum reveals certain mechanics of this bone. The ratio between the posterior limit of the tuber and the anterior end of astragalocalcaneal facet to the distance from the latter to the cuboid facet is the largest in *Protungulatum* (3.4) and the smallest in *Plesiadapis* (2.1). It is relatively small also in *Procerberus* (2.4). This is supported by the consideration of the changing lever arm of the Achilles tendon during plantar flexion of the foot. Figure 11 shows, in spite of the unavoidable inaccuracies of a two-dimensional illustration, that the difference, and hence the leverage from a dorsiflexed to a plantarflexed position is largest in *Protungulatum* and smallest in *Procerberus* and *Plesiadapis*. The upper ankle joint axis is shown to be perpendicular to the paper and the arrow on the tibial trochlea represents the inferred long axis of the tibia. In order to standardize the unknown differences in the place of origin of the plantar flexors, the line representing the tendo calcaneum was drawn parallel to the inferred long axis of the tibia. The lever arm of *Protungulatum* is actually longer than is shown in Fig. 12B. In the everted position the calcaneum is oriented medially when the upper ankle joint axis is perpendicular to the paper; therefore, y on Fig. 12 is distinctly shorter than it would be on Fig. 8 if the lever arm was drawn in the latter.

In summary, although the necessary anterior parts of the tarsus and metatarsals are lacking to support this, one might suggest that the calcaneum of *Protungulatum* provided more leverage to the triceps surae than that of *Plesiadapis* or *Procerberus*.

Tarsal Function and Substrate Preference among the Ancestral Eutheria and Early Primates

A study of the earliest primate tarsal remains cannot ignore the over a century old controversial problem of whether the earliest "mammals" (widely differing concepts), therians, and then either the metatherians or eutherians were arboreal or terrestrial (for a brief review see Haines, 1958 and Jenkins, this volume). The problems stated or answered by numerous students probably cannot as yet be solved in reference to the Theria. Although it is likely that the metatherian-eutherian dichotomy

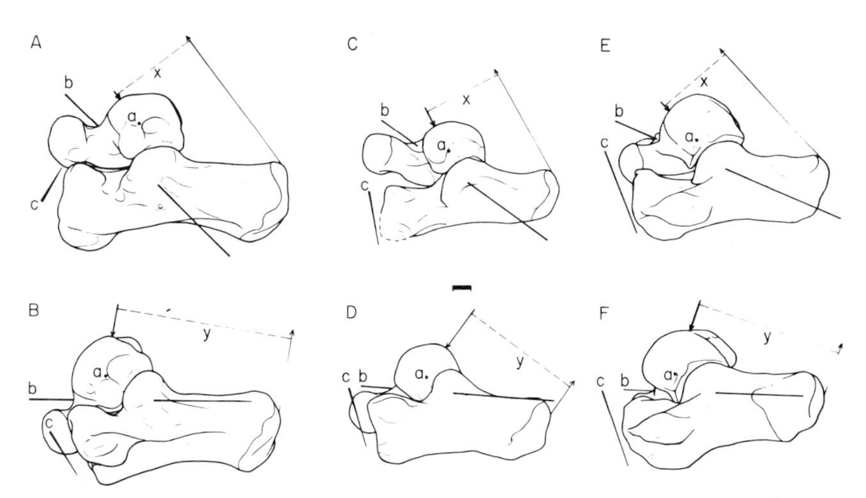

Fig. 12. Lateral view of the left astragalocalcaneal complex, the axis of rotation of the upper ankle (a), lower ankle (b), and calcaneocuboid (c) joints, and the lever arm of the Achilles tendon in *Protungulatum* (A, B), *Procerberus* (C, D), and *Plesiadapis* (E, F) in inverted (top row) and everted (bottom row) positions. The arrow on the tibial trochlea represents the inferred long axis of the tibia. In order to standardize the unknown differences in the points of origin of the triceps surae the line representing the tendo calcaneum is drawn parallel to the inferred long axis of the tibia. The lever arm of *Protungulatum* in the everted position (B, y) appears shorter in this orientation (with the upper ankle joint axis perpendicular to the paper) because the tuber is swung relatively more medially than in *Procerberus* or *Plesiadapis*. The scale represents 1 mm. x, lever arm of the calcaneum during extreme inversion; y, lever arm of calcaneum during extreme eversion.

was already in existence in early Cretaceous Albian times (Slaughter, 1965, 1968; approximately 100–120 million years ago), postcranial remains of these early forms have not been recovered. Even if individual species were known by their postcranials it would not reveal needed information on the adaptations of their close contemporary relatives or descendants. Individual species of "ancestral" metatherians or eutherians did not occur by themselves but were parts of faunas and adaptive radiations, which, presumably were comprised of a variety of lineages with diverse adaptations. It is hoped, however, that as collections improve and consequently inferred phylogenies become based on better evidence it will become possible to adduce substrate preference to the common ancestors of various supraordinal categories of the Theria.

Although the beginning of the ungulate orders, fifteen of them, which originated from the Condylarthra, have a good structural ancestor in

Protungulatum, the late Cretaceous leptictids and palaeoryctids, as stated above, are not ancestors to either the Primates, Carnivora, Hyaenodonta, or the Anagalida. One can ask, however, whether the postulated ancestral stock of eutherians, best considered now as Insectivora, which probably gave rise to the latest Cretaceous Insectivora, was adapted best for arboreal or terrestrial existence, and to what extent. It appears that although a limited ability to climb cannot be precluded for the ancestral early to medial Cretaceous eutherian stock, adaptations for the arboreal habitus were not present in the common ancestors of Insectivora, Condylarthra, Carnivora, Hyaenodonta, and Primates. The ancestral eutherians were adapted in their pes for a plantigrade type of locomotion on a flat to irregular substrate, presumably a forest litter environment. Our tentative conclusion agrees with that of Haines (1958) who suggested, based on the convergent structural-functional type of the manus of the known early eutherians (based largely on the evidence of Paleocene condylarths in Matthew's, 1937, outstanding monograph), that a terrestrial, rather than an arboreal adaptive zone was the element of the earliest Eutheria.

We do not support Haines' similar conclusions for the Metatheria for which the relevant fossil and extant evidence indicates that the primitive members had a tarsus adapted for varying orientations, particularly habitual inversion, and hence probably for habitual travel on an arboreal substrate.

It is significant that tarsals of Cretaceous and Paleocene metatherians show a little calcaneofibular contact, an extremely well-developed calcaneocuboid pivot, and some other features of the articular facets recalling the derived characters of such arboreal eutherians as the advanced primates which do not appear in the fossil record until the Late Paleocene–Early Eocene. Although a calcaneofibular contact is probably primitive in the Theria, and persists in early Tertiary representatives of the Condylarthra, Carnivora, and Hyaenodonta, a survey of the fossil evidence and of the most primitive living metatherians suggests that this was reduced very early in the Metatheria. Our preliminary evaluation suggests that a high degree of inversion characterized marsupial feet when contemporary eutherians were probably primarily adapted to level orientations of the pes.

The absence of adequate intermediates has been repeatedly emphasized in the literature dealing with the origin of higher level taxonomic characters. Although no representative intermediates are known between the feet of the known Cretaceous eutherians and those of paromomyiforms, the morphological gap between a *Protungulatum*-like tarsus and that of a paromomyiform is small indeed, yet highly significant in terms of habitual orientations and hence of adaptations to different substrates. Virtually all the noted functional differences in the proximal tarsus of paromomyiforms from those of the latest Cretaceous eutherians are characters shared with Eocene and later primates. Although in respect to the peroneal tubercle and the conformation of the calcaneocuboid joint in paromomyiforms there are differences from later representatives of the order, these features show intermediate characters between the primitive eutherians, on the one hand, and early representatives of other primate higher categories, on the other.

Our interpretations of the morphology and of joint function indicate that whereas the very primitive condylarth astragalocalcaneal complex, that of *Protungulatum*, was primarily adapted for a limited range of level orientations, the astragalocalcaneal complex of the paromomyiform primates was adapted for habitual inversion. We believe that these functions were caused by selection for habitual terrestrial locomotion in the former and primarily arboreal existence in the latter. *Procerberus*, as the late Cretaceous *Cimolestes* or Paleocene leptictids, appears to have been closer to *Protungulatum*.

What are the musculoskeletal features of the tarsus suggestive of the adaptive zone in which the palaeoryctoid *Procerberus*, the condylarth *Protungulatum*, and the paromomyiform primates functioned? Before attempting to answer this question, another one might be necessary to answer: why was the tarsus, or elements of it, chosen rather than the total available evidence? The total evidence, of course, will be used in the final evaluation, when all known elements will have been studied. This analysis was undertaken partly because the best available identifiable comparative sample of Late Cretaceous and Early Tertiary eutherians is that of the astragalocalcaneal complex, and because we believe that the foot is a sensitive indicator of the substrate on which it is used. The hallmarks of an inverted tarsus, associated with an arboreal existence, can be differentiated from those of the habitually level, terrestrial one. When the pes assumes a level orientation we would expect those aspects of the joint surfaces in articular contact to provide more advantage to the transmission of forces than those aspects of the articular surfaces in contact during inverted rotation of the pes.

Proceeding on the assumption that in equilibrium the resultant of forces (the forces exerted by muscles and tendons and the mass of the animal) acting on a joint surface must be perpendicular to the surface of the joint, a comparison of *Protungulatum*, *Procerberus*, and paromomyiforms is most revealing. Figure 10 demonstrates that the plantar surface of the foot was distinctly more medially oriented in paromomyiforms and *Procerberus* than in *Protungulatum.*

The calcaneofibular articulation of the primitive eutherian ancestry and its restrictive influence disappear in early primates, but for reasons of flexibility of the ankle the fibula plays its important role of opposing forces from the lower ankle joint, as shown by the large fibular facet on the astragalus. Unlike some of the living tupaiine tupaiids, some soricoids, talpids, and many rodents, and a few lineages of highly specialized primates with a saltatory form of locomotion, such as some of the tarsiiform vertical clingers and leapers, primates do not usually fuse the fibula with the tibia and retain the independent force-opposing role of this element of the crus.

The change in the position of the groove and the anterior plantar tubercle for the attachment of the plantar calcaneocuboid ligament points to a migration of this ligament from a primitively transversely wide, plantar position to an increasingly medial one in paromomyiforms and this, in turn, is correlated with the changed orientation of the calcaneocuboid joint axis. These features appear to reflect the necessity to brace the tarsus in a changed habitual orientation of the foot, suggesting changes in the direction of the substrate reaction. From a habitually everted, level, plantigrade position to a habitually inverted one, the forces that would tend to place the greatest bending moment on the rigid calcaneocuboid articulation during the propulsive phase shifted more medially as the relative orientation of the substrate changed from a horizontal to an inclined one.

The upper ankle and lower ankle joints of *Protungulatum, Procerberus,* and paromomyiforms are all different from each other, yet those of the primates differ significantly from some shared characters of the former genera. As the astragalocalcaneal and calcaneoastragalar facets in paromomyiforms respond to the structural needs of an increased range of inversion, their posterior parts enlarge and their movements become more screwlike (helical) from the primitive eutherian one in which the movements were simple rotations. Helical movements increase the degree of anteroposterior change in astragalocalcaneal relationships during inversion and eversion of the foot. This accommodates the astragalonavicular articulation which is concurrently brought closer to the calcaneal side of the lower ankle joint during inversion and further away during eversion. Therefore rather greater tarsal mobility is achieved without loss of the close association among the tarsals necessary for receiving stresses during different orientations of the foot. In *Protungulatum,* and to a lesser degree in *Procerberus,* the transversely broad trochlea reflects the transverse orientation of these facets facilitating simple rotations There are several correlates of these changes in the lower ankle joint for increased inversion from a system evolved probably primarily for habitual eversion. To oppose these new forces efficiently the tibial trochlea of paromomyiforms enlarges laterally and elongates, extending the trochlea slightly forward and back above the calcaneoastragalar articulation.

When the lemuriforms are compared with paromomyiforms for indications of mobility in the upper ankle joint it is obvious that the former are more mobile as to extension of this joint. Continuing in the arc of the trochlea, the tibial facet often extends far past the superior position of the astragalar foramen whereas the position is insignificant in paromomyiforms. However, unlike in palaeoryctoid insectivorans, for example, wherein the foramen is eliminated, in lemuriforms the tibioastragalar articulation bypass it medially. Extension is limited at the posterior astragalar shelf. Stresses thus produced on the shelf are apparently resisted by the posterior astragalocalcaneal articulation beneath extending the length of the shelf, and by the flexor fibularis tendon which angles here. Potential shear is therefore transformed into compression.

The articular facets of the lower ankle joint show important functional differences concerning the extent of medial or lateral rotation in *Protungulatum, Procerberus,* and *Plesiadapis.* A perpendicular line drawn from the point where the lower ankle joint axis pierces the calcaneum in the sulcus astragali to the posterior and anterior limits of the astragalocalcaneal facet will delineate the posteromedial and anterolateral areas which are traversed by the astragalus during inversion and eversion, respectively. As shown on Fig. 13 the anterolateral diameter, indicative of the relative importance of eversion or stability in this position, is larger in *Protungulatum* than the posteromedial diameter of the facet, depicting the relative importance of inversion. In *Plesiadapis* the opposite of the condition in *Protungulatum* can be seen, whereas in *Procerberus* the relative sizes of the two respective areas are intermediate between *Protungulatum* and *Plesiadapis.*

It was noted before that in *Protungulatum* the long axis of the astragalar calcaneoastragalar facet is transverse, and it forms a large angle with the long axis of the astragalar body. In *Plesiadapis,* however, the orientation of this posterior articulation of the lower ankle joint changes and the long axis of the facets of this posterior articulation forms a much smaller angle with the long axis of the astragalar body. This is in part the cause for the changes in the facets noted above. This trend is further accentuated in such highly arboreal forms as living lemuroids or *Notharctus* (Decker and Szalay, this volume). On the astragalar sustentacular facet the *Protungulatum*-like condition in which the anterolateral area is larger

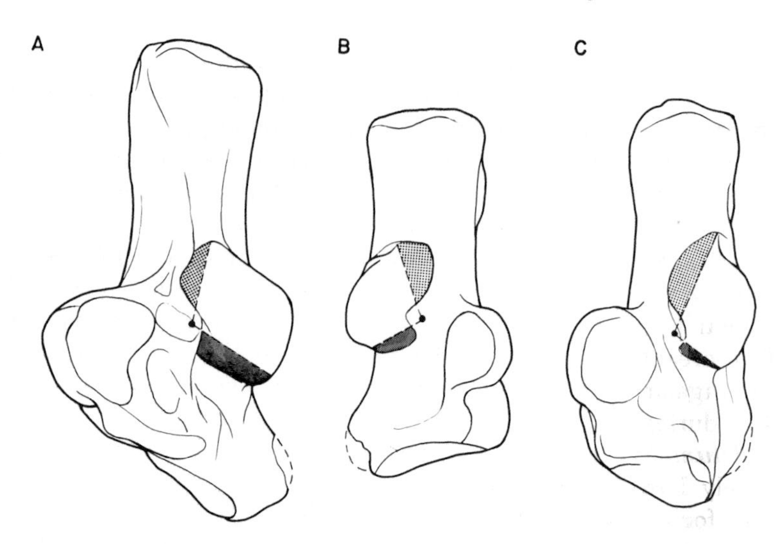

Fig. 13. Calcanea of *Protungulatum* (A), *Procerberus* (B), and *Plesiadapis* (C) oriented so the largest area of the astragalocalcaneal facet is visible. The large black dot represents the point where the lower ankle joint axis pierces the calcaneum. The broken lines are drawn from the axis to the most anterior and most posterior limits of the astragalocalcaneal facets; the stippled areas represent the position of the facet covered during inversion by the dorsiflexed astragalus and the hatched areas the portions covered during eversion by the plantarflexed astragalus. The area covered during eversion is relatively larger than that covered in inversion in *Protungulatum,* whereas it is reversed in *Procerberus* and *Plesiadapis.* This trend is distinctly more developed in the latter. The relative sizes of the diameters of the posteromedial and anterolateral parts delineated on the facets reflect the relative stability during inversion and eversion rather than the mobility during these positions. Thus, it is during eversion or level position of the foot that the astragalocalcaneal articulation was most stable in *Protungulatum,* whereas in *Plesiadapis* adaptations for the greatest stability appear to be for the inverted position of the pes. The scale represents 1 mm.

than the posteromedial one is altered in *Plesiadapis* in which these surfaces become nearly equal in size.

As in most primates (Barnett, 1970) the axis of the lower ankle joint more or less retains the primitive eutherian oblique orientation. In a number of mammals the joint is suited by these conditions for adjusting the foot to uneven surfaces. Again, both the Paromomyiformes and Lemuriformes show the recurved posterior astragalocalcaneal facet on the calcaneum suggesting screwlike rotation at the lower ankle joint. The significance of this in increasing the potential range of lower ankle joint rotation effecting inversion and eversion of the foot has been previously pointed out (Hall-Craggs, 1965). Other arboreal eutherians, such as the procyonid *Potos flavus* (the kinkajou) and the viverrid *Arctictis bintu-*

rong, show recurved posterior astragalocalcaneal facets suggesting possible screwlike movements of the lower ankle joint.

These changes of the lower ankle joint resulted in order to cope with the increasingly complicated mechanical requirements of support of a habitually inverted paromomyiform pes derived from a *Protungulatum*-like ancestral condition.

The evidence from the astragalonavicular joint has been fully described above. The differences in the lateral and medial dimensions of the astragalar head also show evidence for terrestriality of *Protungulatum* and arboreality in paromomyiforms. The increased medial thickness of the astragalar head in *Plesiadapis* clearly shows that the forces transmitted during inversion were relatively more significant than in *Protungulatum.* Similarly, the fact that the facet for the spring ligament is relatively larger in *Plesiadapis* than in *Protungulatum* supports the inference for a habitually level foot in the latter and an inverted one in the former. One might ask why the extensive lateral area of the astragalar head does not disappear in arboreal mammals. At this time it is adequate to state that the ability for level orientations also remains at a premium. What does remain significant is the relative thickness of the pertinent medial and lateral parts of the astragalar head. The conformation of the astragalonavicular articulation depends to a large degree on the movements at the lower ankle and calcaneocuboid joints especially if the association between the navicular and cuboid is a relatively stable one.

Although the number of articulations between the cuneiforms and navicular and cuboid may add effectively to the total potential for rotation of the pes the individual contribution of each is probably small, in most cases, compared to the more proximal articulations.

Movements at the calcaneocuboid joints are of particular importance in the mammalian foot. Unless it is a specialized condition in which the navicular moves independently from the cuboid, movements of the cuboid are transferred to the navicular and, consequently, to the cuneiforms. Thus, movements of the cuboid on the calcaneocuboid joint axis, passing through the cuboid but not the calcaneum, affect the motions of the tarsals in the primitive eutherian tarsus. One of the most significant features for both the earliest eutherians and the primates lies in the clear differentiation of the direction of rotation of the calcaneocuboid axis. Unlike the mediolaterally oriented near hinge-type articulation of *Protungulatum,* the paromomyiform cone-shaped articulation has been modified clearly in the direction of later primates. The primarily extension–flexion movements, rotations in an anteroposterior direction at this joint in the inferred primitive eutherian condition has become modified in primates for rotation in a more mediolateral direction (Fig. 14).

In *Protungulatum* the cuboidocalcaneal facet is nearly parallel to the axis of rotation and the direction of motion is perpendicular to the long axis of the calcaneum; this is nearly so in *Procerberus*. The paromomyiforms, however, display the beginnings of a very characteristic primate condition. The axis of rotation of the cuboid is very oblique anterior to the calcaneocuboid facet, and it is therefore intermediate between the condition present in *Protungulatum* and that in most later primates. In the latter, the facets are nearly parallel to the direction of rotation of the

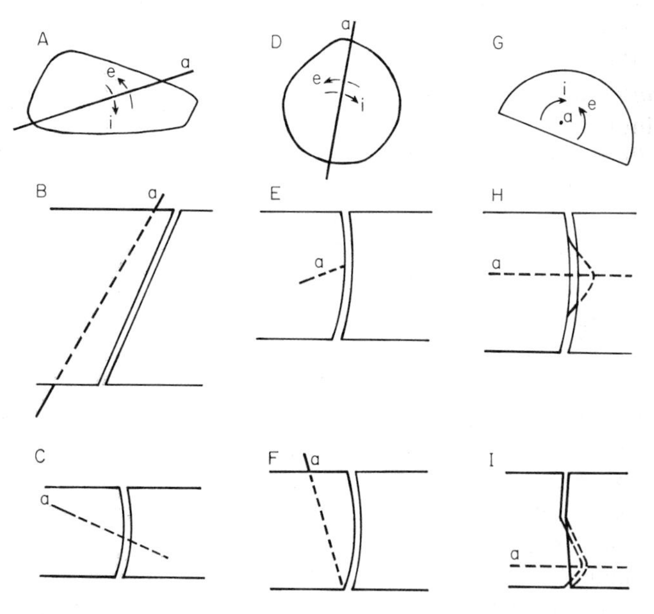

Fig. 14. Schematized stages in the evolution of the calcaneocuboid joint and its axis of rotation from a primitive eutherian stage (A, B, C) to the early primates, the paromomyiforms (D, E, F), and from the latter to the pivotal joint of the more advanced primates (G, H, I). The top row shows the distal view of the cuboid facet of the left calcaneum, and the middle and lower rows depict the dorsal and lateral views of left calcanea and cuboids in articulation. The calcaneum is on the right and the cuboid is on the left in each pair. The axes of rotation of the calcaneocuboid joints (a) are shown, and the arrows indicate the direction of rotation during inversion (i) and (e).

In the primitive eutherian and paromomyiform stages the diameter of the area stressed during either inversion or eversion is larger to ensure joint stability. In more advanced primates, due to the nature of the cuboid pivot, requirements for stability manifest themselves in the differential development of the base surrounding the pivot on the calcaneal and cuboid facets.

cuboid whereas the axis itself is almost perpendicular to the facets. This changed relationship between the calcaneocuboid facet and the axis of the calcaneocuboid joint results in some changes in the differential enlargement of parts of the calcaneocuboid facet in forms such as *Protungulatum,* on one extreme, and later primates, on the other. In *Protungulatum,* during inversion, it is the ventrolateral portion of the calcaneocuboid facet that is traversed by the cuboid partly leaving the mediodorsal part, as proximally it rotates around the axis of the joint in a ventrolateral direction (Fig. 14 A, B, C). During eversion the rotation is reversed, and the cuboid proximally rotates mediodorsally. During this phase it is the dorsomedial side of the calcaneocuboid facet that is traversed by the cuboid and the ventrolateral side is not stressed strongly. In *Protungulatum* the mediodorsal facet is distinctly larger, and thus seems to indicate that a greater resultant of forces was habitually transmitted while the foot was held horizontally, in a level position.

Comparison of the derived paromomyiform condition of both the axis of rotation of the calcaneocuboid joint and the facet on the calcaneum with that of *Protungulatum* and with lemuriform, platyrrhine, and catarrhine primates is most revealing. As the nearly parallel joint axis-joint facet relationship of a *Protungulatum*-like ancestral condition is altered into the oblique axis-facet relationship the positions of the calcaneocuboid facet which are traversed by the cuboid during inversion and eversion remain the same. The relative enlargement of the areas traversed by the cuboid during inversion in *Protungulatum* and *Plesiadapis* is clearly shown on the distal views of the calcaneum (Figs. 5 and 11). The differential enlargement of the respective surfaces can be related to compromise in joint stability and mobility.

The paromomyiform cone-shaped articulation has clearly been modified in the direction of later primates. In the advanced primate condition represented in *Adapis, Notharctus,* or *Propithecus* the axis lies more nearly parallel to the long axis of the pes and is therefore almost completely disposed for longitudinal rotation of the foot alone. Movements of this joint adjust the pes to the substrate in the most economical manner. Opposability of the hallux aside, it is probably the most significant pedal adaptation of primates to the arboreal substrate.

The differences which occur in advanced primates to produce this set of events involves the change of the calcaneocuboid joint to a pivotal nature from its gliding predecessor. On the calcaneal element the cuboid facet receives in a depression a projection from the cuboid. Both are generally surrounded by a plane margin. The axis of rotation intersects the projection at its tip.

Evolution of the calcaneocuboid pivot joint in primates from a more primitive eutherian condition is important in at least two respects regarding the ability of the foot to adjust to uneven surfaces of the arboreal substrates. First, because the rotation of the calcaneocuboid axis is oriented nearly parallel to the long axis of the foot, all movements of the joint are disposed almost completely for inversion and eversion of the foot. This increases the range of these movements of the distal portion on the proximal parts of the tarsus or vice versa. Second, imagine the calcaneum–cuboid–navicular complex to be a single rigid body. Were this true, because the tendency of the astragalar head is to be displaced from the navicular concavity during rotations of the lower ankle joint, the range of movements of the lower ankle would be limited to the small displacement possible while the astragalonavicular joint remained in contact to take up the stresses during inversion and eversion. As it is, the complex is not a rigid body. Through its association with the cuboid the navicular is displaced among a curvilinear path by movements of the calcaneocuboid joint. Obviously the larger the component of movement of this joint in a frontal plane, the greater are the dorsolateral and ventromedial displacement of the navicular necessary to retain the astragalonavicular articulation receiving stresses in inversion and eversion. Rather ultimate conditions are obtained to this effect as the calcaneocuboid joint becomes a pivot with its axis of rotation parallel to the long axis of the foot. If the range is increased during which the astragalonavicular contact is retained, it follows then that potential rotation of the lower ankle joint may be increased also without loss of stability. Suggestions that this has occurred were previously indictated.

It is of utmost significance for understanding late Mesozoic and early Tertiary Theria that the cuboid pivot was well established in all metatherian fossils known to us. Thus, very significant adaptations correlated with arboreal substrate preference were entrenched in metatherians when most eutherian representatives appear to have been terrestrial.

Summary

An evolutionary analysis of the astragalocalcaneal complex of the late Cretaceous condylarth *Protungulatum*, the palaeoryctoid insectivoran *Procerberus*, and the Paleocene primate *Plesiadapis* utilized phylogenetic and functional features from the upper ankle, lower ankle, calcaneocuboid, and astragalonavicular joints. Emphasis was placed on the axis of rotation of the various joints and the differential enlargement of the articular facets indicative of either habitual inversion or level orientation of the pes.

Judged from the astragalocalcaneal complex, the foot of *Protungulatum* appears to be the most primitive eutherian one known. Derivation of most orders of the Eutheria from a palaeoryctoid stock represented by *Cimolestes* or *Procerberus* is unlikely; a postulated medial Cretaceous group of "Insectivora" with a tribosphenic dentition derived from dentally *Pappotherium*-like ancestors with an astragalocalcaneal complex similar to that of *Protungulatum* probably gave rise to palaeoryctoid and erinaceotan-derived Insectivora, the Condylarthra, Primates, Carnivora, Hyaenodonta, and the Anagalida. The palaeoryctid–leptictid group of Insectivora is probably ancestral to the Pantolestidae, Microsyopidae, Rodentia, Taeniodonta, and Tupaiidae, and the palaeoryctoid derived Pantolestidate is probably ancestral to the Metacheiromyidae and Epoicotheriidae, neither of the Pholidata. Derivation of the Rodentia from the Primates is not likely.

There is a rather consistent correlation between the existence on a terrestrial substrate and a level, or everted, orientation of the pes, whereas in order to habitually grasp or place the pes on a rounded, arboreal substrate, such as a branch or tree trunk, the foot is usually adapted for inversion. *Protungulatum,* a structural ancestor of the Condylarthra and the latter ancestral to approximately fifteen orders, possessed a pes which appears to have been adapted for habitual level orientation and eversion rather than inversion. The pes of *Plesiadapis,* in most ways representative of the earliest major primate radiation, appears to be derived from a primitive eutherian (*Protungulatum*-like) ancestry and it is modified for habitual inversion like the pes of other Primates. The astragalocalcaneal complex of *Procerberus,* and other known Cretaceous palaeoryctoids, derived from a complex not unlike that of *Protungulatum,* has a highly modified astragalar tibial trochlea permitting extremes of plantarflexion.

It is suggested that the ancestry of the eutherian orders was terrestrial and that the differentiation of primate postcranial modifications are causally correlated with the invasion of the arboreal adaptive zone from an erinaceotan or condylarth ancestry occupying a terrestrial one.

Acknowledgments

We are grateful to Dr. Robert E. Sloan of the University of Minnesota for the loan of late Cretaceous eutherian postcranials and to Dr. Donald E. Russell for the loan of some paromomyiform material. We thank Miss A. J. Cleary, who skillfully prepared our illustrations and Miss M. Siroky for expert technical assistance.

This work was supported by NSF Grant GS 32315 and a CUNY doctoral faculty grant, both to Szalay.

References

Barnett, C. H. (1955). Some factors influencing angulation of the neck of the mammalian talus. *J. Anat.* **89**, 225–230.

Barnett, C. H. (1970). Talocalcaneal movements in mammals. *J. Zool.* **160**, 1–7.

Barnett, C. H., and Napier, J. R. (1952). The axis of rotation at the ankle joint. *J. Anat.* **86**, 1–9.

Barnett, C. H., and Napier, J. R. (1953). The rotary mobility of the fibula in eutherian mammals. *J. Anat.* **87**, 11–21.

Bock, W. J., and Wahlert, G. V. (1965). Adaptation and the form-function complex. *Evolution* **19**, 269–299.

Cartmill, M. (1972). Arboreal adaptations and the origin of the order Primates. *In* "The Functional and Evolutionary Biology of Primates" (R. Tuttle, ed.), pp. 97–122. Aldine-Atherton, Chicago and New York.

Charles-Dominique, P., and Martin, R. D. (1970). Evolution of lorises and lemurs. *Nature (London)* **227**, 257–260.

Clemens, W. A., Jr. (1963). Fossil mammals of the type Lance Formation, Wyoming. Part I, Introduction and multituberculata. *Univ. Calif. Publ. Geol. Sci.* **48**, 1–105.

Clemens, W. A., Jr. (1966). Fossil mammals of the type Lance Formation, Wyoming. Part II, Marsupialia. *Univ. Calif. Publ. Geol. Sci.* **62**, 1–122.

Clemens, W. A., Jr. (1970). Mesozoic mammalian evolution. *Annu. Rev. Ecol. Syst.* **1**, 357–390.

Elftman, H., and Manter, J. (1935). The evolution of the human foot with especial reference to the joints. *J. Anat.* **70**, 56–67.

Emry, R. J. (1970). A North American Oligocene pangolin and other additions to the Pholidota. *Bull. Amer. Mus. Natur. Hist.* **142**, 455–510.

Haines, R. W. (1958). Arboreal or terrestrial ancestry of placental mammals. *Quart. Rev. Biol.* **33**, 1–23.

Hall-Craggs, E. C. B. (1965). An analysis of the jump of the lesser galago (*Galago senegalensis*). *J. Zool.* **147**, 20–29.

Lillegraven, J. (1969). The latest Cretaceous mammals of the upper part of the Emonton Formation of Alberta, Canada, and a review of the marsupial placental dichotomy in mammalian evolution. *Univ. Kans. Paleontol. Contrib. Vert.* **12**, 1–122.

McKenna, M. C. (1961). A note on the origin of rodents. *Amer. Mus. Nov. No.* **2037**, 1–5.

McKenna, M. C. (1969). The origin and early differentiation of therian mammals. *Ann. N. Y. Acad. Sci.* **167**, 217–240.

Manter, J. T. (1941). Movements of the subtalar and transverse tarsal joints. *Anat. Rec.* **80**, 397–410.

Martin, R. D. (1972). Adaptive radiation and behavior of the Malagasy lemurs. *Phil. Trans. Roy. Soc. London* **B264**, 295–352.

Matthew, W. D. (1937). Paleocene faunas of the San Jaun Basin, New Mexico. *Trans. Amer. Phil. Soc.* **30**, 1–510.

Russell, D. E. (1964). Les Mammifères Paléocène d'Europe. *Mem. Mus. Nat. Hist. Natur.* (*Paris*) **C13**, 1–324.

Schaeffer, B. (1947). Notes on the origin and function of the artiodactyl tarsus. *Amer. Mus. Nov. No.* **1356**, 1–24.

Simpson, G. G. (1931). A new insectivore from the Oligocene, Ulan Gochu horizon, of Mongolia. *Amer. Mus. Nov.* No. **505**, 1–22.

Simpson, G. G. (1935). The Tiffany fauna, upper Paleocene. Part II, Structure and relationships of *Plesiadapis*. *Amer. Mus. Nov. No.* **816**, 1–30.

Slaughter, B. H. (1965). A therian from the lower Cretaceous (Albian) of Texas. *Postilla* **93**, 1–18.

Slaughter, B. H. (1968). Earliest known marsupials. *Science* **162**, 254–255.

Sloan, R. E. (1969). Cretaceous and Paleocene terrestrial communities of western North America. *Proc N. Amer. Paleontol. Conv.* **E**, 427–453.

Sloan, R. E., and Van Valen, L. (1965). Cretaceous mammals from Montana. *Science* **148**, 220–227.

Szalay, F. S. (1968). The beginnings of primates. *Evolution* **22**, 19–36.

Szalay, F. S. (1969). Mixodectidae, Microsyopidae, and the insectivore-primate transition. *Bull. Amer. Mus. Natur. Hist.* **140**, 193–330.

Szalay, F. S., and McKenna, M. C. (1971). Beginning of the age of mammals in Asia: the late Paleocene Gashato fauna, Mongolia. *Bull. Amer. Mus. Natur. Hist.* **144**, 269–318.

Sulimsky, A. (1968). Paleocene genus *Pseudictops* Matthew, Granger, and Simpson, 1929 (Mammalia) and its revision. *In* "Results of the Polish-Mongolian Palaeontological Expeditions" (Z. Kielan-Jaworowska, ed.) Part I, pp. 101–129.

Van Valen, L. (1966). Deltatheridia, a new order of mammals. *Bull. Amer. Mus. Natur. Hist.* **132**, 1–126.

Van Valen, L. (1967). New Paleocene insectivores and insectivore classification. *Bull. Amer. Mus. Natur. Hist.* **135**, 217–284.

Van Valen, L. (1970). Adaptive zones and the orders of mammals. *Evolution* **25**, 420–428.

Van Valen, L., and Sloan, R. E. (1966). The extinction of the multituberculates. *Syst. Zool.* **15**, 261–278.

Wood, A. E. (1962). The early Tertiary rodents of the family Paramyidae. *Trans. Amer. Phil. Soc.* **52**, 1–261.

9

Origins and Function of the Pes in the Eocene Adapidae (Lemuriformes, Primates)

RICHARD LEE DECKER
and
FREDERICK S. SZALAY

Introduction

In this study of the pes in the adapids, the most ancient known family of the Lemuriformes, we present a functional analysis of most of the foot skeleton. We believe that the foot structure of *Notharctus, Pelycodus,* and *Adapis,* and *Leptadapis* although the latter three genera are less well known, clearly show that the adapids were derived from a hitherto unknown primate stock possessing both morphological and functional features displayed by the paromomyiform primates of the Paleocene (Szalay and Decker, this volume). There are a sufficient number of features characterizing lemuriforms to indicate that their foot structure is a derived, rather than a primitive one in the order.

Lemuriforms in many ways were one of the most varied higher categories of primates, and particularly among the species of the Madagascan radiation, they display an impressive array of adaptive diversity. Acquaintance with the Eocene level of organization, however, is necessary to facilitate examination of the living and subfossil lemuriforms. Only a knowledge of the Eocene forms allows a keener appreciation of the hitherto unknown, colonizing, first Lemuriformes of Madagascar. The study of Madagascan lemurs, on the other hand, offers one of the most unusual opportunities among mammals to study a major radiation within an order, stemming probably from a single ancestor, a species or several species of a genus, that evolved into animals as diverse in their ways of life as *Lemur, Indri, Daubentonia, Palaeopropithecus, Archaeoindris,* or *Hadropithecus.*

The Pes in the Eocene Adapidae: Morphology and Axes of Rotation

The postcranial anatomy of the European adapines, the tarsal morphology in particular, is poorly known (see Gregory, 1920). Stehlin (1916, Fig. 302) figured an astragalus as that of *Caenopithecus lemuroides?* This, however, appears to be not of a primate but either that of a carnivoran or hyaenodontan. Astragali of *Leptadapis magnus* were reported by Filhol (1883) and by Stehlin (1916, Fig. 302). The only calcaneum of an adapine previously reported in the literature (Gregory, 1920) is a specimen in the American Museum of Natural History collection (A.M.N.H. No. 10016) from the Phosphorites of France. First, second, and third metatarsals, and a phalange of an adapine from the Phosphorite were also figured by Gregory (1920, Fig. 18). The pes of *Adapis parisiensis* and *Leptadapis magnus,* are known to us by several specimens of astragali and calcanea (Figs. 1 and 2) and by metatarsals.

The North American early Eocene notharctine *Pelycodus* is poorly known postcranially; a calcaneum, astragalus, and an entocuneiform (A.M.N.H. No. 16852) from the Gray Bull beds of the Big Horn Basin, Wyoming, have been illustrated by Matthew and Granger (1915). A few additional fragmentary calcanea of *Pelycodus* are known to us.

The remaining account of the notharctine pes is based largely on the skeletal remains in the American Museum of Natural History of *Notharctus tenebrosus,* previously described by Gregory (1920), and additional specimens (A.M.N.H. No. 91663) which can be identified as *Smilodectes gracilis.*

Our comparison of the adapid foot morphology has been with extensive collections of paromomyiforms, and with feet of most living genera of Madagascan lemuriforms. Through the facilities of the Academie Malagache in Tananarive we made comparisons with the tarsal remains of the subfossils *Varecia, Megaladapis,* and *Archaeolemur.*

Fig. 1. Dorsal view of a left astragalus (Basel No. QE 496) and right calcaneum (Basel No. QE 505) of *Leptadapis magnus* from the late (?) Eocene Phosphorites of Quercy. Subdivisions on the scale are 0.5 mm.

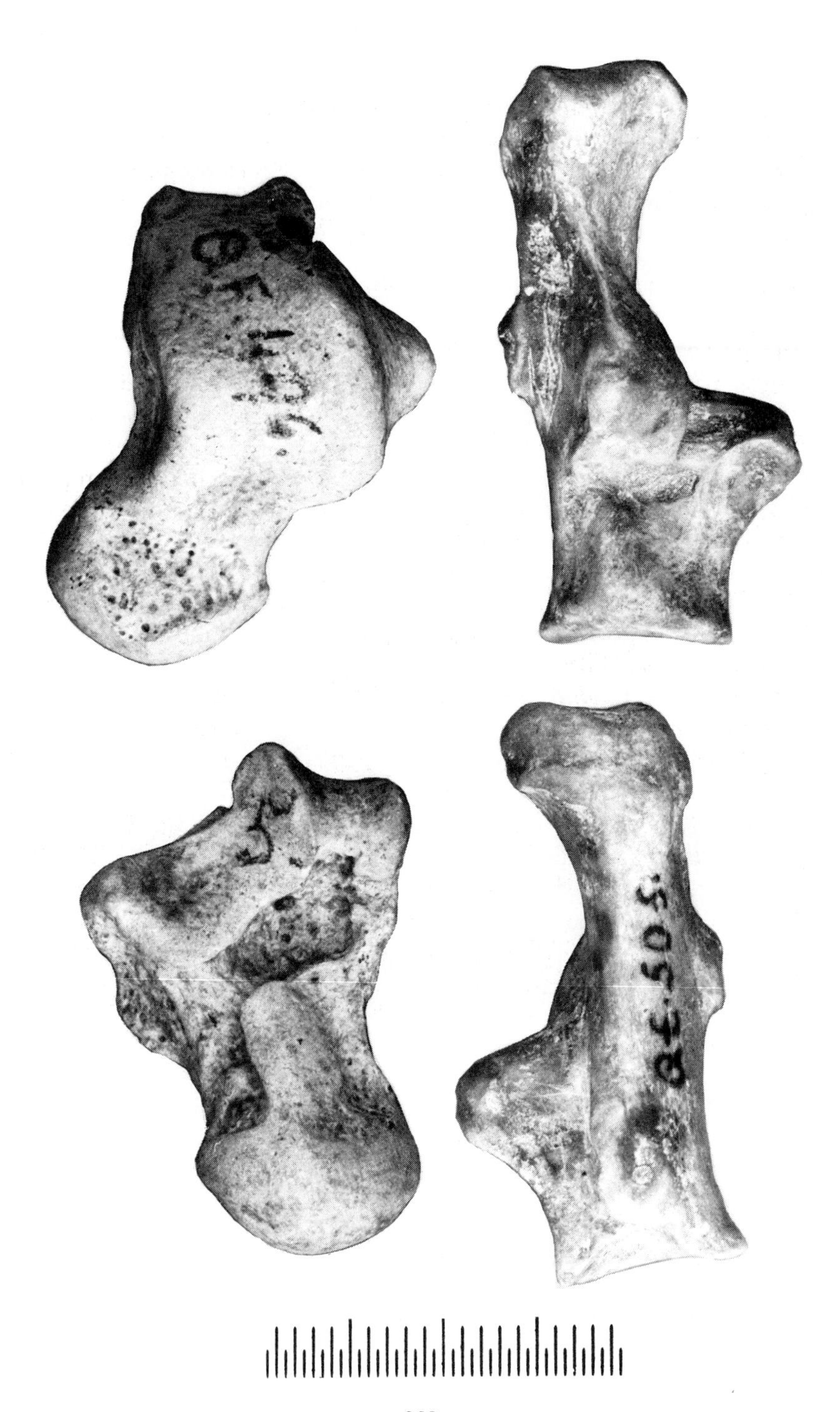

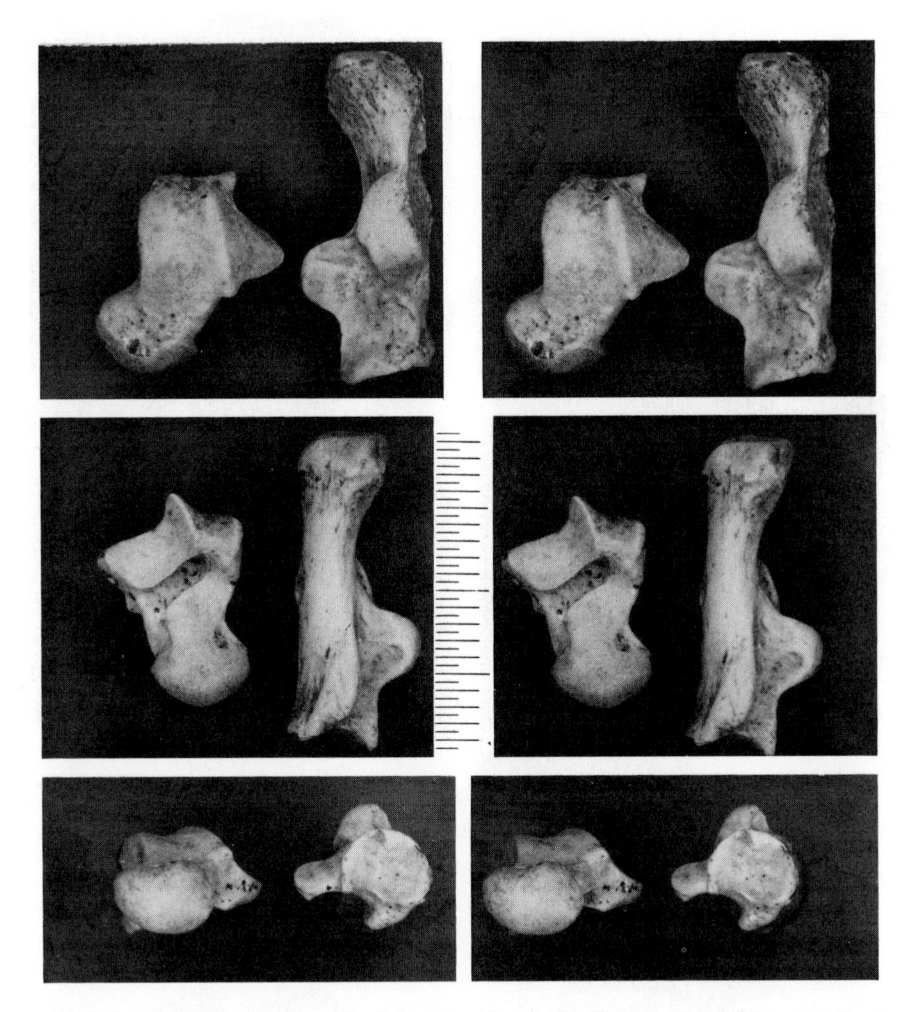

Fig. 2. Stereoscopic pairs of a left astragalus and calcaneum of *Adapis parisiensis* from the late (?) Eocene Phosphorites of Quercy, France (Montauban collection). Top, dorsal view; middle, ventral view; bottom, distal view. Subdivisions on the scale represent 0.5 mm.

THE ADAPINE PES

Adapis parisiensis and *Leptadapis magnus*, share a uniformly rounded naviculoastragalar facet and recurved posterior astragalocalcaneal facet with other known Paleocene and Eocene primates (see Fig. 3). As in other Eocene primates the cuboid facet of the calcaneum shows a subconical

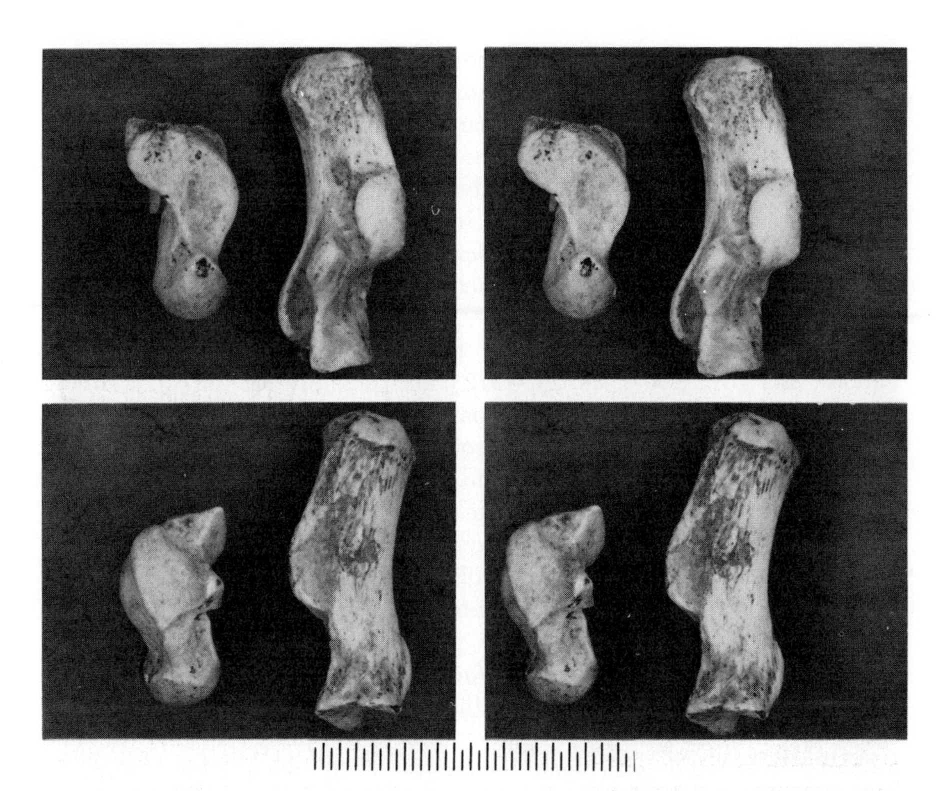

Fig. 3. Stereoscopic pairs of a left astragalus and calcaneum (same as in Fig. 2) of *Adapis parisiensis* from the Phosphorites of Quercy, France (Montauban collection). Top, medial view; bottom, lateral view. Subdivisions on the scale represent 0.5 mm.

depression of variable depth. The calcaneal groove for the tendon of the flexor fibularis is again deep.

As in paromomyiforms and notharctines, in which the form of the trochlea is similar, the axis of upper ankle joint rotation probably lay obliquely in the body of the astragalus.

In the astragalus, conspicuous differences from notharctine adapids include the presence of an extensive "squatting facet," relatively smaller secondary fibular facet, and a deeper groove for the tendon of the flexor fibularis. Probably affected by these latter two aspects, and by the more limited posterior extent of the superior tibial facet, the trochlear shelf is less well developed, especially in *A. parisiensis.* As the trochlear shelf is well developed in *L. magnus,* a dentally slightly more primitive adapine, it appears to us that the reduced condition of the trochlear shelf in *A. parisiensis* is a derived adapine condition.

In the calcaneum, the most remarkable difference from notharctines is the relatively greater length of the tuber calcaneus or heel. The calcaneal index may be taken as the tuber length up to the posterior margin of the sustentaculum relative to the length anterior to this margin. Comparable values are obtained in *A. parisiensis* and *L. magnus* despite the considerable size differences between these two forms (mean = 144, range 112–163, $n = 12$; and mean = 143, range 116–171, $n = 12$, respectively). The index exceeds values for other strepsirhines which may seldom vary over 100 (Martin, 1972) and possibly for all other, living and known fossil, primates except paromomyiforms to which the values are comparable. In both *Adapis* species, unlike in *Notharctus,* the distal astragalocalcaneal articulation extends to the end of the calcaneum and the proximal tarsals do not alternate.

A further difference concerns the disposition of the lateral tubercle. Homologies here are, however, problematical. The lateral tubercle in notharctines, lying even with the anterior margin of the posterior astragalocalcaneal facet, was identified as homologous with the peroneal tubercle of paromomyiforms. In *Pelycodus* (A.M.N.H. No. 16852) there is a second smaller tubercle behind this level. Beneath a pit for calcaneofibular and astragalocalcaneal ligaments there is a third very tiny tubercle probably associated with such ligaments, as in living lemuriforms. In *Adapis* and *Leptadapis* there is only a single tubercle present distinctly behind the anterior margin of the facet. Whether this represents the peroneal tubercle in a more posterior position, or whether this has disappeared and another tubercle became more pronounced, we cannot say.

The first metatarsal is similar to those of notharctines and other lemuriformes and the hallux was undoubtedly opposable. The remaining metatarsals differ specifically from *Notharctus* in the comparative bluntness of the heads. In *Notharctus* these have well-defined keels bordered by grooves for the sesamoid bones. The phalanges are curved.

In summary, *L. magnus* and *A. parisiensis* resemble other Paleocene and Eocene primates in characters of the pes we have associated with arboreal adaptation. However, calcaneal proportions differ from the proportions usually associated with leaping adaptation. For example, these dimensions suggest great leverage for the triceps surae.

The Notharctine Pes

The head of the astragalus is slightly broader than the neck which is distinct, elongate, and constricted at its junction with the body. The body is high and narrow in contrast to the living indriids or *Lemur catta*, in which it is low and broad (it is higher in *Lemur fulvus*). In notharctines, the lateral side of the body is particularly high relative to its length, and higher than the medial side, resembling paromomyiforms, in which this is more marked. As in later lemuriforms, the facet for the fibula slopes steeply down a shelf projecting from the lateral side of the body. In paromomyiforms this shelf was small or absent, and the fibula articulated with the vertical face of the lateral side, which by comparison was very high and broad to transmit the forces from the tibia and fibula. Development of the shelf in adapids might have relieved the area directly beneath the lateral crest of the trochlea from much of its burden and consequently resulted in a reduction in the height of the lateral side. Correlated with these developments is the reorientation of the calcaneoastragalar facet on the plantar side of the body (from a position directly beneath the lateral border of the trochlea in paromomyiforms) to the more oblique one of adapids and later lemuriforms. Thus, anteriorly this facet extends beneath the fibular shelf transmitting stresses from the shelf for the fibula directly to the calcaneum; in other words, these are compressive rather than shearing ones.

In notharctines, the trochlea is largely restricted to the body, whereas in paromomyiforms, adapines, and later lemuriforms the medial and superior facets continue onto the neck, forming a "squatting facet" abruptly limiting anterior travel of the tibia. The notharctine complex probably represents the derived condition. The medial facet is slightly depressed for the tibial malleolus and the lateral facet is high and narrow, and convex anteroposteriorly. Although the anterior limits of the tibia do not seem well developed, the crests of the trochlea, where the medial and lateral facets are continuous with the superior facet, diverge gradually anteriorly, as in *Daubentonia*. Perhaps this form of the astragalus is resilient to the crus during dorsiflexion. The superior facet in notharctines is strongly arched and extensive posteriorly as in extant lemuriforms, and, in contrast to paromomyiforms indicates an increased hingelike mobility of the upper ankle joint, particularly during plantarflexion. In *Daubentonia*, however, the superior facet is less extensive posteriorly. The groove on this facet in specimens of notharctines is neither as marked nor as medially disposed as in *Propithecus* or *Indri*, nor as shallow or absent as in *Daubentonia madagascariensis*. It is rather shallow to vaguely distinct.

A major characteristic of the lemuriforms, developed early and most strikingly in the notharctines and *Leptadapis magnus*, is a bony embankment protruding posteroventrally from the body of the astragalus. This structure is conveniently termed the trochlear shelf. In notharctines it is robust dorsoventrally and bears a complex of features which on the whole are better developed than in later lemuriforms. On the dorsum of the shelf the facet is broad and flat but curves upward at its posterior margin, limiting posterior travel of the tibia. Posterior on the shelf is a shallow groove (it may be deeper in *Daubentonia* than in *Notharctus* or *Smilodectes*) for the tendon of the flexor fibularis. The tubercle lateral to this is enlarged laterally to support a separate posterior fibular facet. This facet and its bony support are poorly developed or absent in later lemuriforms. On the plantar surface of the astragalus, the calcaneo-astragalar facet is extended under the posterolateral corner of the shelf. This facet is unequally developed beneath the enlarged lateral tubercle and posterior fibular facet, giving the former facet a unique notched appearance. Thus, in contrast to the anterior limits of the tibial facet, the posterior limits of the facets for both the tibia and fibular are well developed. The trochlear shelf is absent in paromomyiforms and, clearly as a secondary modification from the usual lemuroid condition, in *Megaladapis*. Its absence in *Megaladapis* is probably causally related to the limited arc and the posterior extent of the superior facet and the enormous development of the groove for the tendon of the flexor fibularis. *Megaladapis* also has the relatively largest superior astragalar foramen of any of the known primates.

On the plantar surface of the neck and head the extensive convex facets for the calcaneum and the plantar calcaneonavicular (spring) ligament are continuous with one another and with the distal spheroid-shaped facet of the head for the navicular. This characterizes most lemuriforms (except *Archaeolemur;* see below). The tendency for continuity of these facets occurs in paromomyiforms, particularly through an enlargement of the spring ligament facet.

The sulcus calcanei separates the calcaneoastragalar and sustentacular facets and may be pierced by one or more plantar astragalar foramina, the largest lying in the most posterior corner of the sulcus. The latter may have been continuous with a smaller superior astragalar foramen via an astragalar canal. This is more apparent in the condylarth *Protungulatum* and in *Plesiadapis* (Szalay and Decker, this volume) and in some of the subfossil lemuriforms of Madagascar such as *Varecia insignis.*

The calcaneum anterior to the astragalocalcaneal facet is about 40% of the length of the bone in notharctines, *Daubentonia,* or *Propithecus,* or longer by a few percent in *Lemur* compared to about 32% in *Plesiadapis.* The astragalar sustentacular facet is disposed somewhat more forward in relation to the astragalocalcaneal facet in notharctines or

Daubentonia than in *Propithecus* or *Lemur* so that from this point the apparent anterior elongation of the calcaneum is less while that of the heel is more. Walker (1970) has shown that there is an allometric relationship between the calcaneum elongating anteriorly and a decrease in its size.

A peroneal (?) tubercle is present, disposed slightly less anterior to the astragalocalcaneal facet than the primitive large one in paromomyiforms. In living lemuriforms this tubercle is largely absent. Certain of the lemuriform subfossils show enlargement of a lateral tubercle, but this is probably not homologous with the peroneal tubercle of notharctines. In paromomyiforms the large peroneal tubercle probably served in part as a shelf over which the tendon of the peroneus brevis angulated downward to its insertion into the base of the fifth metatarsal, as this occurs in living Mammalia which retain this condition of the tubercle (Reed, 1951). In living primates, the inferior peroneal retinaculum is attached to the tubercle, binding the peroneal muscles to the calcaneum. We do not believe that the changes in the tubercle and its relationships to the tendons of peroneus muscles are entirely related to changes in the calcaneocuboid joint (see below). Rather, the reduction of the peroneal tubercle is correlated with elongation of the elements. This, in addition to changes in the calcaneocuboid joint, would have changed the requirements in the directions of muscle forces of everted movements in the calcaneocuboid and lower ankle joints, both of which are produced by the same peroneal muscles.

The anterior plantar tuberosity is low and depressed anteriorly for the calcaneocuboid ligaments, most important of which is the long plantar calcaneocuboid ligament which is attached somewhat distant from the calcaneocuboid joint. Medially the tuberosity is grooved for the tendon of the flexor fibularis. The tuberosity may be more prominent in extant indriids.

The long axis of the convex astragalocalcaneal facet lies longitudinally as in later lemuriforms and as in paromomyiforms (Szalay and Decker, this volume). As mentioned above the concave calcaneal sustentacular facet is disposed slightly more anteriorly to the posterior facet in notharctines and *Daubentonia* than in living Madagascan lemuroids. This facet is almost continuous with the distal astragalocalcaneal facet anteriorly. A moderately developed navicular facet is present and convex on the most distomedial corner of the calcaneum while the presence, shape, and size of this facet may vary through the genera of Lemuriformes. For example, although the tarsus alternates (differently, however, from the primitive eutherian condition in which the astragalus and cuboid might be in contact), i.e., the tarsal joints are offset from one another, calcaneonavicular contact may be avoided in *Lemur catta* but is well developed in *Lemur fulvus*.

The cuboid facet of the calcaneum is semilunar, flat, but depressed ventromedially to receive the projection of the cuboid. The calcaneocuboid articulation shown by notharctines and adapines, or a modified version of this, is common in varying degrees in most primates except the paromomyiforms. The pivotal articulation between the cuboid and calcaneum is a major advance in primates. Although it is later modified in *Tarsius* (Hafferl, 1929; Halls-Craggs, 1966), in *Homo*, and others, it is undoubtedly one of the most important modifications for mediolateral rotation of the pes anterior to the astragalocalcaneal complex.

There are few distinctions of the cuboid in notharctines from those of other lemuriforms. It is similar in proportion to those of living Lemuridae of comparable size. It is slightly larger in length than breadth, but smaller in dorsoventral diameter than in the extant indriids. The plantar groove for the peroneus longus tendon is directed obliquely distally and medially. Laterally a moderately developed dorsal lip or wall of this groove bounds the tendon as in most Lemuridae, but is better developed in *Lemur catta* whereas it is normally absent in living Indriidae. Plantar eminences for the calcaneocuboid and tarsometatarsal ligaments are relatively low in notharctines (Fig. 4).

The calcaneal facet is flat on the margin surrounding the subconically shaped cuboid pivot. The base tends to be larger and the projection smaller in the living indriids (Fig. 11E, F, and G, p. 284) so that this articulation is more stable. However, in notharctines the projection and the reciprocal depression on the calcaneocuboid facet are often slightly broader relative to the base than is found in living Lemuridae although not as much as in *Daubentonia*. One can understand the need for this stability in living indriids in which the tarsus does not alternate to the degree that it does in the lemurids. Neither is the pes as alternating in notharctines as in the lemurids, nor the calcaneocuboid joint as advantageous to stability.

On the medial side there is a flat facet proximally for the navicular; this facet may be convex in living indriids, *Lepilemur*, or *Cheirogaleus*. Distal to this is a slightly convex facet for the ectocuneiform, flat in the indriids and *Daubentonia*. Still farther distally there is a second flat facet for the ectocuneiform. There are probably differences in the extent of the limited mobility between the cuboid and the navicular and ectocuneiform in *Notharctus*, the extant Indriidae, and *Daubentonia*. *Notharc-*

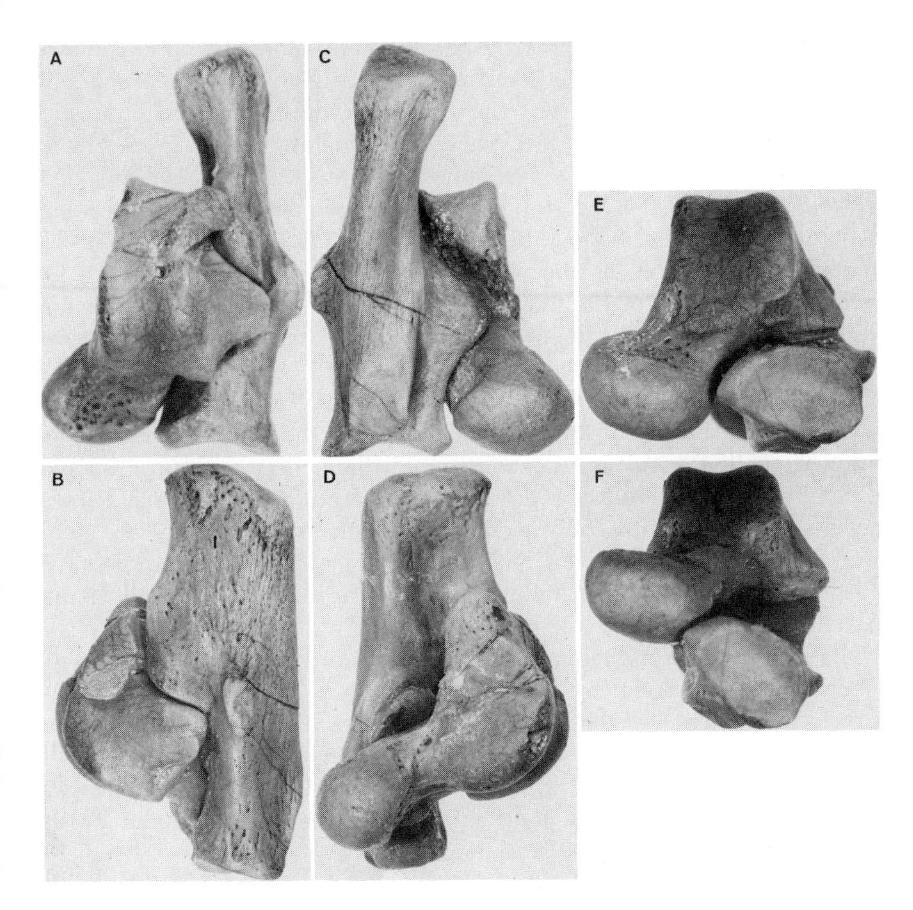

Fig. 4. The astragalocalcaneal complex of a notharctine, A.M.N.H. No. 91663. (A) Dorsal view, everted; (B) lateral view, everted; (C) plantar view, everted; (D) medial view, everted; (E) distal view, everted; (F) distal view, inverted.

tus appears to have been similar in this respect to the majority of the extant Lemuridae.

Concerning the navicular, Gregory (1920, p. 99) wrote: "The bone slants sharply inward, backward, and downward in such a way that its under surface flanks medially the deep gutter for the flexor tendons on the underside of the tarsus. But the slanting position of the navicular is especially connected with the turned-in position and the thumb-like character of the hallux."

In general this bone largely resembles the navicular of the extant lemurids, although it is slightly broader and less elongate. The navicular articulates with the head of the astragalus proximally via a semispheroid concave facet, a bit enlarged by the moderate development of the posteroinferior process projecting backward, but shallower than in the living lemuroids. It differs from the navicular in the extant Indriidae and *Daubentonia,* in which it is shorter with the process undeveloped. Development of this process is probably correlated with the elongation of the navicular. Elongation of the navicular would increase the distance between the insertions of the muscles of inversion, the tibialis anterior and posterior, and the astragalonavicular joint, decreasing their control over it. Thus, an elongated process on the elongated navicular would allow the effective maintenance of force transmission from the navicular to the astragalus. Laterally the astragalar facet of the navicular is notched to avoid conflict with the calcaneum, for which there may be an actual facet. Distally the concave ectocuneiform and convex entocuneiform facets are bent laterally and medially, respectively, on the convex mesocuneiform facet, forming an S-shaped arrangement. The middle facet faces less dorsad than in living lemurids; the ectocuneiform (lateral) facet is more spheroid than in later lemuriforms.

The navicular abuts against the cuboid ventrolaterally by a flat facet, which borders on the facets for the two lateral cuneiforms. This facet is curved and concave in the living Indriidae, *Lepilemur,* and *Cheirogaleus major.*

The cuneiforms combine with the cuboid and navicular to form the transverse arch of the pes. This arch maintains the alignment of the flexor tendons to the digits during rotation of the pes. The movements among these elements add slightly to the total rotation possible in the pes (see below).

The mesocuneiform of *Notharctus* (Figs. 5 and 6) narrows ventrally into a wedge more markedly than in any of the extant lemuriforms. *Daubentonia* and the living Indriidae resemble *Notharctus* in this respect. Conversely, the ectocuneiform narrows less from the dorsum ventrally; the dorsum is particularly narrow proximally. In *Notharctus* a prominence projects ventrally from the entocuneiform forming the medial wall of the transverse arch and in particular of the long tendon to the first digit. This differs from living lemuroids in which the process deflects laterally beneath this tendon, forming a more complete osseous containment. In *Daubentonia* the prominence is more blunt but is otherwise similar to that of *Notharctus.*

The degree to which the tarsals interlock may also inhibit tendencies

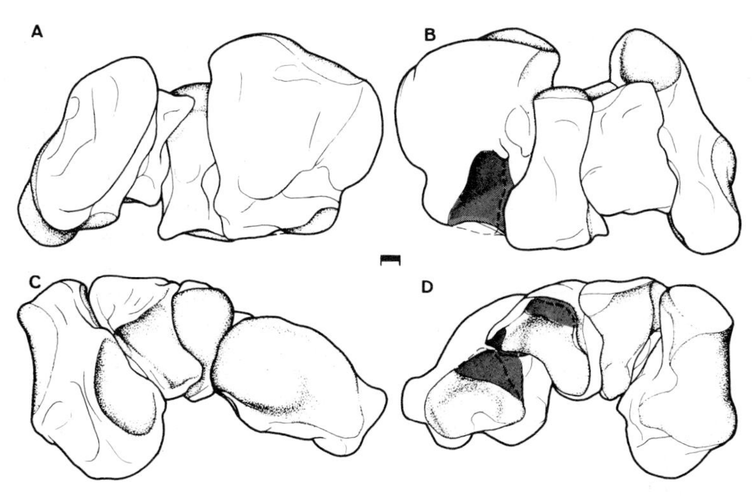

Fig. 5. Composite of the right cuboid and cuneiforms of a notharctine. Cuboid: A.M.N.H. No. 13030; ectocuneiform: A.M.N.H. No. 11484; mesocuneiform: A.M.N.H. No. 13030; entocuneiform: A.M.N.H. No. 11484. Plantar (A), dorsal (B), proximal (C), and distal (D) views. Stippled surfaces are joint facets; hatched areas represent breakage. The scale represents 1 mm.

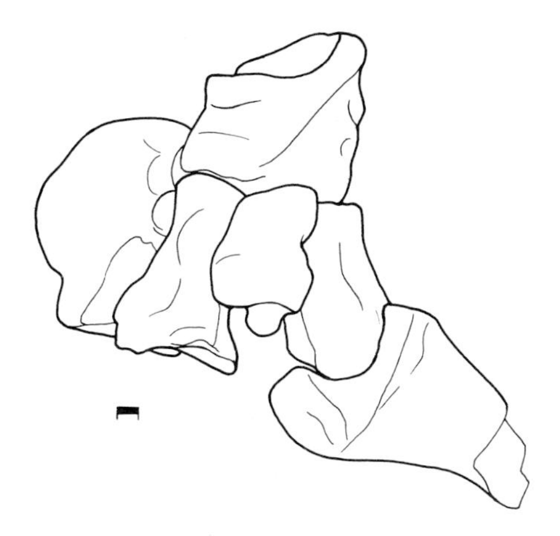

Fig. 6. Composite of the right navicular, cuboid, cuneiforms, and proximal part of first metatarsal of *Notharctus*. The first metatarsal, probably that of *N. tenebrosus*, has been reduced in size to match the tarsus. Navicular: A.M.N.H. No. 11484; cuboid, ectocuneiform, mesocuneiform, and entocuneiform: same as in Fig. 5; proximal fragment of metatarsal I: A.M.N.H. No. 11484. A.M.N.H. No. 11484 probably represents mixed individuals of possibly two species of *Notharctus*. The scale represents 1 mm.

of the lever system to buckle dorsally. In lemuriforms, the mesocuneiform and cuboid may join ventrally to form a seat for the ectocuneiform, in particular, proximally where the cuboid and mesocuneiform actually make contact. In living Lemuridae the mesocuneiform proximally twists ventrolaterally about a small keel of the ectocuneiform to articulate with the cuboid. For different reasons this twisting, or torsion, of the mesocuneiform is insignificant in *Northarctus, Daubentonia,* and extant Indriidae. In living indriids the proximal keel of the ectocuneiform is reduced, and twisting of the mesocuneiform is therefore unnecessary. Instead, a process of the cuboid (poorly developed in the extant Lemuridae) projects for most of the distance beneath the ectocuneiform to articulate with the mesocuneiform. In *Lepilemur,* in addition to less twisting of the mesocuneiform, the process of the cuboid projects somewhat to meet it as in living indriids. In *Daubentonia* and *Notharctus,* the proximoventral keel of the ectocuneiform is well developed and obstructs significant mesocuneiform-cuboid contact. The *Notharctus* mesocuneiform is only slightly twisted around the ectocuneiform, and the process of the cuboid is undeveloped. In *Daubentonia* these features are even more marked by their absence, and for these reasons the mesocuneiforms of *Daubentonia* and *Notharctus* are very similar. Gregory (1920) suggested that these differences in the mesocuneiform of *Notharctus,* i.e., the lack of torsion, were associated with the function of the hallux. This may rather be a mechanism for stabilizing the tarsus, a condition less developed in notharctines than in the other known lemuroids.

According to Gregory (1920) the metatarsals are short and stout and the phalanges long and slender. The first metatarsal is shorter than in *Propithecus* (although Gregory made no relative comparisons), and its ability to abduct was probably less than in the living Lemuroidea. The terminal phalange of the hallux is expanded more than in *Lemur* but not so much as in *Lepilemur.* The degree to which the second metatarsal is interlocked between the cuneiforms is less. We observed on a complete fourth metatarsal the differential robusticity of this element (probably of the other lateral metatarsals). Not only does the dorsoventral diameter of the shaft increase distally but that of the head is larger than the base. The degree to which this occurs in *Notharctus* is unusual among the Lemuroidea, except in *Daubentonia,* and often the reverse may occur. This may indicate distal displacement of some of the major bending moments, perhaps due to the forceful flexors of the digits.

Comparisons of the nearly whole pes of *Notharctus* (A.M.N.H. Nos. 11474 and 11478) to extant lemuriform feet of similar size (e.g., *Daubentonia madagascariensis, Lemur fulvus, Lemur catta,* and *Propithecus verreauxi*) reveal some important differences in proportions. From the heel to the anterior margin of the calcaneal astragalocalcaneal facet is 29–31% of the distance from the latter to the tip of the third metatarsal. This is similar to the proportions of *Daubentonia* (29%, $n = 1$), while in *Lemur* (24–27%, $n = 4$) and *Propithecus* (25%, $n = 3$) this is somewhat less. *Notharctus* differs, however, from *Daubentonia* in that its metatarsus contributes less (50%) to the anterior proportion than in the latter (58%), and thus it is more comparable to *Lemur* (50%). The condition in *Notharctus* is unlike that in *Propithecus,* however, in which the metatarsus is quite long (62%). As previously noted when calcaneal proportions were compared (see Adapine section), *Notharctus* is certainly nearer to the living lemurs in this respect than *Adapis* and *Leptadapis* are. The phalanges in *Notharctus* are long by comparison to the metatarsals (Gregory, 1920) and the additional information indicates that its grasp was probably as efficient in perching or clinging as that of any of the living genera.

An interesting point, noted before, is that due to differential elongation of the navicular and calcaneum the degree to which the tarsus of *Notharctus* alternates is less than is common in living Lemuridae. The extant indriids and *Daubentonia* display an even less alternating condition than that shown by *Notharctus.*

THE PES OF *Archaeolemur majori*

Lamberton (1939) described the known parts of the pes of the subfossil Madagascan lemuriformes, compared these to living representatives, and made certain functional interpretations. In addition, Walker (1967) has described a cuboid and the first four metatarsals of *Archaeolemur;* the fifth metatarsal was described by Lamberton. Walker (1967) has presented evidence that some of these genera, *Archaeolemur* among them, were at least partially terrestrial. Because we have additional evidence from the pes to suggest a terrestrial habitus in *Archaeolemur majori,* a brief account of the astragalocalcaneal complex and the entocuneiforms of this taxon help us to evaluate the notharctine pes by means of contrast.

The characteristic posterior trochlear shelf of the astragalus in lemuriforms is present. The principle differences in the pes from adapids and other lemuriforms are (1) the known tarsals are as large or larger than those in *Indri,* except for the entocuneiform; (2) the calcaneum is very robust; the whole pes is probably broader relative to its length; (3) the astragalocalcaneal and cuboidonavicular complexes are nearly transverse, not alternating, like in the indriids, but unlike the living Lemuridae; (4) facets of the posterior astragalocalcaneal articulation are relatively broad; (5) the facet of the astragalus for the plantar calcaneonavicular ligament is small, resulting in the tendency for the plantar facets of the head and neck of the astragalus to be discontinuous. The medial region of the navicular facet is small in comparison to the lateral region of this facet; (6) the superior facet of the trochlear is small posteriorly and its borders diverge slightly anteriorly; (7) though similar in basic structure, the depression in the cuboid facet of the calcaneum is more spheroid and enlarged at the expense of the base, which is liplike; (8) the facet for the tendo calcaneus on the tuber of the calcaneum is flat rather than arched; (9) there is a large lateral (peroneal?) tubercle on the lateral aspect of the calcaneum inferior to the posterior astragalar facet;* (10) there is a large pit perhaps for an astragalocalcaneal ligament on a shelf-like projection continuous with the lateral tubercle; (11) the navicular is short proximodistally; the astragalar concavity is shallow, the cuboid is flat, and the entocuneiform facet is reduced compared to the other two cuneiform facets which form a ventral peak as they meet; (12) on the very broad cuboid the groove for the tendon of the peroneus longus is deep, nearly transverse, and the lateral lip superior to the groove is a prominent projection as may occur to a lesser degree in *Lemur catta.* It may be correlated with increase in the importance of the peroneal muscles. This might be significant because *L. catta* was recently found to be more terrestrial than previously suspected in comparison to *L. fulvus* (C. R. Sussman, personal communication); (13) the entocuneiform is relatively elongate, not robust, and is compressed dorsoventrally (mediolaterally); the facet for the first metatarsal is small and narrow on the distal end, not very extensive or arched, and the process for the flexor tendon to the first digit is poorly developed; (14) the ectocuneiform bears a flat facet for the cuboid; (15) the metatarsals are robust, larger based, with less curved shafts, and the styloid process on the fifth metatarsal is very large. Walker (1967) described a dorsal facet of the head limiting dorsoflexions of the proximal phalanx. The first metatarsal shows plantarward torsion of the head and the peroneal tubercle is absent.

* This process in *Archaeolemur* should not be referred to as a peroneal tubercle since this is an unusual position for a true peroneal tubercle. Whenever developed in primates, such as in *Lepilemur*, it is in a similar relative position as in notharctines, though never as well developed. The tuberosity developed in *Archaeolemur* is in the precise position of a small tubercle common in the living lemuriforms giving attachment to calcaneofibular and astragalocalcaneal ligaments. Lamberton's (1939) conclusions were similar.

JOINTS AND THEIR AXES OF ROTATION

We have estimated the disposition of the axes of the major movements of certain important tarsal joints in *Notharctus, Propithecus,* and *Archaeolemur* (Figs. 7–10). In this procedure, as outlined in Szalay and Decker (this volume), we used the curvatures of the articular surfaces to estimate the direction of movements. When possible, manipulation of articulated tarsals provided the probable path of movement which could be traced on the articular surfaces.

In all three, the axis of rotation in the upper ankle joint is disposed obliquely through the body of the astragalus. This condition was ac-

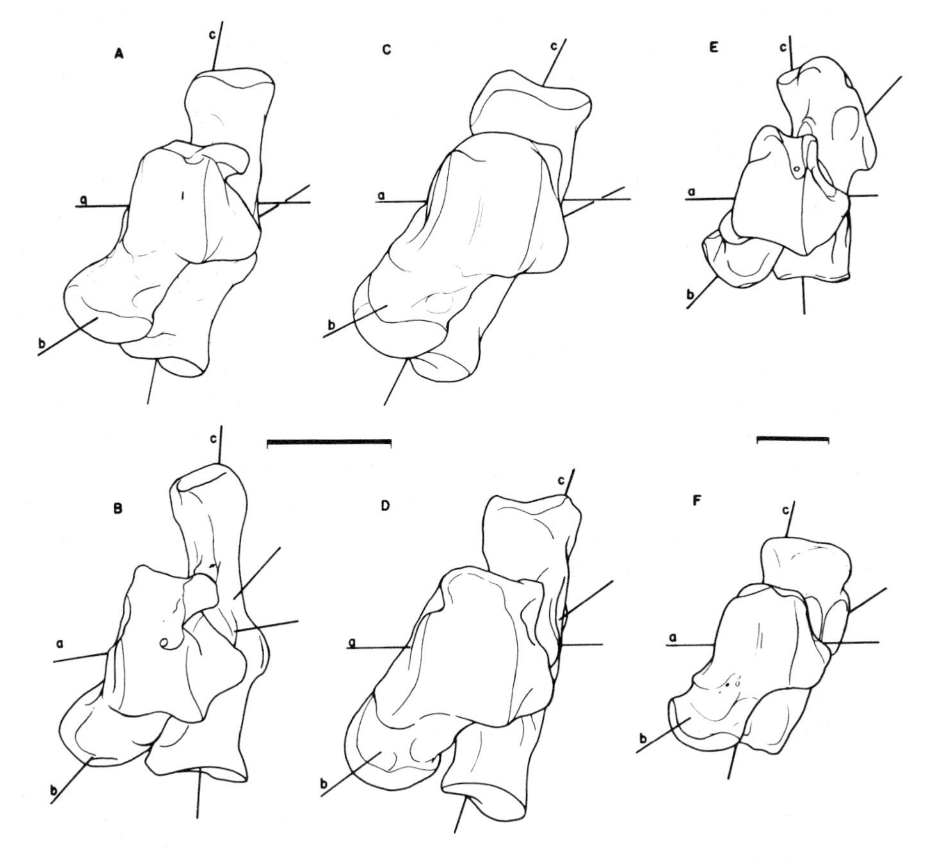

Fig. 7. Dorsal view of the left astragalocalcaneal complex and the axis of rotation of the upper ankle (a), lower ankle (b), and calcaneocuboid (c), joints in *Notharctus* (A,B), *Propithecus* (C,D), and *Archaeolemur* (E,F) in inverted (top row) and everted (bottom row) positions. The scale represents 1 cm.

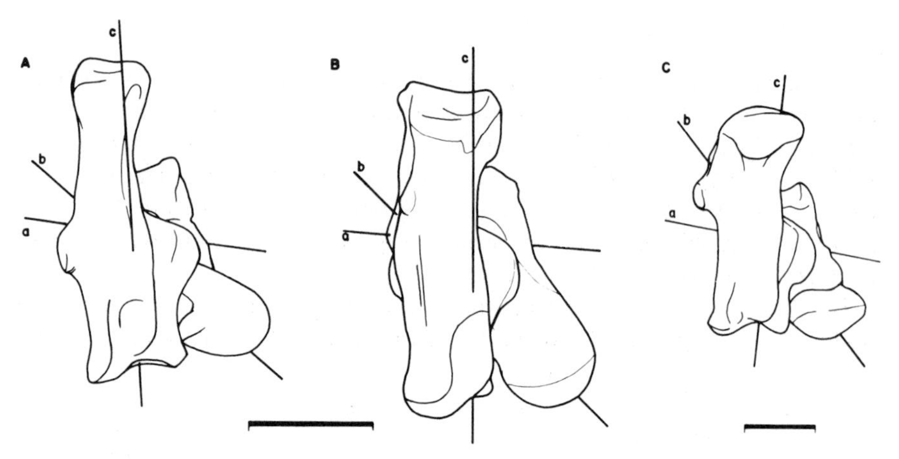

Fig. 8. Ventral view of the left astragalocalcaneal complex and the axis of rotation of the upper ankle (a), lower ankle (b), and calcaneocuboid (c) joints in *Notharctus* (A,B), *Propithecus* (C,D), and *Archaeolemur* (E,F) in an everted position. The scale represents 1 cm.

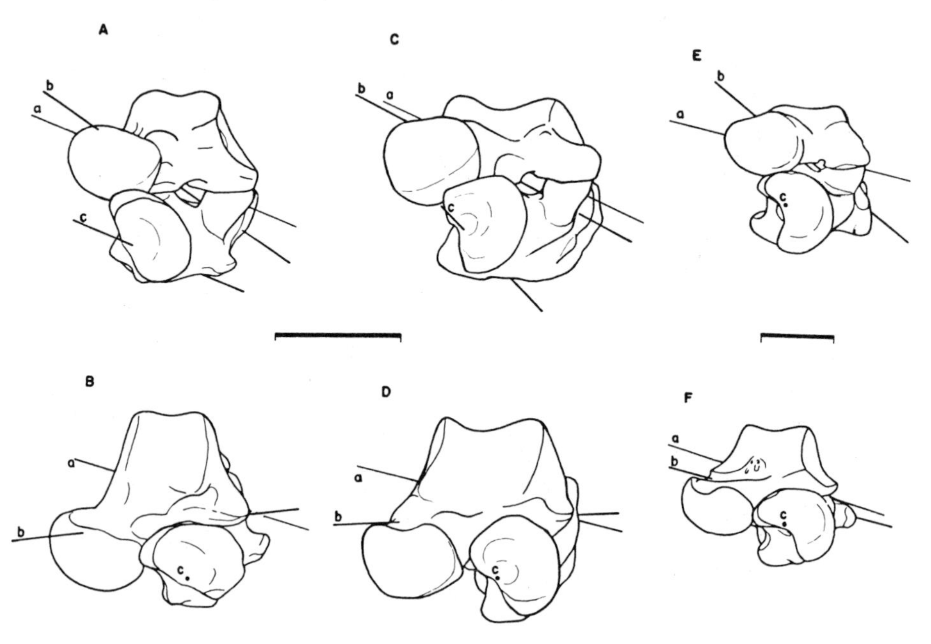

Fig. 9. Distal view of the left astragalocalcaneal complex and the axis of rotation of the upper ankle (a), lower ankle (b), and calcaneocuboid (c) joints in *Notharctus* (A,B), *Propithecus* (C,D), and *Archaeolemur* (E,F) in inverted (top row) and everted (bottom row) positions.

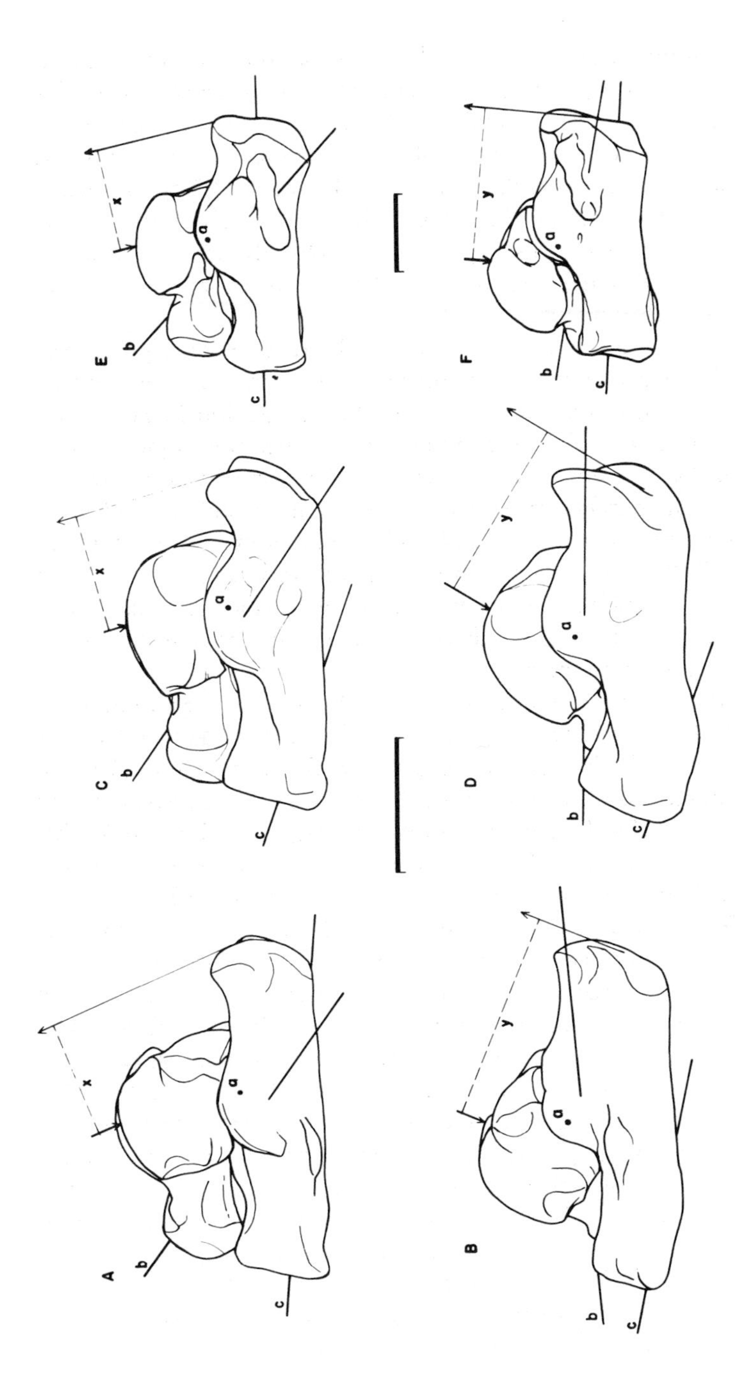

Fig. 10. Lateral view of the left astragalocalcaneal complex, the axis of rotation of the upper ankle (a), lower ankle (b), and calcaneocuboid (c) joints, and the power arm of the Achilles tendon in *Notharctus* (A,B), *Propithecus* (C,D), and *Archaeolemur* (E,F) in inverted (top row) and everted (bottom row) positions. The arrow on the tibial trochlea represents the long axis of the tibia. This is inferred in *Archaeolemur*. In order to standardize the unknown points of origin of the triceps surae the line representing the tendo calcaneum is drawn parallel to the long axis of the tibia. The scale represents 1 cm. x, Power arm of the calcaneum during extreme inversion; y, power arm of calcaneum during extreme eversion.

quired early in primate evolution (Szalay and Decker, this volume). It has been associated with providing efficient lubrication for the astragalofibular articulation and fibular mobility in its capacity to receive the strains of inversion and eversion which are prevalent in mammals adapted for uneven, i.e., arboreal, substrates (Barnett and Napier, 1953). In *Notharctus,* on the medial side of trochlea, the axis appears at the center of the curvature of this aspect while laterally it intersects the posterior astragalocalcaneal articulation. These points are similar in *Archaeolemur* and *Propithecus.*

In all, the axis of the lower ankle joint is directed obliquely through the astragalocalcaneal complex from medially above to a ventral, lateral, and posterior direction. The movements of the astragalocalcaneal complex, therefore, produce oblique "rotations" of the pes not along its long axis. In *Notharctus* the anterior extremity of the axis exits the head of the astragalus immediately superior to the navicular facet, in fact, in a small roughened area between medial and lateral sides of this facet. In *Propithecus* or *Notharctus,* the axis traverses the sinus tarsi and exits on the lateral side of the calcaneum inferior to the posterior astragalar facet. In *Archaeolemur* the axis exits in the same area, but this is obscured by the presence of the lateral process.

The axis of the calcaneocuboid joint of the lemuriforms examined lies more parallel to the long axis of the pes and therefore produces true rotations of the foot. Hall-Craggs (1966) pictured the axis of the calcaneocuboid joint similarly for *Galago.* The axis intersects the calcaneocuboid articulation through the depression and projection, exits the cuboid distally on its medial aspect, and may traverse the ectocuneiform and the base of the third metatarsal. Ultimately the axis lies beneath the metatarsals since these are normally dorsiflexed. In *Notharctus* the axis exits the calcaneum more posteriorly to the anterior plantar tuberosity than in *Propithecus.* In *Archaeolemur* the axis may traverse the length of the calcaneum, exiting the inferomedial corner of the tuber calcis. Even though the calcaneum of *Propithecus* is bent dorsally on the fore pes, the axis of calcaneocuboid movements maintains its alignment to the fore pes through the cuboid. The axis figured for *Archaeolemur* may be artificial. The large, more spheroid depression on the cuboid facet of the calcaneum, as well as the lipped rather than flat nature of the base, suggests that the movements here may resemble conjunct rotation rather than the adjunct rotation in *Notharctus* and *Propithecus.*

There are no great differences in the relationships of the axes of movement. The axis of the calcaneocuboid joint does vary in disposition relative to the calcaneum. The axis remains aligned with the fore pes on which the calcaneum is bent dorsally in *Propithecus*. Here the calcaneum might be disposed in a more advantageous disposition to the plantarflexors for an initial propulsive phase on a vertical support, while the fore pes and its rotations remain aligned upon the substrate. This condition is not pronounced in *Notharctus* or *Archaeolemur*.

The oblique movements of the lower ankle joint appear to have more than one consequence. Aside from permitting a large degree of the total range of inversion and eversion of the pes, they influence the mechanical advantage of the foot with respect to the upper ankle joint (Figs. 7–10). The anteroposterior component of lower ankle joint rotation would inevitably affect the leverage of the triceps surae if the foot is either inverted or everted as it is plantarflexed and dorsiflexed. This is probably even more true here and in earlier primates (Szalay and Decker, this volume) than in many other Eutheria which retain the primitive oblique orientation of this axis, because screwlike movements of the lower ankle joint increase its anteroposterior component. This development is more importantly associated with the more forward tarsus. With increased participation of the latter in inversion and eversion of the foot, as made possible, for example, by the pivot nature of the calcaneocuboid articulation, the deviations of the navicular with which, as we shall see, it is strongly associated, are increased. During inversion the navicular is brought closer to the calcaneal face of the lower ankle joint and during eversion further away. Therefore, to remain in direct contact with the navicular and to receive stresses during inversion and eversion of the foot as may be required, the astragalus must also shift backward and forward, respectively, relative to the calcaneum. This, again, is made possible by the increased anteroposterior component of screwlike movements of the lower ankle joint.

Though a mobile articulation, no axis has been determined for movements of the astragalonavicular joint. Among hominoids this articulation has been functionally tied with the calcaneocuboid articulation through the term "transverse (Chopart's) joint" and with the lower ankle articulation as the talocalcaneonavicular joint, since the movements of these joints are mutually dependent.

Though equally true of nonhominoids, the astragalonavicular and calcaneocuboid joints are often less transversely aligned. Directly and via the remaining tarsals the navicular is morphologically and functionally associated with the cuboid, and it, therefore, revolves about the longitudinal axis of the calcaneocuboid joint. On the other hand, the navicular, through the astragalonavicular articulation, is directly associated with the astragalus which, as previously noted, rotates about a separate axis.

To avoid conflict with the direction of either movement, the astragalonavicular articulation is a suitably adapted semispheroid or ovoid joint enabling each to occur without interference. For example, it may permit conjunct rotation of the navicular concavity on the head of the astragalus around the lower ankle joint axis, an axis lying outside this articulation. In addition, it may also allow the navicular to slide ventrally and dorsally on the head of the astragalus around the axis of the calcaneocuboid joint as this latter movement is required.

Except when substrate contact occurs directly beneath the head of the astragalus the substrate or "ground" reaction may pass forward onto the metatarsal heads as in the event of propulsion. This reaction is passed further forward as the pes is elongated and the strain at the astragalonavicular articulation is increased. In *Notharctus,* and to a great degree in extant lemuroids, the medial wall of the navicular concavity projects proximally into a process to brace the articulation here. It is probably correlated with elongation of the foot anterior to this point, or elongation of the navicular.

Though small, the gliding movements of the naviculo–cuboido–cuneiform complex add to the total potential rotation of the pes. In *Notharctus,* the proximal one of the two cuboidoentocuneiform articulations is curved. The cuneiform facet of the cuboid is the broader of the two articular surfaces and it is convex anteroventrally to dorsoposteriorly. This permits simultaneous proximal sliding and dorsal rotation of the cuneiform complex on the cuboid (during inversion). These movements are accommodated by the curved naviculocuneiform articulations, allowing a combination of some sliding, flexion, and rotations conjunct to the cuboidocuneiform movement. In *Notharctus* the naviculoentocuneiform articulation is more spheroid than in other lemuriforms to permit this.

Because of the differences in the cuneiform construction in the extant Lemuridae, cuboidoentocuneiform rotations are accommodated by naviculocuneiform movements and structure of a slightly different nature. Together the lateral facing navicular facet of the mesocuneiform and part of the navicular facet of the ectocuneiform form a socket for a projection formed by the reciprocal navicular facets. This nonetheless permits some sliding, more flexion, and rotations conjunct to the cuboidcuneiforms rotations.

The living Indriidae and to some degree *Cheirogaleus* and *Lepilemur* differ from the others and *Notharctus.* The cuboidocuneiform articulations are flat. The naviculocuboid articulation is curved to permit rotations of the cuboido–cuneiform complex on the navicular. These again are accommodated by conjunct rotations on the curved naviculocuneiform articulations.

In *Archaeolemur* or *Daubentonia* both cuboidocuneiform and naviculocuboid articulations are flat. In *Daubentonia* the naviculocuneiform articulations are shallow but permit some degree of flexion and little rotation. The rotational mobility of the distal tarsus is no doubt correlated with the differential mechanisms for stabilizing the tarsus discussed earlier.

Curvatures of the entocuneiform-mesocuneiform articulations which often occur in the extant lemuroids and indicate additional mobility of the hallux complex are absent in *Notharctus, Archaeolemur,* and *Daubentonia.* In addition, in *Archaeolemur* the mobility, and therefore opposability, of the hallux at the tarsometatarsal joint definitely appears to have been reduced.

Mechanical Function and Biological Role of the Pes in Lemuriformes

The methods and conceptual framework to interpret the movements responsible for the differential enlargement of joint facets to determine habitual orientation of the pes is that used by Szalay and Decker (this volume). As particular orientations of the pes become important to an animal (i.e., inverted, level everted) aspects of the articular surfaces in contact during these phases enlarge to transmit the forces incurred or to stabilize the joint. On the whole, the parts in contact during eversion of the pes are usually the larger because this orientation is probably common whether the animal is on a tree or on the ground. However, as an inverted orientation of the pes becomes habitual the surfaces in contact during inversion also enlarge. Because an inverted orientation of the pes is correlated with an arboreal substrate and a more level orientation of the foot with a terrestrial substrate, these differences, as discussed in Szalay and Decker (this volume), may distinguish major adaptive zones. However, more level orientation of the pes is also possible on an arboreal substrate if the latter is particularly large or the animal is small relative to its preferred substrate.

In extant lemuriforms the axis of movements at the calcaneocuboid joint is directed perpendicular to the joint surfaces and lies near the anatomical long axis of the pes. The movements of the cuboid on the distal end of the calcaneum are medial rotation during eversion of the pes, and lateral rotation during inversion of the pes (see Fig. 11).

Rotated laterally (inversion) the facets on the cuboid and calcaneum are more nearly in full contact than during medial rotation. Rotated medially (eversion) the facet on the cuboid loses contact with the facet of the calcaneum dorsomedially, thus freeing the latter from ventrolateral contact. It is the base on the perimeter of either facet that is affected most by these differences for it not only transmits much of the resultant forces communicating the calcaneocuboid joint but also stabilizes the joint in

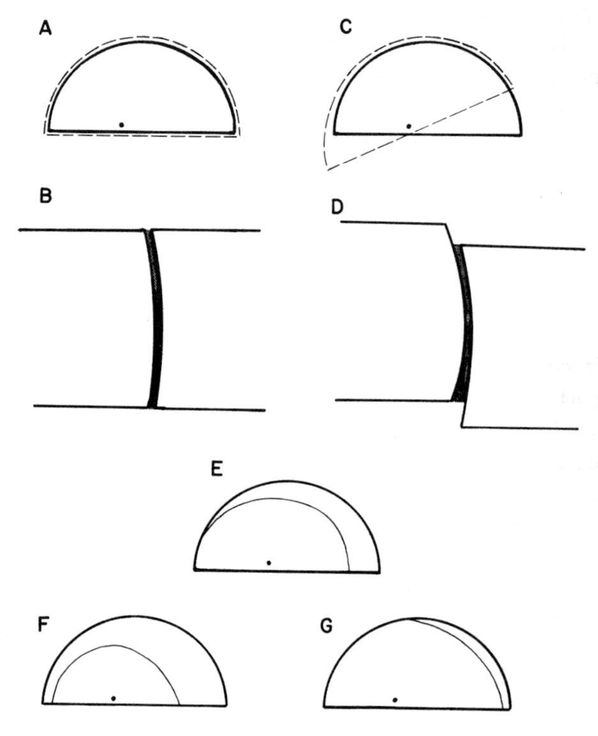

Fig. 11. Schematized views of the left calcaneocuboid joint in lemuriforms (A–D) and the left calcaneal cuboid facet in *Notharctus, Propithecus,* and *Archaeolemur.* A and C represent the distal view of the cuboid facet (solid line) with the calcaneal facet of the cuboid (broken line) superimposed during inversion (A) and eversion (B). B and D represent the same calcaneocuboid contact during inversion and eversion in dorsal view, showing that the cuboid is medially slipped off the calcaneum. E, F, and G represent distal views of the calcaneolcuboid facet in *Notharctus* (E), *Propithecus* (F), and *Archaeolemur* (G). The size differences in the base of the facet indicate relative stability of the calcaneolcuboid joint. In *Propithecus* (F), which has a narrower based cuboid pivot but a broad based calcaneal facet, the joint is more stable than in *Notharctus* (E). This joint is the least stable in *Archaeolemur* judged from the very broad based cuboid pivot and the narrow, dorsolateral base on the rim of the calcaneum.

nonmobile planes of either orientation. We would therefore expect its extra development were the pes more often inverted. In *Notharctus* the development of the base is no more than may be found in a number of the extant Lemuridae, and the dorsomedial aspect is less extensive. *Propithecus* and *Archaeolemur* exhibit two opposing trends compared to the condition in *Notharctus*. In *Propithecus* not only is the base enlarged dorsomedially but the entire base is enlarged at the expense of the size of the projection of the cuboid or the depression of the calcaneum (Fig. 11F). However, in *Archaeolemur* the base is reduced dorsomedially and altogether as the depression in the cuboid facet of the calcaneum enlarges (Fig. 11G). In the extant indriids, noted for their spectacular leaping abilities, there is clearly a trend to stabilize this joint from dorso-buckling; in these, where the tarsus is not very alternating, the advantage of the form of this joint is clear. The constructions in *Notharctus* and, more so, in *Archaeolemur* are less stable than in *Propithecus*. This less stable condition is more evident in *Daubentonia madagascariensis* than in *Notharctus*.

Again, as we noted for the early eutherians (Szalay and Decker, this volume) differences in the astragalonavicular articulation distinguish the relative importance of inversion, eversion, and orientation of the pes. In *Archaeolemur* the lateral portion of the head of the astragalus is proportionately larger than the medial aspect, and in either *Notharctus* or *Propithecus* the head is more globular and either side receives nearly equivalent emphasis. This indicates that inversion and inverted orientations of the pes are more important in the latter two genera (this is a fact for *Propithecus*), and level orientations are more important and inversion of the foot less important in *Archaeolemur*.

During inversion of the pes the lateral side of the naviculoastragalar facet of the astragalus slides off the navicular dorsally while the medial side remains in contact. During eversion the opposite occurs while the lateral side of the naviculoastragalar facet is responsible for a larger share of the support. The degree to which the medial side of this facet is enlarged provides a clue to the role the astragalonavicular joint plays in mobility of the pes during an inverted phase of rotation.

The head of the astragalus in *Notharctus* and in *Propithecus* is globular and the navicular facet is extensive on both medial and lateral sides. This indicates the greater extent to which the navicular traverses up the medial aspect of the astragalus during inversion of the pes when compared to *Archaeolemur* in which the lateral aspect is disproportionately large and the medial aspect smaller. These differences are as great as when paromomyiforms are compared to the latest Cretaceous *Protungulatum*.

In this regard it should also be pointed out that the facet by which the astragalar head is supported by the spring ligament has undergone a secondary reduction in *Archaeolemur*.

The previously shown differences in the construction of the joints of rotation suggest the following in regard to preferred orientations of the pes in *Notharctus* and *Archaeolemur* compared to other lemuriforms. In *Archaeolemur* from the evidence of the astragalonavicular and calcaneocuboid joints the pes would have been less well adapted for a fully inverted position than those of others; this animal clearly preferred a more level orientation of the pes. The shallowness of the astragalonavicular articulation in *Notharctus* and *Daubentonia* is indicative of a stability less developed than in living lemuroids. The degree to which these joints in *Notharctus* contribute to the total potential rotation of the pes while maintaining stability was probably less than in other lemuriforms. Therefore, in some respects the *Notharctus* pes was most stable in a less inverted position.

Inferring the nature of the substrate from evidence concerning the preferred orientations of the pes is less easily accomplished among primates all of which share an arboreal ancestry. Although efficiency of the mechanism might be decreased in the initial propulsive phase in a pes demonstrably less stable in an inverted orientation (*Papio, Archaeolemur*), it should be remembered that there is no apparent disadvantage to the ability for grasping a branch, dependent on other factors.

Some Indications of Locomotor Mode in the Lemuriform Pes

We urge caution concerning the various concepts of primitive primate morphology and locomotor patterns inferred from that in the literature. We believe that it is not only necessary to be acquainted with extant lemurs and lorises, but an appreciation of the entire primate record, including that of the Paleocene, and an equal knowledge of early eutherians are also prerequisites. Repeated opinions and views on locomotor patterns of extinct early primate remains which have not been studied by the workers discussing them are unfortunate. Even when the morphological and inferred functional (mechanical) differences can be noted, e.g., between the condylarth and paromomyiform bony elements, and the biological role of either eversion or inversion can be assessed and probable locomotor environments can be inferred, the third level of inference, that pertaining to actual mode of locomotion, is a very difficult one indeed. The validity, therefore, of an inference of past locomotor patterns can only be tested by the elaboration of muscle and joint mechanics of the known fossils and pertinent living forms, with the necessary careful considerations given also to the phyletic features of the morphology examined. Thus, for example, the studies of Napier (1967), Walker (1967), and Napier and Walker (1967a, b), which raised the view that a vertical clinging and leaping form of locomotion was primitive for the order Primates, have not demonstrated that any tested morphological–functional inferences exist in the actual members of ancestral radiations to support this postulate.

Short comments on the relevance of Eocene and older fossils to vertical clinging and leaping as the ancestral primate locomotor patterns have been published by Szalay (1972). In brief, so far none of the pertinent fossil evidence supports the hypothesis that vertical clinging and leaping is the primitive mode of locomotion of the Primates. In addition to our (Szalay and Decker, this volume) second-level inference that paromomyiforms were adapted for habitual arboreal existence, there are no musculoskeletal features in the known tarsal elements to suggest a particular mode of locomotion other than what might be called quadrupedalism. This probably involved a certain amount of agile hopping and bounding as these are usually performed by most quadrupeds.

Walker (1967) has described at least two characters of the pes correlated with the vertical clinging and leaping mode of locomotion shown by a number of Primates. These are the great elongation of the load arm for the given size of the animal, and an enlarged hallux and prominent peroneal tubercle of the first metatarsal. The first of these has been associated with greater acceleration in the push-off phase of leaping. The second, according to Walker (1967), may be associated with the power of the grasping hallux and in proportion to the force of gravity to be overcome when holding onto vertical stems.

Demonstrating this first feature Walker (1967, p. 164) plotted log femur length versus log calcaneal index (lever arm length × 100)/load arm length. Femur length was used as an indicator of body size. The results show regression lines of different gradients for vertical clingers and leapers, and for arboreal quadrupeds. The apparent conclusions are that for a given length of the femur (body size), vertical clingers and leapers have a lower calcaneal index, i.e., relatively shorter lever arm. However, Walker has also shown that vertical clingers and leapers show low intermembral indices in having long hindlimbs, so it is uncertain that femur length mirrors actual body size, and if the latter were plotted versus the calcaneal index, that it would give the separateness of the regressions. In addition, the regression for vertical clingers and leapers shows the gradient with a great increase in the calcaneal index per increase in the length of the femur compared to the gradient for arboreal quadrupeds. This also might be a factor for the discrepancy noted between femur length and body size.

The load arm of the calcaneum is greatly elongated in various Rodentia, Lagomorpha, the soricid *Suncus,* and the talpid *Uropsolis,* to mention a few forms among terrestrial nonprimates. Load arm elongation correlates with leaping but not with vertical clinging.

From Gregory's (1920) reconstruction of *Notharctus osborni* (A.M.N.H. No. 11474) the appearance of the first digit, in comparison to the remaining ones, may well be an enlarged one. According to his descriptions the distal phalanx is more expanded than in *Lemur* though not as much as in *Lepilemur.* He states that the first metatarsal is shorter than in *Propithecus,* but all the metatarsals are shorter than in *Propithecus.* However, in the pedal elements attributed to *Adapis parisiensis,* certainly the opposite trends are shown in either of the two characters (see above).

In our examinations of the pes we found a character which may correlate with vertical clinging and leaping locomotion in the lemuriforms, or probably more closely with lever functions on a vertical substrate. According to the degree to which the calcaneal cuboid facet is inclined posterodorsally, the calcaneum may be variably bent dorsally on the fore pes. This may be better developed in known lemuriforms of Walker's (1967) vertical clinger and leaper category, in the living Indriidae, and in *Lepilemur* rather than in the lemuriform arboreal quadrupeds. However, to a lesser degree this inclination may also occur in *Daubentonia.* Perhaps, this feature enables the fore pes to remain in contact with the substrate while the plantarflexors are given a maximum heel advantage during a vertical resting phase or in the initial phase of propulsion from a vertical support. Dorsiflexions of the tarsometatarsal joints would also facilitate these functions. Neither *Notharctus, Adapis,* nor *Leptadapis* show the modifications noted. In a number of ways the living lemuriforms show a configuration of the tarsus adapted to stability. These include alternating of the tarsal elements in the Lemuridae, enlargement of the flat surfaces in the calcaneocuboid articulation in the Indriidae, and, in both, depth of the navicular concavity for the head of the astragalus, and one or another mechanism whereby the cuboid and mesocuneiforms are in contact beneath the ectocuneiform. In the development of these characters *Notharctus* apparently has not reached the level of improvement of the living lemuriforms, although apparently the living lemurids are closer to *Notharctus* in this respect than is *Daubentonia madagascariensis.* The tarsus alternates less than in Lemuridae, and the flat surfaces of the calcaneocuboid joint are less enlarged than in the Lemuridae or Indriidae.

The navicular concavity for the head of the astragalus is shallower, and cuboidomesocuneiform contact is largely inhibited by the ectocuneiform. *Daubentonia* exemplifies a different condition. Certain of the subfossil lemuriforms, e.g., *Archaeolemur*, exhibit the *Daubentonia*-like features. In *Notharctus*, to a greater degree in *Daubentonia*, and to some degree in *Archaeolemur*, the trochlear borders of the astragalar body diverge anteriorly. This condition may be resilient during dorsiflexions of the pes.

Summary

An analysis of the structure and function of the adapid lemuriform pes is presented with particular emphasis on the major tarsal joints. Though pertinent observations and tentative interpretations are made regarding elements of the pes of *Adapis* and *Leptadapis*, only the notharctine pes is sufficiently known to treat it as a complex. In the known morphology of the pes of *Pelycodus* no differences from that of *Notharctus* exist which might not be attributed to allometry.

A significantly consistent character of the lemuriforms is the form of the posterior trochlear shelf as well as the presence of a shelflike projection of the astragalar body for the fibular malleolus, largely absent in paromomyiforms. These developments are probably associated with the flexible nature and the power exerted at the upper ankle joint during plantarflexion.

The adapids as well as later lemuriforms possess a pivot type of calcaneocuboid joint, increasing the total potential for rotations in the middle of the tarsus along the long axis of the pes. This represents an advance in comparison to the condition in late Cretaceous eutherians and in the majority of paromomyiform primates, some of which, however, may possess an incipient pivot. The pivot-type of calcaneocuboid joint, developed first in marsupials, is also basic structural innovation in the primates, though it is modified in terrestrial forms and in *Archaeolemur*.

Adapis, Leptadapis, and notharctines differ from later lemuriforms in the presence of some probably primitive characters of primates in having a peroneal tubercle, and a high, narrow lateral side to the trochea. Changes in the peroneal tubercle are assessed as correlates of changing directions of the forces required by the calcaneocuboid and lower ankle joints of the peroneal muscles resulting from elongation of the tarsus, as well as changes in the nature of the calcaneocuboid joint. In respect to one specialization, the extent of the posterior trochlear shelf of the astragalus, *Notharctus* is better developed than later lemuriforms, including the development of a larger posterior fibular facet suggesting a remarkable force during plantarflexion requiring well developed muscles. Compared to living lemuriformes (except *Daubentonia*), the construction of the tarsus of *Notharctus* is less efficiently rigid in the middle and in the cuneiform complex.

The axes of rotation of the major tarsal joints were determined for *Notharctus, Propithecus,* and *Archaeolemur* and were found to be not significantly different except possibly at the calcaneocuboid joint. Evidence from the differential enlargement of the facets suggests a less habitually inverted pes in *Archaeolemur* and this is a possible correlate of habitual existence on a terrestrial substrate.

In light of the present knowledge about the adapid pes doubts are raised concerning some alleged correlates of vertical clinging and leaping suggested for Eocene primates.

Acknowledgments

We thank Dr. R. van Gelder of the American Museum of Natural History for his permission to study skeletons of the extant lemuriforms in his care, Drs. Clayton E. Ray and Robert J. Emry of the National Museum for the loan of several specimens of *Notharctus,* Dr. A. Cavaillé of the Montauban Museum for loan of some specimens of *Adapis,* Dr. J. Hürzeler of Basel for loan of specimens of *Adapis,* and Dr. Pierre Verin of the Musée d'Art et Archaeologie of Tananarive, Madagascar for his gracious aid to Szalay in Madagascar. We thank Miss A. J. Cleary, who skillfully prepared our illustrations and Miss M. Siroky for expert technical assistance.

This research was supported by NSF Grant GS 32315, a grant for African studies by the Wenner-Gren Foundation for Anthropological Research, Grant No. 2686, and a CUNY doctoral Faculty Research Award, all to Szalay.

References

Barnett, C. H. and Napier, J. R. (1953). The rotary mobility of the fibula in eutherian mammals. *J. Anat.* **87**, 11–21.

Filhol, H. (1883). Observations relatives au Memoire de M. Cope intitulé: Relation des horizons renferment des Debris d'Animaux Vertebrés fossiles en Europe et en Amerique. *Ann. Sci. Geol.* **14**, 1–51.

Gregory, W. K. (1920). On the structure and relations of *Notharctus. Mem. Amer. Mus. Natur. Hist* 3 [N.S.], 49–243.

Hafferl, A. (1929). Bau und Function des Affenfusses. Ein Beitrag zur Gelenk und Muskelmechanik. *Z. Anat.* Entwicklungsgesch. **90**, 46–51.

Hall-Craggs, E. C. B. (1966). Rotational movements in the foot of *Galago senegalensis. Anat. Rec.* **145**, 287–294.

Lamberton, C. (1939). Contribution a la connaissance de la faune subfossile de Madagascar. *Mem. Acad. Malagache* **27**, 5–203.

Martin, R. D. (1972). Adaptive radiation and behavior of the Malagasy lemurs. *Phil. Trans. Roy. Soc. London* **B264**, 295–352.

Matthew, W. D., and Granger, W. (1915). A revision of the lower Eocene Wasatch and Wind River faunas. Part IV, Entelonychia, Primates, Insectivora (part). *Bull. Amer. Mus. Natur. Hist.* **34**, 429–483.

Napier, J. R. (1967). Evolutionary aspects of primate locomotion. *Amer. J. Phys. Anthropol.* **27**, 333–342.

Napier, J. R., and Walker, A. C. (1967a). Vertical clinging and leaping—a newly recognized category of locomotor behaviour of primates. *Folia Primatol.* **6**, 204–219.

Napier, J. R., and Walker, A. C. (1967b). Vertical clinging and leaping in living and fossil primates. *In* "Progress in Primatology" (D. Starck, R. Schneider, and H.-J. Kuhn, eds.), pp. 66–69. Gustav Fischer Verlag, Stuttgart.

Reed, C. A. (1951). Locomotion and appendicular anatomy in three soricoid insectivores. *Amer. Midl. Natur.* **45**, 513–671.

Stehlin, H. G. (1916). "Die Säugetiere des schweizerischen Eocaens. Critischer Catalog der Materialien" Vol. II, second half. *Abh. Schweiz. Palaeontol. Gesellsch.* **41**, 1299–1552.

Szalay, F. S. (1972). Paleobiology of the earlies primates. *In* "The Functional and Evolutionary Biology of Primates" (R. Tuttle, ed.), pp. 3–35. Aldine-Atherton, Chicago and New York.

Walker, A. C. (1967). Locomotor adaptation in recent and fossil Madagascan lemurs. Thesis, Univ. of London, London.

Walker, A. C. (1970). Post-cranial remains of the Miocene Lorisidae of East Africa. *Amer. J. Phys. Anthropol.* 33 [N.S.], 249–262.

10

Electromyography of Forearm Musculature in Gorilla and Problems Related to Knuckle-Walking

RUSSELL H. TUTTLE
and
JOHN V. BASMAJIAN

Introduction

Electromyographical techniques are used routinely in clinical medicine and elaborate multichannel electromyographical studies have been conducted on human posture, locomotion, and other motor functions (Basmajian, 1967). By contrast, electromyography has received scant employment in research on the functional and evolutionary biology of man's nearest relatives, the great apes, even though such studies were recommended more than 20 years ago (Washburn, 1950).

The juxtaposition and close collaboration of Yerkes Regional Primate Research Center, which houses the world's largest comprehensive collection of great apes, and an electromyographical research unit at Emory University in Atlanta, Georgia, have made pongid electromyographical studies both practical and productive.

We will first outline our initial attempts to conduct electromyographical studies on gorillas, stressing the uniqueness of our approach by comparison with electromyographical studies on man.

Second, we will discuss preliminary results on selected forelimb muscles, especially as they relate to several questions about the mechanics of knuckle-walking.

Finally, we will review contemporary theories on hominoid phylogeny in light of revised concepts of knuckle-walking and knuckle-walkers.

Electromyographical Studies of *Pan gorilla*

Selection and Habituation of Subjects

In electromyographical studies of normal human activities, investigators obtain cooperative communicative subjects among their colleagues or by hiring other volunteers. The human subject requires, at most, a brief explanation about the prospective, negligibly painful assault that will be performed on him. In electromyographical studies on great apes, by contrast, the selection and habituation of suitable subjects is a much more lengthy and challenging segment of the project.

We elected gorillas, instead of chimpanzees, for our pilot investigations because they were reported to be relatively tractable, even in adulthood (Schaller, 1963; Groves, 1970; Fossey and Campbell, 1970). Further, gorillas probably exhibit the most advanced development of the knuckle-walking complex (Tuttle, 1969b, 1970). From the colony of fifteen subadult and young adult gorillas at Yerkes Regional Primate Research Center we chose five prospective subjects. Our selection was biased toward the smallest animals that were not involved in other projects. One of us (R.T.) conducted feeding-habituation sessions wherein the gorillas were fed from positions outside their home cages. The feeding-habituation period was characterized by the following features and events.

Fifty-one hours were spent in close proximity with the subjects. The average times per day that the investigator spent with them during morning (milk and biscuits), midday (cabbage and carrots) and afternoon (oranges, white or sweet potatoes, onions, and bean cake) feedings are shown in Table I. The investigator offered food bit by bit to the animals, often placing it directly in their mouths. Every opportunity was taken to pet and to talk at the subjects. The investigator did not punish

TABLE I

Approximate Distribution of Time during Forty-One Day Feeding-Habituation Period

Feeding	Number of days	Average minutes per day	Total (hours)
Morning	40	20	13½
Midday	29	23	11
Afternoon[a]	39	41	26½
Corridor	18	34	10¼

[a] Includes time in the corridor with G-26.

them for misdeeds and pranks, and he would exit promptly if fights occurred between cagemates.

On the fourteenth day and on many succeeding days of the habituation period, the investigator entered an outdoor corridor with one 4-year-old female gorilla (G-26). Her afternoon meal was given to her there and she was petted and cuddled whenever she solicited these attentions. Invitations to rough-and-tumble play were refused by the investigator. On two occasions, early in the corridor habituation period, the subject became excessively active and bold. The investigator left the corridor into a room where he was out of the subject's view. Thereupon her vigorous activities ceased, and she remained calm and solicited cuddling when the investigator reentered the corridor.

On the basis of our habituation experiences and discussions with caretakers who had long-term associations with the gorillas in the nursery and large animal wing, we decided to work intensively with subject G-26 until we developed a technology especially adapted to gorillas. Then we would attempt also to work with subjects G-22 and G-24, which were females between 5 and 6 years old. The other two gorillas were judged unsuitable for our study because of potential hazards to the delicate equipment and to persons who would be with them.

In the second phase of habituation, we faced problems of ensuring that the subject would engage in relatively normal knuckle-walking and other behavior in the experimental setting, and that it would not damage equipment on its body and in the recording booth. Our initial experiments on G-26 showed that we had been unduly concerned about such matters. G-26 was introduced into the testing area only once before we conducted electromyographical experiments on her. Although she was clearly intimidated by her new surroundings and curious about the gadgetry and apparel on her body, she soon began to explore and to move freely about the room between bouts of cuddling with the investigator.

While still groggy from anesthesia, she intermittently bit at the apparatus on her right arm and forearm, but she also nibbled at her left forelimb which bore no equipment. In order to keep her from damaging the equipment, we placed a terry cloth towel against her mouth and gently but steadily pushed her face away from it.

By the fourth experiment, G-26 was trained virtually not to dislodge electrodes or to damage equipment. G-26 was very cooperative as we removed equipment from her so that we did not have to anesthetize her at the end of experiments.

Our attempts to habituate subjects G-22 and G-24 to our procedures were unsuccessful. They damaged equipment and threatened the investigator.

Thus, at the end of the first year of pilot studies on gorilla, we report excellent technical success with one 4-and-½-year-old female subject and much less success with two older subjects. We may renew efforts to habituate larger gorillas to our experiments when technical and operating room procedures are refined to the extent that results of the kind that are obtained with G-26 may be anticipated with them.

Experimental Setting, Equipment, and Procedures

Our experiments were conducted in a specially constructed, air-conditioned trailer consisting of an exercise area 3.6 × 3.9 m (Figs. 1–3) and a separate recording booth.

In our pilot studies we employed two somewhat different recording systems. In phase 1, comprising seven experiments on G-26, we recorded the activities of two forearm muscles during each session.

The subject was injected in the hip muscles with 10 mg/kg body weight of Ketelar [*dl*-2-(*O*-chlorophenyl)-2(methylamino)cyclohexanone hydrochloride]. As the drug took effect, the subject was transported to the operating room where Karma fine-wire bipolar indwelling electrodes (Basmajian and Stecko, 1962) were implanted in selected muscles (Fig. 4a). We confirmed placement of the electrodes by manipulating the joints which the muscles cross before withdrawing the implantation needles (Fig. 4b).

After the needles were withdrawn, the free ends of the electrodes were connected by special encased metal springs to separate differential preamplifiers (Fig. 5). The preamplifiers were taped to the shaved surface of the forearm. A brass plate was cemented and taped onto the forearm in order to ground the system. Lead wires from the preamplifiers were taped loosely along the arm and shoulder in a manner that permitted free movement across joints. Then all equipment was covered by loose wrappings of elastic bandage. The subject was dressed in a tailored one-armed denim jacket (Figs. 1–3). The proximal ends of the lead wires and one component of a connector were drawn through a hole into a pocket on the back of the jacket (Figs. 1a and d).

The subject was transported to the trailer where the two components of the connector were joined. The connector was attached to equipment in the recording booth by approximately 8 m of flexible ribbon cable that allowed complete freedom of movement by the subject (Fig. 3a) and slack was drawn quickly into the recording booth by the staff.

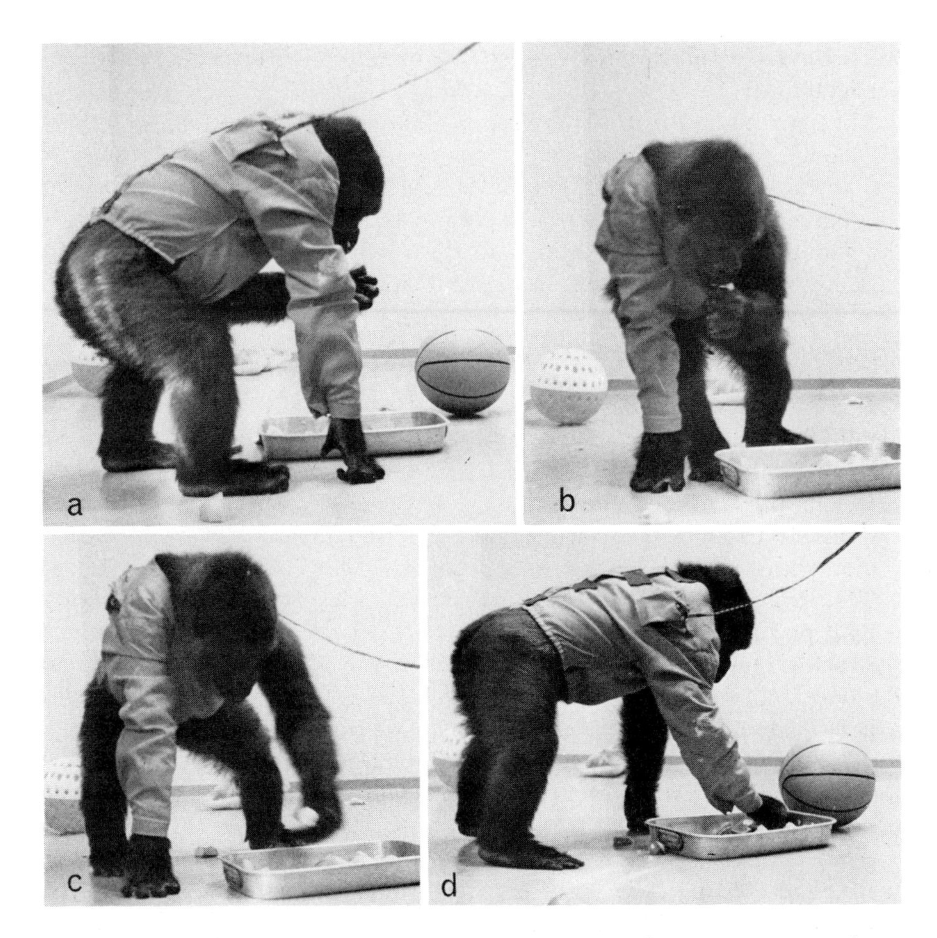

Fig. 1. G-26 executing tripedal stances while feeding during an EMG study on the right forearm. Right digits II–IV contact the floor in (a) whereas right digits II–V contact the floor in (b) and (c). Note pocket on back of jacket (a and d) which contains connector. (F. Keirnan, Yerkes RPRC.)

Two channels of electromyogram, one channel of reference pulse and one channel of narration, were calibrated and monitored on a Model 564B Tectronix oscilloscope and recorded permanently on a Hewlett Packard tape recorder. Narration of the investigator and behavior of the subject were recorded on a Model VR 7000 Ampex Video Recorder. All systems ran continuously from the time the subject was connected to the cable until termination of the experiment, generally after discontinuance of all electrodes.

Subsequent to each experiment, visicorder records were produced from

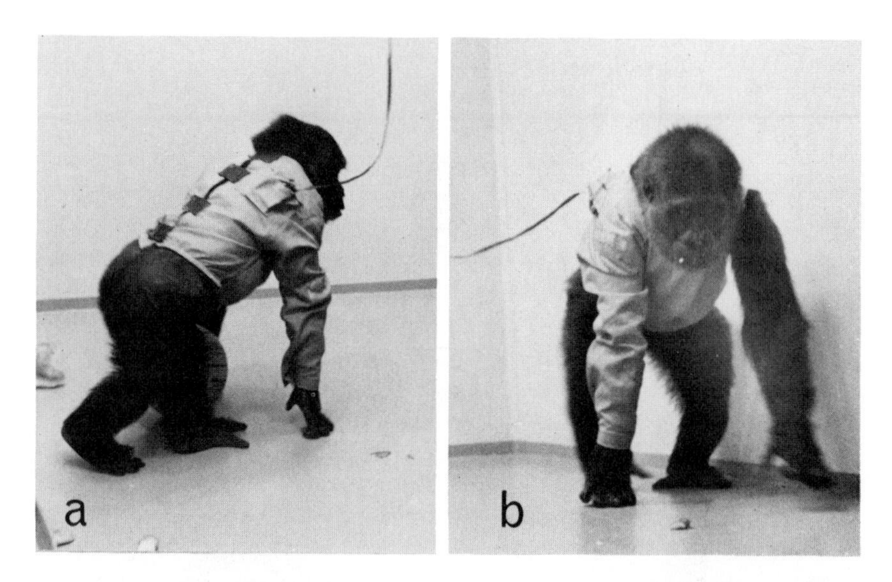

Fig. 2. Knuckle-walking locomotion by G-26. Note that digit V is not in contact with the floor (b). (F. Keirnan, Yerkes RPRC.)

the FM tapes. These were analyzed for details of EMG potentials and graded qualitatively according to the following arbitrary scale: very marked, marked, moderate, slight, negligible, and nil activities. The entire visicorder record of each experiment was examined in order to locate the most prominent activities of each muscle. Unlike experiments with human subjects, we could not command the gorillas to "make a muscle" or otherwise produce maximum contractions at the beginning and at the end of each session. The bursts of EMG potentials were scored on the basis of frequency and magnitude according to methods described by Basmajian (1967, pp. 46–47). Sound channels on the FM and video tapes were synchronized so that particular behavioral episodes could be related to EMG potentials on the oscilloscope and visicorder records.

The technique employed in experiments of phase 1 possessed notable inadequacies. We sought to correct them by developing a more sophisticated technology which we employed in phase 2, comprising (for the purpose of this paper) three experiments on G-26. The springs on the preamplifiers that we used in phase 1 were difficult to manipulate in the time available in the operating room and because the subject sometimes moved unexpectedly. Further, the electrodes sometimes broke at or near their connections with the springs. Therefore, in phase 2, we attached

Fig. 3. Bipedal walking (a) and bipedal stance (b) by G-26. Note subject's tolerance of cable draped across her head (b). (F. Keirnan, Yerkes RPRC.)

the electrodes to contact plates on the preamplifiers by a conducting metallic suspension in quick-drying glue (Fig. 6). The electrodes were capped with plastic tape. This mode of connecting electrodes to preamplifiers has proved very durable and relatively easy to employ.

In order to increase the amount of data collected in a given experiment during phase 2, we increased the number of preamplifiers from two to five and interposed a channel selector into the recording system. This enabled us to record simultaneously the activities of any triad of muscles from among the five in which electrodes were placed and to change the composition of the triad literally "at the flick of a switch."

We reduced the size of the preamplifiers, thereby permitting them to be accommodated comfortably on available surface of the forearm.

Finally, and perhaps most importantly, we employed a splitscreen Panasonic television outfit that recorded the oscillogram of muscle activity in the upper right-hand quadrant of the video tape while ongoing

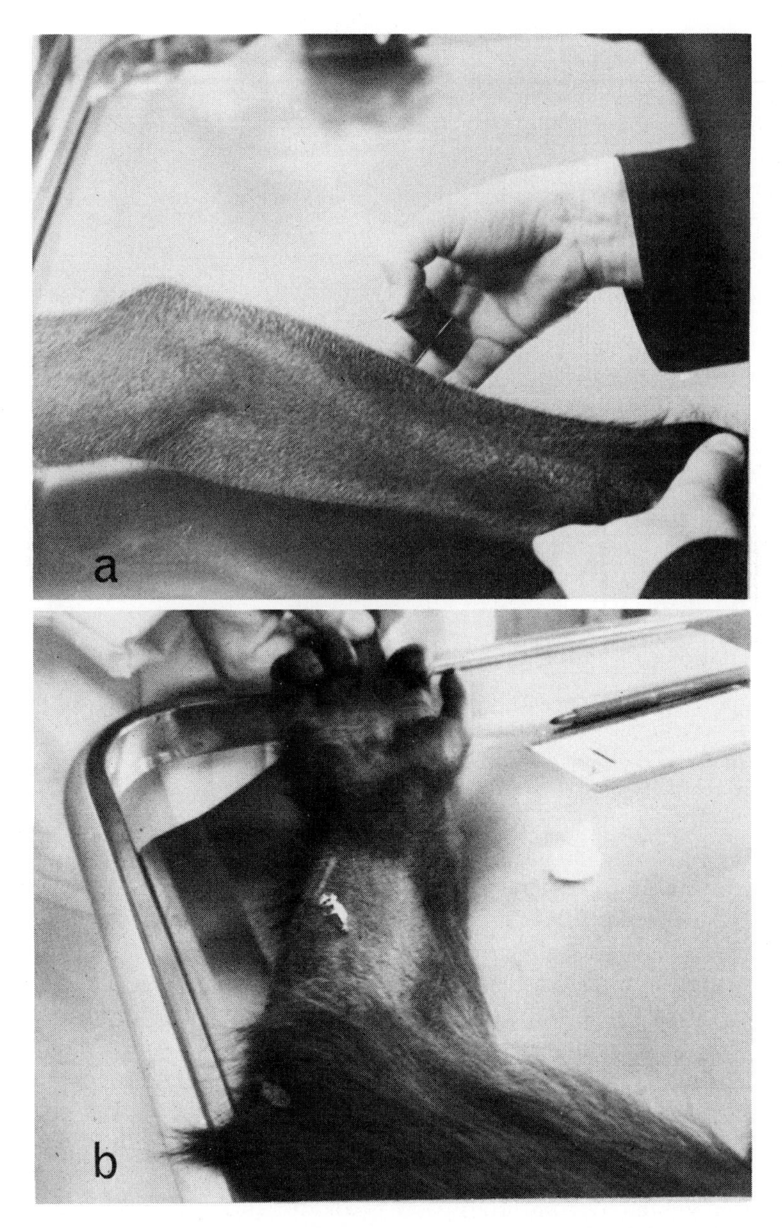

Fig. 4. (a) Insertion of electrode into right forearm. (b) Manipulation of right digit III in order to establish placement of electrode in long digital flexor muscle.

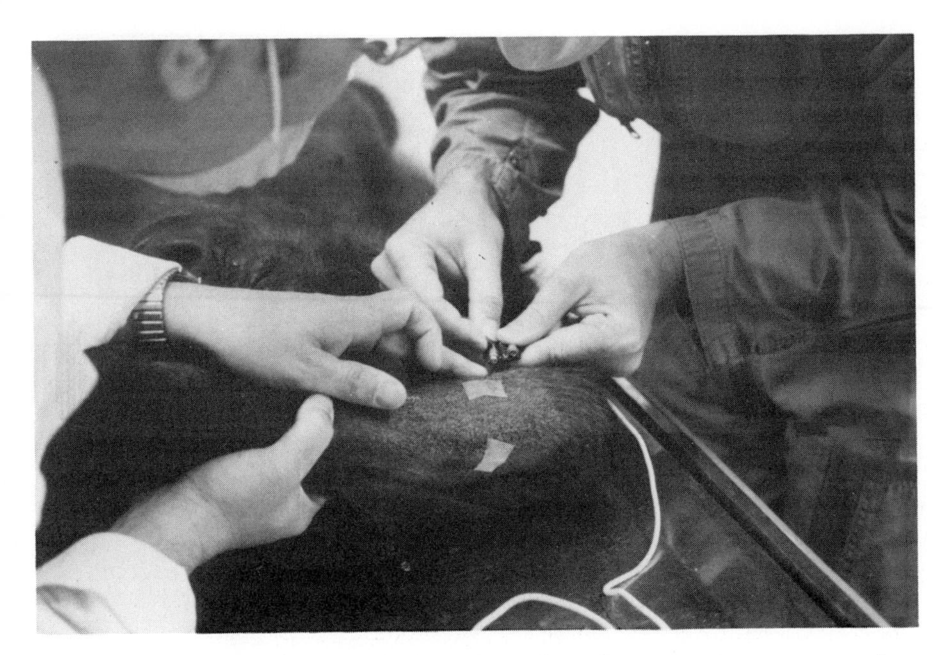

Fig. 5. Attaching the proximal end of an electrode to a spring connector of a differential skin preamplifier in a "phase 1" experiment.

behavior of the subject was recorded on the remaining area of each picture. Thus, we correlated precisely the behavior of the subject and levels of muscle activity as represented on the oscillogram and as scored on the visicorder records.

Preliminary Results

The methods that we employed in phases 1 and 2 can produce EMG records that are as good as the best electromyograms from normal human subjects. However, the frequency with which we achieve such success, especially during periods when the gorilla is fully alert, is less in our experimental situation than in some human electromyographical studies.

One further important qualification that must be kept in mind when considering the inferences and provisional conclusions that we proffer on the basis of our preliminary results is that we have studied selected muscles in one limb of a single subject. Ordinarily in human electromyographical studies no less than six subjects are required to test hypotheses on the principal activities of muscles and, if possible, proper attention must be given to asymmetries and sex and age variations.

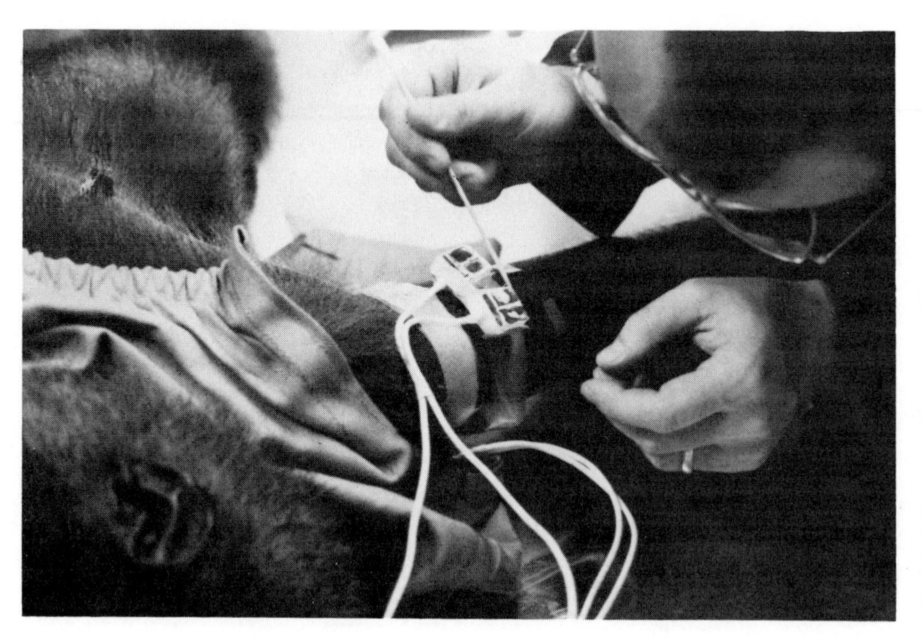

Fig. 6. Attaching the proximal end of an electrode to the contact plate of a differential skin preamplifier with a silver suspension in glue during a "phase 2" experiment.

This report is focused on the activities of five forearm muscles: flexor carpi radialis, flexor carpi ulnaris, extensor carpi ulnaris, flexor digitorum superficialis, and flexor digitorum profundus. We will describe patterns of knuckle-walking in our experimental subject and then discuss EMG results associated with these behaviors.

Behavior

Knuckle-walking by G-26 may be considered arbitrarily under two sets of circumstances—prerecovery and postrecovery from anesthesia—though both are parts of a continuum.

During the initial phase of each experiment, her locomotion was distinctly incoordinated. But once she rose fully into a knuckle-walking posture, she could execute relatively extensive series of short steps with a diagonal-sequence, diagonal-couplets gait before falling down. She employed not only knuckle-walking postures but also modified palmigrade, fist-walking, and other hand postures that are atypical of alert gorillas.

During early and intermediate recovery intervals, the subject typically knuckle-walked with her forelimb outstretched (protracted and abducted at the shoulder) and with her elbow stiffly extended so that her hand was at a sharp angle to the floor. There was notable wobbling, especially in side-to-side directions, at the wrist during early prerecovery knuckle-walking.

As the subject recovered more fully, she employed her forelimb as a well-aligned supporting column under the shoulder (Fig. 2). The elbow now flexed smoothly during the swing phase of progression. The hand was maintained in a position more or less perpendicular to the floor during the support interval of the stance phase of progression (Fig. 2). The wrist was not volarflexed markedly during the swing phase. Elbow flexion apparently provided adequate elevation of the hand so that it cleared the floor in swing phase.

The subject characteristically employed manual digits II–IV during the support interval of the stance phase (Fig. 2). Manual digits I and V generally did not touch the floor during knuckle-walking. The wrist was notably adducted and the load was often carried primarily on digits II and III during support intervals of progression and stance.

Initially the subject moved with very short erratic steps, frequently sliding her feet and sometimes also her hands along the floor. Subsequently, as the effects of anesthesia waned, she took longer strides, increased the tempo of her movements and sat and reclined less frequently between bouts of locomotion. Toward the end of an experiment she ran, walked bipedally (Fig. 3), and display stomped with hands and feet.

Flexor Carpi Radialis

Activities of the flexor carpi radialis muscle were recorded successfully in three experiments. Excellent results were obtained during one experiment in phase 1 (75 minutes) and during one in phase 2 (65 minutes). Good to excellent results were obtained during the initial 20 minutes of a second experiment in phase 2.

In phase 1, moderate and marked potentials were prominent during the first 40 minutes of knuckle-walking, but slight potentials increased in incidence after the first 10 minutes of knuckle-walking progression and stance. Negligible potentials did not occur during the stance phase of progression or during periods of quiescent stance wherein the knuckles of one or both hands were supportive.

In the first experiment in phase 2, marked potentials were prominent during the initial 34 minutes of knuckle-walking progression and stance. The subject was quite groggy at that time. Marked potentials in the flexor carpi radialis muscle were particularly notable during attempts to rise into a knuckle-walking posture from prone and sitting positions, while standing tripedally to feed and manipulate objects with the left hand, and during quadrupedal stances wherein weaving was evidenced or the hands were angled to the floor or both.

Moderate potentials were prominent throughout the experiment when the subject walked at slow and moderate paces, employed tripedal and quadrupedal knuckle-walking stances, and leaned forward between her knuckle-walking hands to reach objects on the floor with her mouth.

Slight potentials began 6 minutes after the onset of knuckle-walking and increased in incidence thereafter. To some extent, especially early in the experiment, slight potentials may be attributed to the fact that the subject leaned against a wall during progression and swayed during stance so that the load on the hand was intermittently lessened. Toward the end of the experiment, one bout of knuckle-walking at a brisk pace was characterized by nine stance phases that were consistently accompanied by slight potentials.

Slight potentials often interspersed with negligible or nil activities in the flexor carpi radialis muscle accompanied sitting–resting or sitting–feeding with the right hand resting in knuckle-walking postures.

Negligible or nil potentials in the flexor carpi radialis muscle cannot be associated with any knuckle-walking progression or stance in which the right hand was load bearing except perhaps during the first strike in a rapid sequence of locomotion and initially in a tripedal stance toward the end of the experiment.

Nil potentials were, however, characteristic of the few fist-walking postures employed during quadrupedal progression and during all modified palmigrade resting postures.

During the second experiment in phase 2, marked potentials were rare. They occurred once, during the first period of knuckle-walking as the hand slid forward on the floor and 15 minutes later during several strides of a knuckle-walking sequence at moderate pace. Moderate potentials were predominant during the entire 20-minute period of intermittent progression and stance. Slight potentials occurred sporadically from the outset of knuckle-walking but they were never especially prominent during progression or stance. Sitting–resting was regularly accompanied by slight potentials, interspersed with negligible potentials. Several rises to knuckle-walking posture from prone and sitting postures were accompanied by slight, as well as moderate potentials. Negligible and nil potentials in the flexor carpi radialis muscle did not occur during knuckle-walking progression or stance, but they always accompanied the few fist-walking stance phases and more frequent modified palmigrade resting postures.

There was no evidence of EMG potentials for the flexor carpi radialis muscle during the swing phase of knuckle-walking progression.

Flexor Carpi Ulnaris

Activities of the flexor carpi ulnaris muscle were recorded in three experiments. Excellent results were obtained during two experiments, each lasting 75 minutes, during phase 1. Poor to good results were obtained during 75 minutes in one experiment in phase 2.

In the first experiment in phase 1, moderate and marked potentials were prominent in the flexor carpi ulnaris muscle during the first 25 minutes of knuckle-walking. No negligible potentials were exhibited during the stance phase of progression, but during relatively long intervals of quiescent stance, including some tripedal ones, the flexor carpi ulnaris showed negligible potentials.

In the second experiment of phase 2, the flexor carpi ulnaris muscle exhibited more prominence of moderate potentials during the inital 25 minutes than subsequently. No marked potentials were evidenced during knuckle-walking.

Negligible potentials occurred in the flexor carpi ulnaris muscle during several cycles of slow, erratic knuckle-walking. Further, more negligible potentials were shown during episodes of quiescent stance in experiment 2 than in experiment 1.

In phase 2, moderate potentials were prominent in the flexor carpi ulnaris muscle during the first 20 minutes of knuckle-walking. The few marked potentials were limited to the first 18 minutes of knuckle-walking and were associated with rising into knuckle-walking postures and with single steps during progressions otherwise characterized by moderate and slight potentials. Slight potentials occurred during the first 15 minutes of knuckle-walking and increased remarkably after about 20 minutes. Likewise, negligible potentials, which were absent during the first 20 minutes of knuckle-walking, increased notably thereafter, not only during quiescent stance but also during the stance phase of progression.

There was no evidence for EMG potentials in the flexor carpi ulnaris muscle during the swing phase of knuckle-walking progression or during fist-walking and modified palmigrade postures.

Extensor Carpi Ulnaris

Activities of the extensor carpi ulnaris muscle were recorded in one experiment of phase 1, lasting 75 minutes and producing excellent results.

Nil, negligible, and occasional short bursts of slight potentials in the extensor carpi ulnaris muscle accompanied the stance phase of progression at slow and moderate paces. No EMG potentials were evidenced during quiescent stances.

Slight or moderate potentials occurred during some swing phases of knuckle-walking, but usually the swing phase was accompanied by negligible potentials.

The generally low level of activity in the extensor carpi ulnaris muscle during knuckle-walking locomotion and stance was notably in contrast to the moderate and marked potentials that it exhibited during certain manipulatory behaviors such as elevating food to mouth with the wrist adducted and extended slightly.

Flexor Digitorum Superficialis

Activities of the flexor digitorum superficialis muscle were recorded in three experiments. Excellent results (during 55 minutes) and good results (during 60 minutes) were obtained in two experiments of phase 1. Poor to good results (over 65 minutes) were obtained in one experiment of phase 2.

During the initial 25 minutes in the first experiment of phase 1, some marked potentials occurred in fasciculus III of the flexor digitorum superficialis muscle. Marked potentials were exhibited not only during knuckle-walking progression but also during stances in which the subject was still groggy from anesthesia.

Moderate potentials were predominant throughout the experiment. Feeding and resting stances generally were accompanied by moderate or slight potentials during the final 30 minutes of the experiment.

In the second experiment of phase 1, fasciculus IV of the flexor digitorum superficialis muscle chiefly exhibited negligible potentials during knuckle-walking progression and stance, but slight, moderate, and marked potentials also occurred during some bouts of brisk locomotion. On several occasions, prehensile actions of the fingers were accompanied by larger potentials than during immediately antecedent bouts of knuckle-walking. During one sequence in which moderate and marked activities were prevalent, there was evidence for notable propellant flexion of the proximal phalanges of manual digits II–IV prior to lift off.

In phase 2, marked, moderate, and even slight potentials occurred very rarely in the flexor digitorum superficialis muscle during knuckle-walking progression. Two bouts of rapid locomotion were accompanied by slight-to-moderate, moderate, and moderate-to-marked potentials, but several other bouts of rapid locomotion were accompanied by negligible potentials.

Moderate and marked potentials never accompanied tripedal and quadrupedal stances. Slight potentials intermittently accompanied one tripedal stance wherein the load shifted back and forth over the supportive hand.

Thus, throughout the experiment, knuckle-walking progression and stance were characterized by negligible potentials in the flexor digitorum superficialis muscle. This set of results contrasts remarkably with the marked and moderate potentials that occurred during the few instances when the subject elevated and suspended itself from the sill of the observation slot. Further, marked, moderate, and slight potentials accompanied certain scratching actions of the hand and manual retrieval of stuffed toys from the floor.

Flexor Digitorum Profundus

Activities of the flexor digitorum profundus muscle were recorded in three experiments. Good results were obtained during 55 minutes of one experiment and poor results were obtained over 60 minutes of a second experiment in phase 1. Good results were recorded during 60 minutes of one experiment in phase 2.

In the first experiment of phase 1, the fasciculus of the flexor digitorum profundus muscle to digit III exhibited slight and negligible potentials during all episodes of knuckle-walking except one bout in which two successive steps were accompanied by moderate potentials. The only other moderate or greater potentials in the flexor digitorum profundus were produced during manipulatory behavior, e.g., as the subject tightly grasped cloth.

In the second experiment of phase 1, despite certain technical difficulties, the fasciculus of the flexor digitorum profundus muscle to digit III or IV exhibited negligible or nil potentials during all instances of knuckle-walking progression and stance.

In phase 2, the fasciculus of the flexor digitorum profundus muscle to digit III predominantly exhibited slight and negligible potentials during knuckle-walking progression. Moderate potentials also occurred with notable frequency during several bouts of progression at brisk and moderate paces.

All quadrupedal and tripedal stances were accompanied by slight or negligible potentials. Moderate potentials occurred in the right flexor digitorum profundus muscle during one long stance in which the right foot and left hand were employed for manipulation such that only two limbs were supportive.

The relatively low levels of activity in the flexor digitorum profundus muscle during knuckle-walking were dramatically highlighted by several episodes of suspensory behavior, scratching, and manipulation in which strong flexion of the distal phalanges is requisite and which were accompanied by marked or greater potentials.

Summary and Discussion

In the initial phases of our experiments, the proper flexor muscles of the wrist (flexor carpi radialis and flexor carpi ulnaris) generally showed high levels of activity during knuckle-walking progression and stance. As the subject recovered from the effects of anesthesia, they evidenced lower levels of activity during knuckle-walking. Diminution in the level of activity occurred sooner in the flexor carpi ulnaris muscle than in the flexor carpi radialis muscle. Further, whereas the flexor carpi ulnaris muscle eventually exhibited many negligible and nil potentials during knuckle-walking stance and progression, the flexor carpi radialis muscle continued to evidence more prominent activities.

Fist-walking, modified palmigrade postures, and the swing phase of knuckle-walking were not accompanied by noticeable EMG potentials in either the flexor carpi radialis or the flexor carpi ulnaris muscle.

The extensor carpi ulnaris muscle was relatively inactive during knuckle-walking. It sometimes evidenced short bursts of low level activity during the stance phase of progression. Occasionally it exhibited somewhat more prominent bursts of activity during the swing phase of knuckle-walking. No notable activities were observed in the extensor carpi ulnaris muscle during quiescent stance, modified palmigrade postures, or fist-walking.

Less consistent results were exhibited by the flexor digitorum superficialis muscle than by the three muscles summarized above. Fasciculus III of the flexor digitorum superficialis muscle exhibited predominantly moderate and higher potentials in one experiment but mainly negligible potentials in a second experiment. Fasciculus IV of the flexor digitorum superficialis muscle chiefly evidenced negligible potentials in one experiment. We will offer here optional explanations that might account for this inconsistency of results. We defer choosing among them until further studies on the flexor digitorum superficialis muscle are analyzed.

In *Pan gorilla,* the long digital flexor muscles (flexor digitorum superficialis and flexor digitorum profundus) are remarkably subdivided into separate fasciculi for each of the manual digits exclusive of the pollex. These fasciculi are arranged in a complex fashion in the forearm. Thus, the exact location of initial placement of an electrode and particularly whether it remained there throughout an experiment were subject to greater ambiguity in case of the long digital flexor muscles than of simpler, more superficial muscles like the proper flexors of the wrist.

Several factors argue that the experiment in phase 1 on fasciculus III of the flexor digitorum superficialis muscle may be more representative of its action during knuckle-walking than the experiment in phase 2. The quality of results was much better in the experiment of phase 1 than in phase 2 which may indicate improper placement of the electrode in phase 2. Further, in phase 1, fasciculus III of the flexor digitorum superficialis muscle behaved differently from fasciculus III of the flexor digitorum profundus muscle, while in phase 2 its action was reminiscent of the flexor digitorum profundus muscle. Thus, it is possible that in phase 2, the electrode actually was placed in the flexor digitorum profundus muscle.

The fact that results on fasciculus IV of the flexor digitorum superficialis muscle resembled more closely "fasciculus III" during phase 2 cannot be employed to argue with certainty that the action of fasciculus III in phase 1 is atypical of the muscle. Once again the quality of results on fasciculus IV was not as good as that on fasciculus III in phase 1. Further, it is possible that the electrode to "fasciculus IV" of the flexor digitorum superficialis muscle also penetrated to the flexor digitorum profundus muscle or that digit IV is not as prominently supportive as digit III is during knuckle-walking.

The flexor digitorum profundus muscle (fasciculi to digits III and IV) consistently produced relatively low levels of activity during nearly all bouts of knuckle-walking.

Placement of electrodes was more certain in the flexor digitorum profundus muscle than in the flexor digitorum superficialis muscle because in experiments on the former we often penetrated the muscle so deeply that the implantation needle struck the anterior surface of the ulna.

Biomechanical Inferences

Introduction

Despite certain arbitrary features inherent to our method of assessing levels of muscle activity and concomitant difficulties in comparing different muscles within an experiment or the same muscle between experiments, we will proffer some inferences from available results so that they may serve as novel (if not long-lasting) bases for hypotheses on the mechanics and phylogeny of knuckle-walking in *Pan gorilla*. First, however, we will briefly review aspects of knuckle-walking in the African apes (gorilla and chimpanzee).

Modes of Knuckle-Walking

Knuckle-walking is the characteristic hand posture employed by gorillas and chimpanzees during quadrupedal progression and quiescent stance. Digits II–V are flexed so that the dorsal aspect of their middle phalangeal segments contact the ground. These regions are covered by friction skin, forming knuckle-pads which exhibit histological features that are similar to the skin on palms and soles (Fig. 7).

During knuckle-walking the metacarpus of the load-bearing digital rays is nearly aligned with the forearm. However, the wrist frequently evidences a convex dorsal curvature or notable adduction or a combination of both postures during progression and stance (Fig. 8).

The proximal phalanges of digits II–V are hyperextended at the metacarpophalangeal joints so that often a sharp angle is formed between them and the metacarpus during load-bearing activities (Fig. 8). The middle phalanges are flexed at the proximal interphalangeal joints and the distal phalanges are flexed at the distal interphalangeal joints. The pulps on the distal phalanges may be apposed against adjacent proximal segments of the hand, but frequently the finger tips are not in contact with the palmar surface of the hand or the substratum during knuckle-walking. The thumb does not touch the ground.

Many details of knuckle-walking behaviors remain to be studied systematically and quantitatively. For instance, we do not know how age, sex, rates of locomotion, individual biomechanical factors, and features of the substratum are associated with many variable characteristics of the knuckle-walking complex, such as (a) which digits are involved fundamentally in supportive functions, (b) how the hand is angled to the line of progression, (c) how the swing phase of progression is executed, (d) the degree of adduction and dorsiflexion in the wrist, (e) the degree of hyperextension of the metacarpophalangeal joints, or (f) the amount of propellant force that is exerted by the hand. Information on these topics and many more besides are imperative to the future delimitation of distinctions between modes of knuckle-walking in chimpanzees and gorillas.

Fig. 7. Knuckle-walking by a juvenile gorilla. Note the fully pronated posture of the left hand and the hyperextended position of metacarpophalangeal joints II–IV of the left fingers. *Inset:* Knuckle pads over the dorsal surfaces of middle phalanges II–V in a gorilla hand. (Yerkes RPRC; Copyright 1969 by the American Association for the Advancement of Science.)

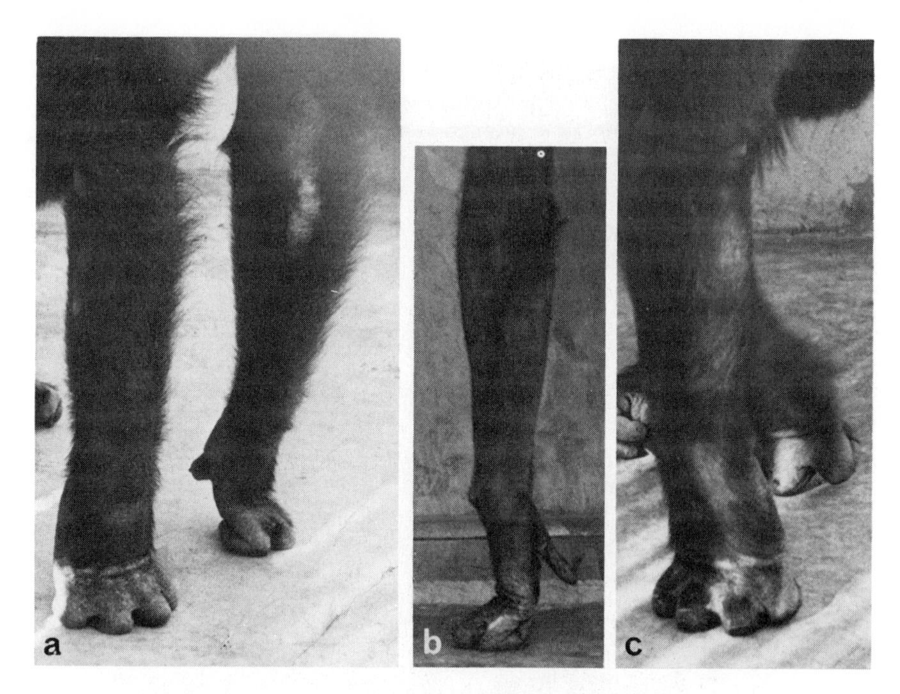

Fig. 8. Quiescent knuckle-walking postures in *Pan gorilla* (a) and *Pan troglodytes* (b) and locomotive knuckle-walking postures in *Pan gorilla* (c). The same subject is in (a) and (c). In (c), the left hand is approximately at maximum stage of load bearing while right hand is in prerelease phase. (Yerkes RPRC; from Tuttle, 1970, pp. 174, 178.)

Role of Muscles in Knuckle-Walking

The prevalence of relatively low levels of activity in the proper wrist flexors and long digital flexor muscles during quiescent stance and progression indicates that close-packed positioning mechanisms may be fundamentally responsible for the integrity of the wrist joint, especially to preserve it against traumatic stresses induced by movement into positions of extreme dorsiflexion during knuckle-walking.

The flexor carpi radialis muscle appears to be particularly important as a factor preventing extreme adduction of the wrist during knuckle-walking behaviors though it also may act concurrently to resist dorsiflexion. We emphasize the activity of the flexor carpi radialis muscle as a safeguard against positioning into extreme adduction since it is strategically placed to serve this function. Marked adduction is possible because of the proximal positioning (termed "retreat" and "withdrawal" by Lewis, 1965) of the distal end of the ulna relative to the distal radius in *Pan gorilla.*

On the basis of available evidence it is impossible to ascertain definitively the extent to which muscles, in contradistinction to or in synergy with bone-ligament mechanisms, are responsible for the integrity of the metacarpophalangeal joints during knuckle-walking. The flexor digitorum profundus muscle, which constitutes approximately 44% of total flexor musculature in the forearm of *Pan gorilla,* is relatively unimportant during the knuckle-walking behavior that we observed in our alert subject. As mentioned above, the role of the flexor digitorum superficialis muscle as a principal support of the metacarpophalangeal joints during knuckle-walking remains ambiguous.

The only other muscles available to prevent traumatic stressing of the hyperextended metacarpophalangeal joints are the intrinsic muscles of the palm, particularly the interossei. These have not been studied electromyographically in *Pan gorilla.*

We are relatively certain, however, from studies of the distal ends and articular surfaces of metacarpal bones II–V in *Pan gorilla* (Tuttle, 1967, 1969a,b, 1970) that special close-packed positioning mechanisms are available to safeguard against traumatic stressing of the hyperextended metacarpophalangeal joints, at least during quiescent stance.

The extensor carpi ulnaris muscle is relatively inactive during knuckle-walking because the same basic posture of the wrist is employed in swing and stance phases. Frequently, the hand is elevated clear of the substrate by flexion of the elbow joint and it seems to be carried forward by actions of the shoulder muscles.

Fist-walking and modified palmigrade resting postures seem to be much less demanding of certain aspects of the wrist and metacarpophalangeal joints than knuckle-walking as evidenced by the virtual absence of noticeable activities in the proper wrist and long digital flexor muscles during these activities.

Evolutionary Inferences

Introduction

Electromyographical and other studies on the muscles of extant apes and man constitute *indirect* sources for inference on the phylogeny of the Hominoidea. Such studies are conducted in order to establish a base for inferring biomechanically feasible models from fossils. Once one or more functional models are inferred from plausible biomechanical explanations of selected fossil structures, they must be tested for compatibility with available information on other bits of the fossil's morphology and contemporaneous environment. Multidisciplinary research and perspectives are imperative if we are to reason from selected activities and structural concomitants of recent apes and man to the fossils and hypothetical intermediaries between them (Tuttle, 1972a).

In order to consider the role that knuckle-walking may have played in hominoid phylogeny, we will establish fundamentally which extant primates engage in knuckle-walking behaviors and the extent to which knuckle-walking is expressed morphologically, especially in bony structures. We will then discuss aspects of selected current theories in light of basic biomechanical inferences that may be gleaned from available electromyographical studies of knuckle-walking in *Pan gorilla* and from other new studies on the behavior and morphology of the Pongidae. We will devote particular attention to theories on hominoid phylogeny advanced by Simons, Pilbeam and their associates, Washburn, Sarich, Tuttle, and Lewis.

Knuckle-Walking and Knuckle-Walkers

Traditionally, knuckle-walking was treated secondarily to brachiation if it were discussed at all. Thus, prior to the 1960's most morphologists devoted only passing attention to specific features that might be related primarily to knuckle-walking. This probably was in large part attributable to the lack of clear recognition that all three great apes (i.e., gorilla, chimpanzee, and orangutan) do not share the knuckle-walking complex or engage in remarkably similar patterns of arboreal locomotion.

Sir Richard Owen (1859, pp. 74–75) designated orangutans, as well as the African apes, "knuckle-walkers" in order to distinguish them from "brachiators" (viz., gibbons). Several other nineteenth-century primate morphologists (Mayer, 1856, p. 285; Gratiolet and Alix, 1866, p. 28; Broca, 1869, p. 26) reported per contra that captive orangutans place their hands on the ground in postures distinct from those of chimpanzees, but their observations evidently did not impress the principal theorists of hominoid evolution. In 1957, Kallner reported and illustrated captive orangutans engaged in nonknuckle-walking hand postures, but again, this observation had no notable effect on models of hominoid phylogeny.

In 1967, Tuttle (pp. 179–186) described the hand postures employed by 26 orangutans that had newly arrived at the Yerkes Regional Primate Research Center in Florida. Many were immature animals that had been captured a relatively short time before shipment to the United States. They exhibited a variety of flexed-finger postures during stance and progression but none of them knuckle-walked. The most common hand postures that they employed were "fist-walking" with the hand aligned with the forearm and "fist-walking" with the wrist remarkably adducted (Figs. 9, 10, and 11). Some individuals, including several young adults (the oldest members of the colony), used fully palmigrade (Fig. 12) and modified palmigrade postures. Repeated cursory surveys of orang-

Fig. 9. Juvenile orangutan exhibiting fist-walking postures in both hands. Note remarkable adduction (ulnar deviation) of right wrist. (Yerkes RPRC.)

utans at Yerkes Regional Primate Research Center and in zoological parks in Europe and the United States did not reveal any instances of knuckle-walking (Tuttle, 1969b) that were reminiscent of gorillas and chimpanzees.

In 1971, Benjamin Beck, Curator of Research at the Brookfield Zoo, observed that "Felix," a 16-year-old orangutan placed his hands in knuckle-walking postures. The following summary of "knuckle-walking" behaviors by Felix is based primarily on pilot studies by Tuttle and Beck (1972). Fuller descriptions on the frequency and patterning of "knuckle-walking" in this singular subject and possible explanations of its development are forthcoming (Susman, in press).

Felix consistently placed his hands in knuckle-walking posture while sitting–resting, sitting–feeding, and engaging in a distinctive mode of squatting progression (Tuttle and Beck, 1972). During sitting–feeding (Fig. 13) and sitting–resting (Fig. 14), Felix characteristically placed his hands with the dorsal aspects of the middle segments of two or more fingers in contact with the substrate. Hyperextension at metacarpophalangeal joints II–V was negligible or very slight. The wrist joints were adducted notably on many occasions.

Felix executed squatting progression (Fig. 15) as follows. He flexed

Fig. 10. Adult male orangutan ("Felix") fist-walking on wet floor. (L. LaFrance, Chicago Zoological Park.)

his thighs markedly and abducted and rotated them laterally so that they were tucked under and often against his abdomen. His knees remained flexed. His toes were flexed and his feet were supinated remarkably such that only their lateral aspects contacted the floor. He maintained his back in a rigid, fully erect posture. Most of the load seemed to have been borne by the feet and ischial tuberosities. Felix moved in this posture by leaning sideways and slightly forward onto one foot and haunch such that the other foot could be elevated a few inches to swing free of the ground. The hand ipsilateral with the swinging foot concurrently swung forward, although their releases and contacts were not precisely coincident. The

Fig. 11. Fist-walking locomotion in a juvenile orangutan compared with knuckle-walking in a juvenile chimpanzee. (1) Right hand just prior to elevation from the ground. (2) Right hand during swing phase of forelimb stride. (3) Right hand of orangutan in contact with the ground. Right hand of chimpanzee just prior to contact with the ground. (4) Support phase of right forelimb. (Drawing by P. Murray; Copyright 1969 by the American Association for the Advancement of Science.)

other hand rested in a knuckle-walking posture in front of the supporting foot. It generally did not bear much weight in excess of the forelimb itself. As the load shifted onto the supporting foot and haunch, the elbow and wrist of the knuckle-walking hand flexed and adducted instead of bracing to support the load.

Sometimes Felix waddled forward with his hands upraised providing little or no augmentary support. Thus, the forelimbs were considered not to constitute essential load-bearing supports during squatting bipedalism (Tuttle and Beck, 1972).

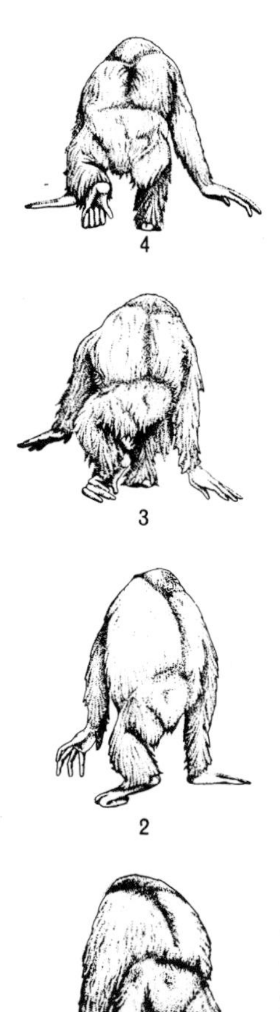

Fig. 12. Posterior view of orangutan engaged in rapid palmigrade locomotion. (1) Right forelimb and left hindlimb serve as supports while left hand is elevated and right foot is swung forward. (2) Right forelimb serves as chief supporting prop just prior to contact by right foot. Left forelimb is swung forward and left hindlimb is elevated. (3) Right hindlimb serves as major supporting prop as left forelimb and left hindlimb are swung forward. Right forelimb is elevated. (4) Left hand contacts the ground. Left forelimb and right hindlimb are principal supports. Right hand and left foot are clear of the ground. (Drawing by P. Murray; Copyright 1969 by the American Association for the Advancement of Science.)

Felix progressed rapidly on dry substrates and sometimes on wet substrates in a semierect quadrupedal fist-walking posture. A diagonal-sequence, diagonal-couplets gait was often employed during fist-walking. He progressed at other times on a wet floor with alternate steps of the hindlimbs but without raising his fists from the substrate (Fig. 10). We once observed him to slide his right hand forward briefly in a "knuckle-walking" posture while his left hand remained a fist (Fig. 16). The load

Fig. 13. Adult male orangutan ("Felix") sitting–feeding with left hand in knuckle-walking posture. Note that digit V is not in contact with the floor. (L. LaFrance, Chicago Zoological Park.)

did not pass over the knuckle-walking hand since his forelimbs were held protracted and abducted throughout the locomotive bout (Tuttle and Beck, 1972).

Felix consistently placed his hands in fist-walking postures when he leaned forward to investigate or to orally pick up objects from the floor, and especially when he descended head first from low elevations to the floor. The most remarkable evidence of Felix bearing notable loads on his knuckle-walking hands appeared in several still photographs (Fig. 17) that were taken when the investigators were not with the photographer. There is no information on events that preceded and followed this sequence. Here we will call attention only to the fact that Felix appeared

Fig. 14. "Felix" sitting–resting with both hands in knuckle-walking postures. Note that digits II and V of right hand and digit V of left hand are not in contact with the floor. (L. LaFrance, Chicago Zoological Park.)

to be sliding his knuckle-walking hands forward such that the load did not pass directly over them. The total impression derived from these pictures is in many ways strikingly reminiscent of knuckle-walking by our gorilla as she attempted locomotion while in early recovery from anesthesia.

In summary, only the African apes are known to engage consistently in knuckle-walking. Although some authors have referred to orangutans as "knuckle-walkers," only one explicitly documented case of facultative knuckle-walking in an orangutan—that reported recently by Tuttle and Beck (1972)—exists in the literature known to us.

Available evidence indicates that "Felix" rarely supported the major portion of his body weight on knuckle-walking hands. He had not been observed to employ his forelimbs as supports directly under his body during knuckle-walking or to place his hands in knuckle-walking postures during crutch-walking as chimpanzees and gorillas customarily do. Instead he placed his hands in fist-walking postures when support was required for loads that were markedly greater than the weight of the

Fig. 15. "Felix" engaged in "bipedal squatting progression." Note how lightly the knuckle-walking left hand seems to rest on the floor even though weight is shifted toward the left side. (L. LaFrance, Chicago Zoological Park.)

Fig. 16. "Felix" progressing on wet floor by sliding his right hand forward in a knuckle-walking posture and his left hand in a fist. (L. LaFrance, Chicago Zoological Park.)

forelimbs themselves. By fist-walking he probably avoided overly stressing metacarpophalangeal joints II–V, which in orangutans lack special morphological adaptations for knuckle-walking of the kind possessed by gorillas and chimpanzees (Tuttle and Beck, 1972).

We believe that in order for locomotor categories to be employed heuristically, the locomotor modes of *Pongo* must be clearly distinguished from those of *Pan*.

Morphological Correlates of Knuckle-Walking

Knuckle-pads, comprised of histologically distinctive friction skin, are the most immediately obvious structures related to knuckle-walking in the African apes (Biegart, 1971; Ellis and Montagna, 1962; Montagna and Yun, 1963; Tuttle, 1967, 1969b, 1970). Knuckle-pads are not developed in orangutans though the dorsal skin over the middle segments of their fingers may sport only scant hair (Schultz, 1933). Despite his facultative knuckle-walking, "Felix" still possesses a notable growth of hair on the areas where knuckle-pads occur in African apes (Fig. 18).

In the present state of research, particular features of the bones, ligaments, and muscles often are more ambiguously associated with knuckle-walking than are knuckle-pads. Major morphological regions that seem to merit special attention in the great apes are the wrist and the metacarpophalangeal joints of digits II–V (Tuttle, 1967 et seq.).

Tuttle (1967) speculated on possible morphological mechanisms for the maintenance of knuckle-walking postures in African apes on the basis

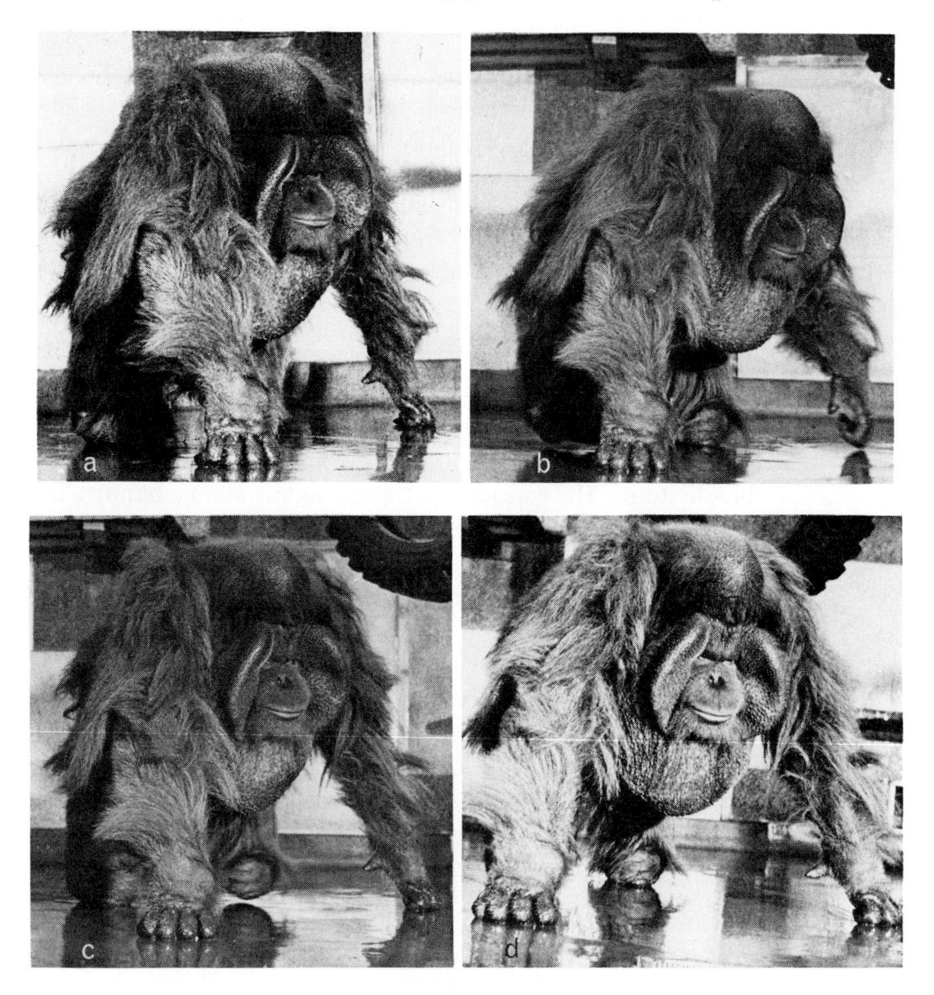

Fig. 17. Knuckle-walking quadrupedal progression by "Felix" on a wet floor. (a) Stance with both hands in knuckle-walking postures. (b) Left forelimb and right hindlimb swinging forward while knuckle-walking right hand and left hindlimb are supportive. (c) Both knuckle-walking hands and right hindlimb supportive as left foot is swung forward. (d) Sliding both knuckle-walking hands forward while both feet are on the floor. (L. LaFrance, Chicago Zoological Park.)

of Schreiber's (1936) description of hominoid wrist joints, and his own studies on (a) hand muscles in *Pan troglodytes* and *Homo sapiens*, (b) disarticulated hand bones of *Pan troglodytes* that he had dissected at Yerkes RPRC and a few museum specimens of *Pan gorilla* and *Pongo pygmaeus*, and (c) passive wrist movements in thirty hands of *Pan troglodytes* and fifty hands of *Pongo pygmaeus*. Tuttle concluded that although synergetic action of flexor and extensor muscles across the wrist joint may provide some support during knuckle-walking, the interaction of bones and ligaments within the wrist are probably the primary factors which prevent it from buckling dorsally, medially, or laterally (Fig. 19). Tuttle (1967, pp. 190–191) suggested that certain features on the ventrolateral aspect of the wrist exemplify one such bone-ligament mechanism. Lewis (1969, 1972a,b) noted that one part of Tuttle's rendering of this particular structural complex is erroneous. Tuttle confused Schreiber's (1936) description of the capsular and intracapsular ligaments in the wrists of *Pan troglodytes* and overgeneralized on the basis of a singular photograph of the lateral aspect of the wrist from one specimen.

Lewis considered the limitations of wrist extension (dorsiflexion) in the African apes to be a "logical consequence of aligning hand and forearm" (1972a, p. 211), presumably coincidental with their heritage and practice as "brachiators." He considered the knuckle-walking habitus to be the result of limited wrist extension which precludes palmigrade

Fig. 18. Left hand of "Felix" showing hairy skin over dorsal aspects of the middle phalanges of digits II–V. (L. LaFrance, Chicago Zoological Park.)

weight bearing. He cited "the relative shortness of the long flexors" as "a potent factor in limitation of wrist extension" in the great apes. Finally, Lewis (1972a,b) confessed an inability to confirm the suggestions of Tuttle (1967, 1969a) that there are other special modifications of the wrist associated with knuckle-walking. Thus, Lewis (1972a, p. 212) concluded that "knuckle-walking requires no especially striking modifications of the wrist joint." Accordingly, clear evidence for the evolution of knuckle-walking should not be expected from the wrist bones of fossil hominoids.

We caution here that neither Tuttle's muddle of the radial collateral and palmar radiocarpal ligaments in *Pan troglodytes* nor the scope and emphasis of Lewis' studies on the hominoid wrist provide sufficient grounds to deny categorically the existence of bony particularities in the wrists of the African apes that may be related to knuckle-walking and that might be used to infer the phylogeny of knuckle-walking in the

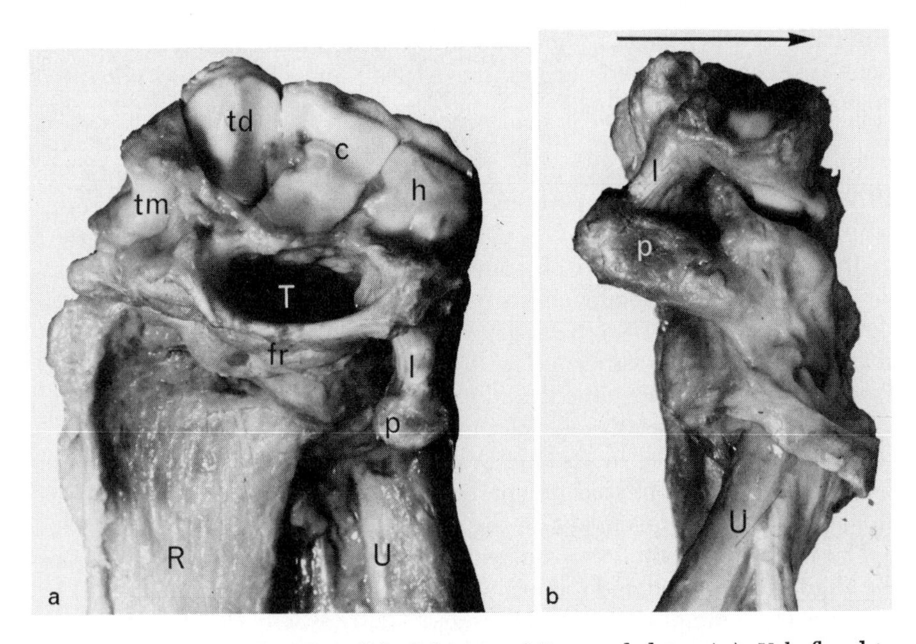

Fig. 19. Fresh preparation of the left wrist of *Pan troglodytes.* (a). Volarflexed to show distal surfaces of the proximal row of carpal bones (tm, td, c, h), and the carpal tunnel (T). (b). Medial (ulnar) aspect. Note the powerful development of the pisohamate ligament (1) which attaches proximally to the elongated pisiform bone (p) and distally to the hook of the hamate bone (h). Direction of dorsiflexion is indicated in (b) by an arrow. (Other symbols: c, capitate; fr, flexor retinaculum; R, radius; td, trapezoid; tm, trapezium; U, ulna.) (From Tuttle, 1970, pp. 200, 212.)

Hominoidea. Clearly, more comprehensive and systematic studies must be conducted in order to identify features of the distal radius, distal ulna, and carpal bones that might be related to knuckle-walking. An obvious research design that would permit distinctions of special knuckle-walking features in the Pongidae would include comprehensive and detailed functional-morphological comparisons between *Pan* (including *Gorilla*) and *Pongo*.

We believe that Lewis passed over somewhat too cursorily certain features in the wrists of knuckle-walking *Pan* that contrast (in some instances quite strikingly) with their counterparts in the wrists of *Pongo*, the arboreal climber and arm-swinger par excellence among the Pongidae (i.e., great apes). For instance, Lewis (1972b) paradoxically concluded that certain features of the carpal bones (viz., the "waisted" appearance of the capitate and early fusion of the os centrale to the scaphoid bone) which are more advanced in the African apes and man than in orangutans, are special adaptations for "brachiation." We should expect that orangutans, which among the great apes engage most frequently in suspensory behavior, would surpass the African apes (which do not often engage in suspensory behavior) in development of these and other structures in the wrist that are purportedly associated with "brachiation." Evidence from the hands (Tuttle, 1969b, 1970), feet (Tuttle, 1968, 1969b, 1970, 1972a,b), bodily proportions (Schultz, 1956; Erikson, 1963), vertebral column (Schultz, 1961), and shoulder girdle (Oxnard, 1967) attest to the preeminent position of orangutans as large-bodied "brachiators." Even exclusion of the ulna from participation in the wrist joint, which Lewis (1969) cited as the hallmark which permitted other morphological changes in the wrist associated with "brachiation," is more advanced in *Pongo* than in other Hominoidea.

The shapes and relative development of yet other carpal bones, e.g., the lunate, also differ in striking ways between *Pongo* and *Pan*. These features might be related to particular suspensory behaviors versus knuckle-walking, respectively, in the two genera.

The logicality of limitations of wrist extension being consequent upon the alignment of hand and forearm in "brachiators" (Lewis, 1972a, p. 211) is based on assumptions that this aligned posture is customary and that there is no need for employment of the hand in a position of notable dorsiflexion in arboreal contexts. Again the orangutan may be cited to caution against such overgeneralization.

Tuttle (1967, 1969a,b, 1970) demonstrated that the wrist of *Pongo* is capable of considerable displacement into dorsiflexion ($\bar{x} = 85°$; 90% fiducial limits, 65–105; $n = 27$). These figures are based on a sample of predominantly subadult (3 to 7 years of age) captive animals (Tuttle, 1969a). Somewhat lower figures might be obtained from free-ranging adult orangutans and from animals that have been caged for considerable periods with little opportunity to climb freely and that customarily fist-walk on the floor.

Tuttle (1967) observed that most hands of *Pongo* were easily moved at least 70° and sometimes more than 90° into dorsiflexion (Fig. 20). One 7- or 8-year-old male orangutan in the Yerkes colony exhibited only 50 (right) and 60° (left) of passive dorsiflexion in its wrists. Several other orangutans that evidenced values below 70° were not deeply anesthetized and actively resisted attempts to manipulate their wrists. Limitation of passive dorsiflexion generally was not correlated with increasing age in the Yerkes colony (Tuttle, 1967). Although the extreme upper value (110°) of dorsiflexion was observed in one of the youngest and the extreme lower value (50°) of dorsiflexion was observed in one of the oldest subjects, more than 80% of the orangutans had wrists that could easily be moved 80–90° into dorsiflexion even when their fingers and elbows were fully extended (Tuttle, 1967). Tuttle (1969a) speculated that free dorsiflexion (that is exhibited by most adult Asian pongids) of the wrist may be of special advantage to orangutans when they climb in peripheral foliage of the canopy. They must draw in from many directions enough branches to support their prodigious weight.

Although studies on captive orangutans may not be completely representative of conditions in free-ranging animals, we doubt that the latter could approach the extreme degree of restriction of dorsiflexion (that is clearly exhibited by African apes) until advanced age or other factors force them to the ground or otherwise restrict their repertoire of arboreal activities.

The potency of the relative shortness of "the long flexors" (i.e., the extrinsic flexor muscles of the manual digits) as factors limiting wrist extension in large-bodied "brachiators" (Lewis, 1972a) may be questioned after careful consideration of conditions exhibited by the Asian apes. Evidence from *Pongo* (Fig. 20) and *Hylobates* (Tuttle, 1967, et seq.), in fact, precludes the assumption that marked "shortness" of the extrinsic flexor muscles of the manual digits is characteristic of all "brachiators" and invalidates the inference that they are related to restriction of extension of the wrist in order to ensure alignment of hand with forearm.

Orangutans and gibbons generally do not exhibit the degree of shortening of the long digital flexor tendons that in African apes prevents full extension of the fingers when their hands are maximally dorsiflexed. In many of the Yerkes orangutans, the fingers were extended fully while the

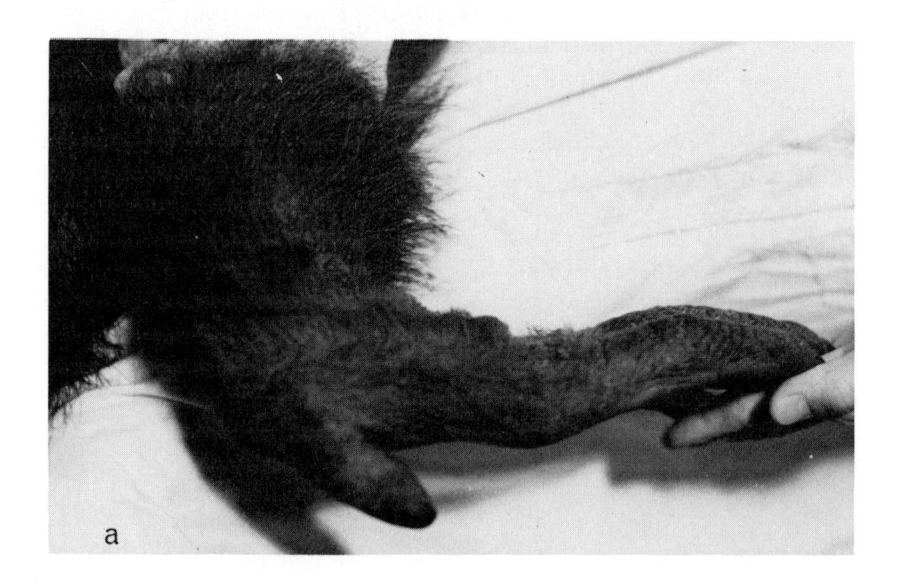

Fig. 20. (a) Left hand of an anesthetized subadult male *Pongo pygmaeus* in maximum passive dorsiflexion with fingers fully extended. (b) Palmigrade resting posture in the right hand of an alert adult *Pongo pygmaeus,* showing its natural capacity for marked dorsiflexion of the wrist and full extension of the fingers. (c) Juvenile orangutan supporting the forepart of its body on a palmigrade right hand.

wrist was held in 80 to 90° of dorsiflexion (Tuttle, 1967, 1969a, 1970; Fig. 20). One would not expect "brachiators" to possess remarkably shortened long digital flexor tendons since it might interfere with disengagement of branches during arm-swinging (Tuttle, 1967). Indeed, the capacity to freely extend the fingers, even when the wrist is markedly dorsiflexed, is probably especially advantageous for orangutans as they climb, forage, and feed in small branch settings (Tuttle, 1969a). Thus, the statement that "the fist-walking of orangutans is doubtless consequent upon the shortness of their long flexors . . ." (Lewis, 1972a, p. 212) is probably incorrect. Instead we suggest that the shortness of the long digital flexor muscles (and of the proper flexor muscles of the wrist) in captive and large free-ranging orangutans is probably a natural outcome of fist-walking as they are induced to engage consistently in such activity.

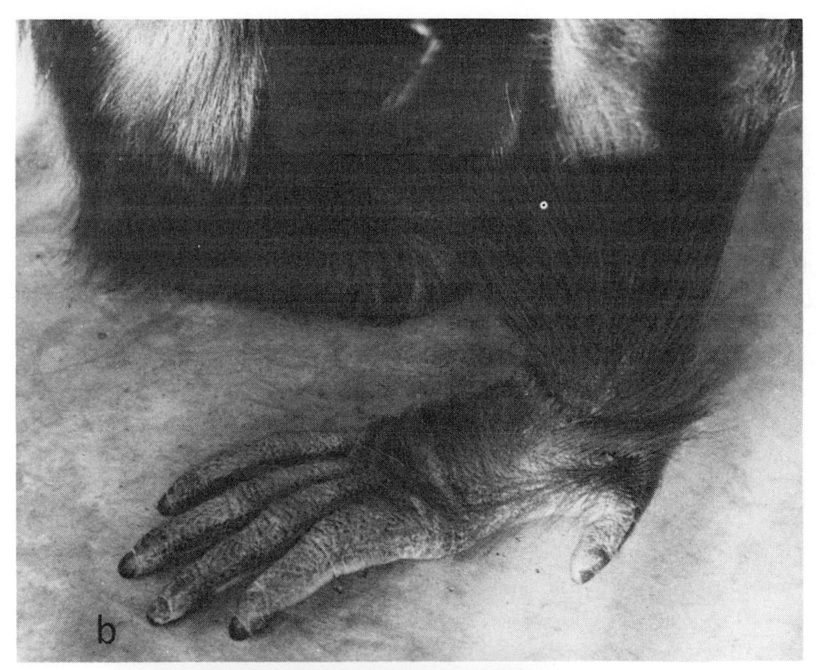
b

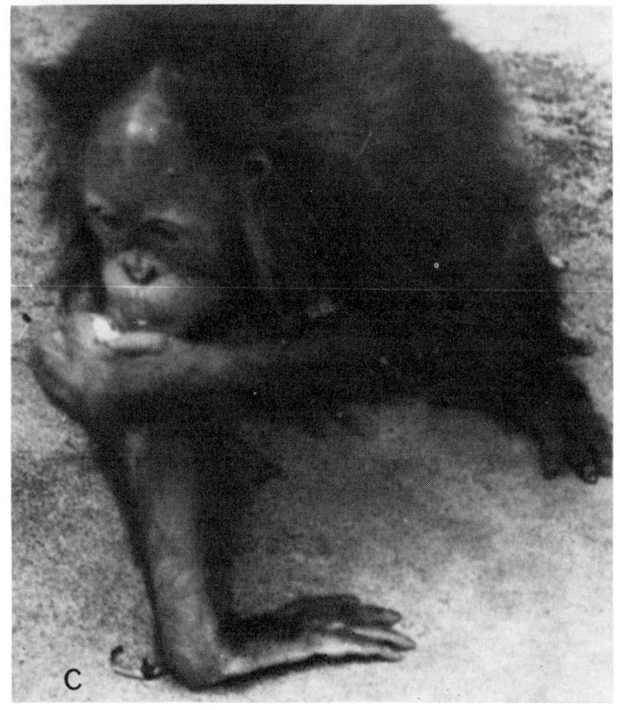
c

Relative shortness of the proper flexor muscles of the wrist also has been cited as a major factor in restriction of dorsiflexion in the chimpanzee (Virchow, 1929). However, Schreiber (1936), Tuttle (1967) and, to a lesser extent, Straus (1940, p. 203) variously demonstrated that Virchow might have overemphasized this feature as a safeguard against wrist dorsiflexion in *Pan*. Further, the fact that the wrists of many orangutans and gibbons (Tuttle, 1967, 1969a,b, 1970) may be moved into a remarkable degree of dorsiflexion precludes the inference that shortened wrist flexors are structural concomitants of "brachiation." The degree of shortening in the wrist flexors that may be exhibited by individual chimpanzees and gorillas probably should be considered a consequence of knuckle-walking.

Evidence from a comparative suite of primates that is much more comprehensive than the apes also may be cited in argument against the notion that alignment of wrist and hand and "shortness" of long digital flexors are primary structural concomitants of "brachiation." Tuttle (1969c) found that among Old World monkeys, the highly terrestrial gelada, baboons, and patas possess remarkable restriction of dorsiflexion by comparison with their nearest arboreal and part-time terrestrial relatives (Fig. 21). This feature is probably better considered a development *de novo* for terrestrial digitigrady instead of a heritage feature related to "brachiation."

Similarly, though not as remarkably developed as in the African apes, the long digital flexor tendons are noticeably shortened in the gelada, patas monkey, baboons, bear macaque, sooty mangabey, and vervet monkey by comparison with their more consistently arboreal nearest relatives (Tuttle, 1969c). This feature too is probably more reasonably associated with their particular patterns of digitigrady than with a history of "brachiation."

Finally, we are compelled to mention that a major feature—"free pronation-supination" of the hand to approximately 180°—that Lewis employs as a universal characteristic related to "true brachiation" in hominoid wrists is not adequately documented in the source (Avis, 1962) that he cites. Indeed, such employment may be judged partly erroneous on the basis of other evidence (Tuttle, 1969b, 1970). It would appear that the value that Avis (1962) claimed for pronation-supination is not based on measurements. It is simply an assumption of what should occur at the radio-ulnar joints when the subject's body rotates markedly relative to the supporting hand. A considerable amount of the total rotation probably occurs at the shoulder. Without precise measurements of joints in moving or (perhaps more feasibly) passive subjects, it is virtually impossible to assess reliably the degrees of potential rotatory movement at any one

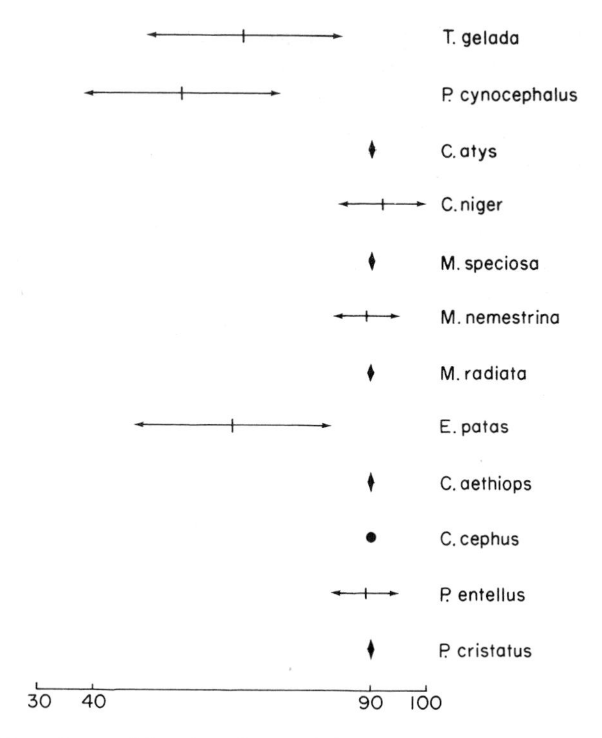

Fig. 21. Mean (vertical bars) and 90% fiducial limits (double-tipped arrows and diamonds) for degrees of passive dorsiflexion in the wrists of selected species of the Cercopithecidae. Dot indicates single specimen. (See Tuttle, 1969c, p. 198, for numbers of specimens.)

joint in a composite series like the forelimb of an arm-swinging ape. Avis' study consisted primarily of relatively brief observations on juvenile chimpanzees and orangutans and adult gibbons in novel settings. Gorillas and adult chimpanzees and orangutans were not tested in the experimental cages. The subjects were not given an opportunity to accommodate to the experimental cages before the investigator began to record data. Clearly the data resulting from such limited materials and procedures does not constitute a proper base for the universal inference that "all [apes] are capable of 180° total trunk rotation and 180° forearm supination (both of which may be slightly greater in the gibbon) . . ." (Avis, 1962, p. 128).

Tuttle's (1969a,b) casual observations on potential rotatory capacities of the hand in anesthetized apes indicate that in adult gibbons and subadult gorillas maximum passive pronation–supination may be considerably less than 180°. If future systematic studies demonstrate that limited pronation–supination is also characteristic of free-ranging hylobatid apes, such information could upset the tendency of some "brachiationists" to overgeneralize features of "the brachiators" despite behavioral and morphological evidence that clearly distinguish the locomotive complexes of ricochetal arm-swinging hylobatid apes (Tuttle, 1969a, 1972b) from orangutans and from the knuckle-walking African apes.

Limitation of pronation-supination in hylobatid apes might be viewed as (a) part of the ricochetal arm-swinging complex, (b) a stabilizing feature associated with hanging-feeding wherein the tendency to revolve too freely beneath the single supporting hand might be disadvantageous, or (c) some combination of these and other factors. Limitation of radio-ulnar rotation in gorillas may be considered most reasonably as part of the knuckle-walking complex wherein hands that are pronated in line with the direction of travel may be best disposed to serve efficiently as supporting props and perhaps also to supply supplemental propulsive force during rapid locomotion.

We offer these counterpoints and criticisms of Lewis' position on the problem of knuckle-walking in order to preserve an atmosphere of open-mindedness in future research that might confirm, deny or otherwise require revision of hypotheses on the existence of bony particularities in the wrists of extant African apes that might be traced into the fossil record.

In African apes certain features of metacarpophalangeal joints II–V can be related less equivocally than features of the wrist to knuckle-walking. This might reflect simply that criticisms of the sort that Lewis applied to Tuttle's inferences about the hominoid wrist have not been generated by more detailed research than that of Tuttle (1967, et seq.) on hominoid metacarpophalangeal joints. We will attempt here to reconfirm Tuttle's suggestions that knuckle-walking indeed has left its mark on metacarpals II–V and probably also on certain muscles to manual rays II–V in the African apes.

In a study on a metacarpal IV from Swartkrans, Republic of South Africa, Napier (1959, p. 9) commented as follows:

> The articular shelf seen on the posterior aspect of the metacarpal heads in chimpanzee and gorilla is absent. This shelf is presumably related in the Pongidae to the hyperextension of the digits during quadrupedal knuckle-walking when the weight of the upper trunk is borne on the dorsum of the middle phalanges. In this respect it is interesting to note that the articular shelf is most prominent in the most terrestrial of the three genera and least well marked in the most arboreal.

Napier, like other contemporary primatologists, apparently considered all three great apes to be knuckle-walkers, differing from one another on the basis of frequency that they engaged in knuckle-walking rather than discretely in terms of the customary hand postures employed on wide horizontal supports.

Once Tuttle (1965, et seq.) recognized that orangutans ordinarily do not knuckle-walk, he reexamined the problem of potential hyperextension at metacarpophalangeal joint II–V in nonhuman catarrhine primates in order to discover features that might be particularly related to knuckle-walking in the African apes. Tuttle (1969c) measured degrees of maximal passive metacarpophalangeal hyperextension in the hands of 154 cercopithecoid monkeys (representing 17 species) and 10 hands in each of the following apes: *Pan gorilla, Pan troglodytes, Pongo pygmaeus,* and *Hylobates lar.* The apes were sharply distinguished from the catarrhine monkeys by their limited capacity for hyperextension at metacarpophalangeal joints II-V. Further, orangutans, and especially gibbons exhibited remarkable restriction of metacarpophalangeal hyperextension by comparison with the African apes (Fig. 22).

Tuttle (1967, 1969b, 1970) observed that the articular surface extends farther onto the dorsal aspect of the heads of metacarpals II–V in *Pan*

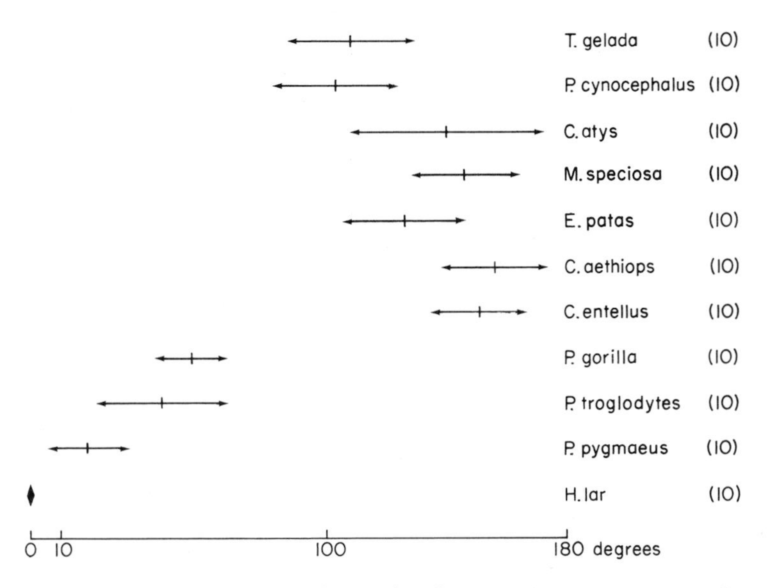

Fig. 22. Mean and 90% fiducial limits for degrees of hyperextension of metacarpophalangeal joints II–V in selected species of the Anthropoidea. Symbols as in Fig. 21. Ten specimens of each species were employed.

than in *Pongo* (Fig. 23). The bases of the proximal phalanges appear to articulate in close conformity with the dorsal "articular shelf" in *Pan* (Fig. 24). Thus, a special close-packed position of hyperextension may characterize metacarpophalangeal joints II–V in the African apes in contradistinction to the condition that maintains in orangutans whereby hyperextension is somewhat more limited and not particularly accommodated by special bony structures.

In addition to bony features related to the close-packed position, other factors have been suggested to safeguard (in varying degrees) the seemingly precarious hyperextended positions of the metacarpophalangeal joints during knuckle-walking (Tuttle, 1967, et seq.). Some of Tuttle's speculations, especially on precise functions of individual digital flexor muscles may require revision in light of electromyographical studies (see above). However, we reconfirm here the viewpoint that shortening of the long digital flexor muscles is considered most parsimoniously as a primary adaptation for knuckle-walking in the African apes (Fig. 25). The "shortening" is probably especially related to the arrangement of the metacarpo-

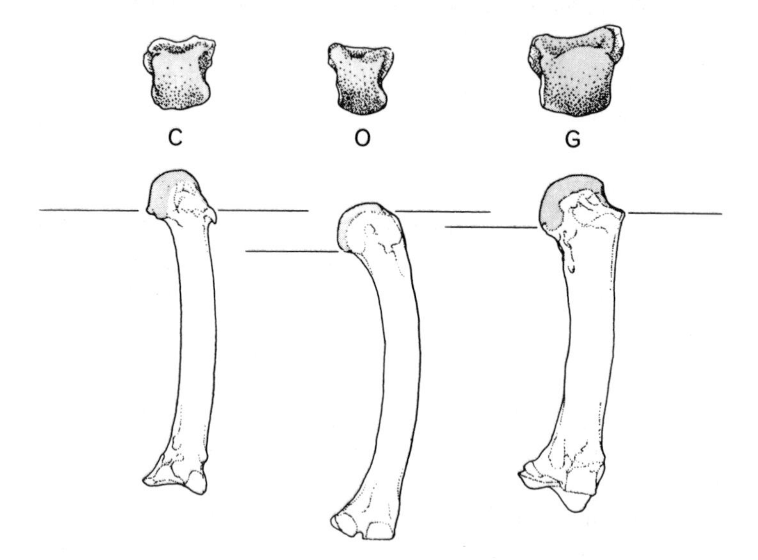

Fig. 23. Medial aspect (bottom row) and distal end (top row) of the third metacarpal bone of chimpanzee (C), orangutan (0), and gorilla (G). The distal articular surfaces are indicated by shading. Horizontal lines indicate the probable extent of the anterior and posterior articular areas. Note the limited posterior extension of the distal articular surface and the greater curvature of the entire metacarpal bone in orangutan. (Drawing by P. Murray; Copyright 1969 by the American Association for the Advancement of Science.)

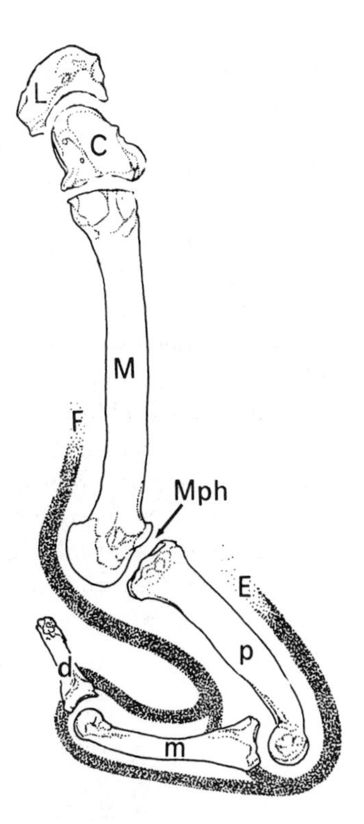

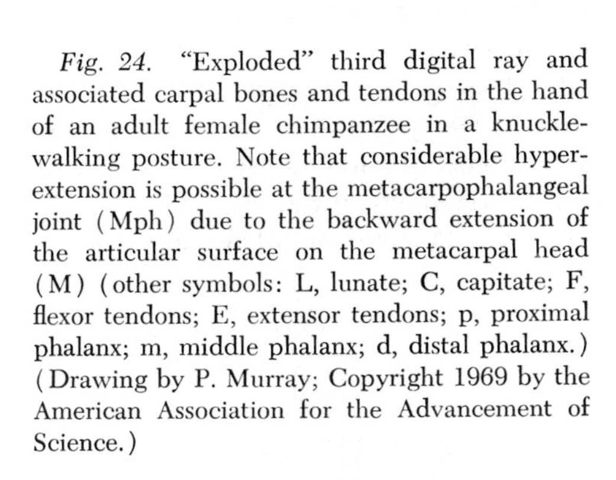
Fig. 24. "Exploded" third digital ray and associated carpal bones and tendons in the hand of an adult female chimpanzee in a knuckle-walking posture. Note that considerable hyperextension is possible at the metacarpophalangeal joint (Mph) due to the backward extension of the articular surface on the metacarpal head (M) (other symbols: L, lunate; C, capitate; F, flexor tendons; E, extensor tendons; p, proximal phalanx; m, middle phalanx; d, distal phalanx.) (Drawing by P. Murray; Copyright 1969 by the American Association for the Advancement of Science.)

phalangeal joints. It might even limit certain sorts of arboreal locomotive activities in African apes by comparison with Asian apes which generally lack remarkable digital flexor shortening (Tuttle, 1970). Unfortunately, pertinent behavioral observations on apes in their natural habitats are not available to permit testing this suggestion.

Phylogenetic and functional inferences from metacarpophalangeal joints II–V in the Hominoidea also would be rendered more secure by detailed biomechanical studies on the palmar plates and collateral ligaments since these structures may play important roles in safeguarding the integrity of the hyperextended joints.

In summary, much research remains to be conducted in order to elucidate fully the morphological bases and biomechanical premises of knuckle-walking in the African apes. Strategic comparisons of African apes with orangutans, hylobatid apes, and man, that focus on the wrist and metacarpophalangeal joints II–V, are requisite to resolve current differences of opinion about morphological aspects of the knuckle-walking complex and its relationship to a heritage of "brachiation."

We reiterate that our electromyographical research on *Pan gorilla,*

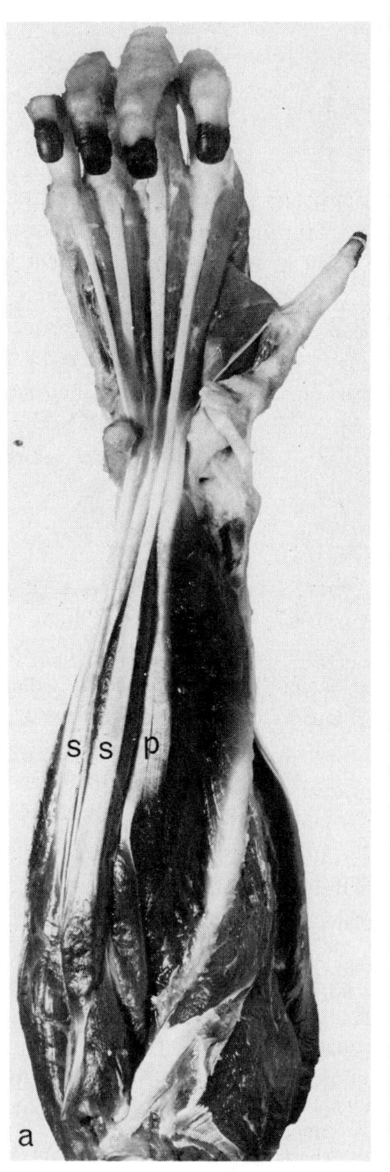
s
s
p
a

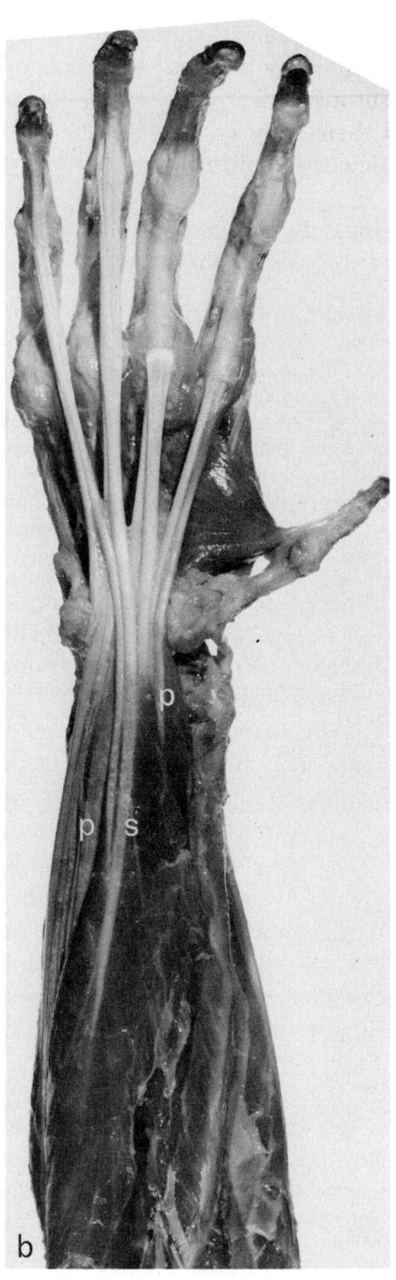
p
p
s
b

Tuttle's previous experiences with the morphology of hominoid hands, and especially his studies on movements of joints in passive apes indicate persuasively that the search for close-packed positioning mechanisms and their bony concomitants in hands of knuckle-walking apes may be particularly fruitful for future studies of hominoid phylogeny.

Possible Pathways of Pongid and Early Hominid Phylogeny

Inferences and theories on the phylogeny of the Pongidae are trellised intimately with those on pre-Pleistocene Hominidae, but the Hominidae have received a preponderance of attention in the writings of most theoretic evolutionary anthropologists. Many points of controversy between members of the "brachiationist," "anti-brachiationist," and "prebrachiationist" schools exist despite overwhelming evidence supporting the close taxonomic relationships of man, chimpanzee, gorilla, and, to a lesser extent, orangutan (Tuttle, 1969b; Pilbeam, 1967; Washburn, 1971).

Contemporary theorists subscribe to a variety of models on pongid and hominid phylogeny, ranging from ones expressing early divergence and remarkable parallel evolution in several hominoid lineages (Pilbeam, 1967, 1968, 1969, 1970, 1972; Uzzell and Pilbeam, 1971; Simons and Pilbeam, 1972) to ones depicting a long common heritage and very late divergence of the great apes and man (Washburn, 1967, 1968a,b, 1971; Sarich and Wilson, 1967; Sarich, 1971). Yet other authors (Tuttle, 1967, 1969a,b,d, 1970; Lewis, 1969, 1971, 1972a,b) present models somewhat intermediate between those of the nonbrachiationists and those of the extreme troglodytophilic "brachiationists."

Unfortunately available fossil evidence, though often informative and provocative, is not complete or otherwise representative enough to allay marked divergences of opinion among leading theorists on hominoid evolution. Consequently, we will summarize several current theories on

---

Fig. 25. Fresh preparations of the right forearm and hand of an adult male *Pan troglodytes* (a) and a juvenile *Pongo pygmaeus* (b). The flexor carpi radialis, flexor carpi ulnaris, palmaris longus, and pronator teres muscles have been removed to expose the flexor digitorum superficialis muscle (s) and parts of the flexor digitorum profundus muscle (p) that are not covered by it. Note the remarkable "tendonization" of the flexor digitorum superficialis and indicial component of the flexor digitorum profundus muscle in *Pan*. Contrast it with the condition in *Pongo*. Such tendonization might be partly responsible for the permanent "shortness" of the long flexor muscles in knuckle-walkers.

the subject and express our preference among available options. We await results of ongoing and future paleontological, comparative biomechanical, behavioral, and other studies on the Hominoidea that will require revision of our preference.

Simons, Pilbeam, and their associates argued emphatically from available "direct" evidence that populations ancestral to chimpanzees, gorillas, man, and perhaps also orangutans, may be recognized as discrete entities in Miocene times. *Ramapithecus wickeri* (cf., *Kenyapithecus wickeri*) and *Ramapithecus punjabicus* from late Miocene and/or early Pliocene deposits of Kenya and South Asia, respectively, are the earliest known hominids in a recent Simons-Pilbeam model (1972). *Dryopithecus* (*Proconsul*) *major* from Kenya and Uganda represents the earlier Miocene ancestry of gorillas. Although in previous works (Simons and Pilbeam, 1965; Pilbeam, 1967), they evidenced uncertainty about which of the earlier Miocene dryopithecine apes might be ancestral to chimpanzees, Simons and Pilbeam (1972) settled on *Dryopithecus* (*Proconsul*) *africanus* for that component in their model. They confessed that the fossil record provides poorer documentation for the ancestry of orangutans than of African apes but submitted that *Dryopithecus sivalensis,* a later Miocene Asian ape, is a likely possibility (1972). Thus, according to their 1972 scheme, Simons and Pilbeam would have lineages leading to chimpanzee and gorilla already distinct in the latest early or early middle Miocene of East Africa [i.e., 20–18 million years BP (before present)]. Since the Eurasian dryopithecine apes are postulated to be descendants of emigrant African dryopithecine apes, perhaps 15 or 16 million years BP, orangutans might be later evolutionary products than their African confamiliar relatives. Finally, Simons and Pilbeam (1972, p. 60) suggested that since *Ramapithecus wickeri* possibly "represents the very earliest detectable stages of hominid divergence from Miocene Pongidae," the date of pongid/hominid divergence might be set at 15 to 20 million years BP. This date, like that for divergence of ancestral orangutans, is somewhat later than their presumptive date of chimpanzee/gorilla divergence.

The 1972 model of Simons and Pilbeam is rather starkly contrasted with that presented as recently as 1970 by Pilbeam though both are based on much the same suite of fossils. According to Pilbeam's model (1970, p. 83), *Pongo* and the Hominidae diverged from basal Hominoidea in Oligocene times and before the divergence of *Pan* and *Gorilla* in early Miocene times.

In 1972, Pilbeam (p. 46) presented yet another model of hominoid evolution in which ancestral chimpanzees and gorillas are distinct "as long ago as 15 million or 20 million years," "ancestral orangutans probably diverged even earlier," and "the hominids seem to have diverged from ancestors of the other apes at least 10 million and probably 15 million years ago." Thus, by comparison with the Simons-Pilbeam model of 1972, he reduced the earliest hypothetical date of hominid divergence another 5 million years and apparently would have ancestral orangutans living in Africa during approximately 10 million years before they began the long migration to South Asia in middle Miocene or later times.

Simons and Pilbeam (1972) suggested that *Dryopithecus major* and perhaps *Dryopithecus africanus* were knuckle-walkers. Accordingly, knuckle-walking developed in parallel and persisted over a long time in lineages culminating in extant gorillas and chimpanzees. Their major support for this model consists of reiterations of personal communications from Alan Walker and a few published accounts on postcranial remains presumably of *Dryopithecus major* (Walker and Rose, 1968; Day and Wood, 1969). None of the published fossils is from an anatomical region especially implicated in knuckle-walking versus other hypothetical locomotive habituses possible to large-bodied apes.

Again following personal communications from Walker, Pilbeam (1972, pp. 39–42) pointed out that in a series comprised of three types of African ape—gorilla, common chimpanzee, and bonobo (cf., "pygmy chimpanzee")—the intermembral index decreases directly with body size. Thus, while the bonobo possesses noncheiridial segments of forelimbs and hindlimbs that are approximately equal in length, the gorilla has relatively elongate forelimbs and short hindlimbs. The common chimpanzee is intermediate between the large and small members of the series. Apparently, the hindlimbs became shorter relative to trunk length as body size increased through the series. Pilbeam reasonably related hindlimb "reduction" in gorillas to their particular mode of terrestrial sitting–feeding instead of postulating that their limb/trunk proportions are solely attributable to a heritage of "brachiation." He drew a provocative analogy between geladas and gorillas regarding this functional–morphological complex (Pilbeam, 1972, p. 41).

He concluded as follows:

> . . . the knuckle-walking ancestor of the gorilla and the other African pongids (if that ancestor was a knuckle walker) would have been a relatively small-bodied animal (around 50 to 70 pounds in weight) with arms and legs of approximately equal length. Knuckle-walking probably developed when the ancestral African apes became relatively terrestrial . . . . The pre-knuckle-walking African apes were probably arm swingers, having arms relatively long compared with trunk length as well as other adaptations of shoulder girdle, thorax, and trunk . . . . Such an animal could have evolved equally well into the knuckle-walking apes, the long-armed orangutan, and

bipedal hominids . . . . Alternatively, orangutan and human ancestors may have been knuckle-walkers.

Subsequently, Pilbeam (1972, p. 99) offered the following detailed model of hypothetical close predecessors of *Ramapithecus:*

> The earliest hominids presumably evolved toward the end of the Miocene from forest-dwelling apes that were relatively large-bodied [weighing 30 to 50 pounds—p. 48] arm swingers or knuckle-walkers. It is possible that the hominids had evolved many of their distinctive dental adaptations before they became habitual bipeds. Initially, they may not have indulged in bipedal behaviors any more than do the living apes. Indeed, the period during which they retained arboreal hanging and swinging or knuckle-walking capabilities may well have lasted several million years. However, we can assume that during the Pliocene hominid bipedal behavior became much more efficient, perhaps as our early ancestors were developing into hunters.

While Pilbeam merely entertained the possibility that Miocene protohominids may have been "at least incipiently" knuckle-walkers (1972, p. 48), Washburn (1967, 1968a,b, 1971) and Sarich (1971) speculated that knuckle-walking constituted an important stage in hominid evolution between phases of advanced brachiation and habitual bipedalism. Washburn and Sarich did not propose detailed models on the evolution of gorillas, chimpanzees, and orangutans, but they clearly favored divergence of orangutans from large-bodied "brachiators" prior to the development of the knuckle-walkers from which gorillas, chimpanzees, and man subsequently diverged. Sarich (1971, p. 78) would have orangutans divergent from the large-bodied hominoid stock approximately 7 million years BP, and the radiation of ancestral chimpanzees, gorillas, and hominids occurring between 3.5 and 5 million years BP.

Washburn (1968a, p. 14) suggested that "the later the separation and the closer the relationship between man and the African apes, the more likely it is that our ancestors went through a stage of knuckle-walking." In Washburn's model (1967, p. 23), knuckle-walking "provides a kind of intermediate condition in which, if selection were for more bipedalism, the arms might be used less and less in locomotion." He suggested further that knuckle-walking would allow prehominids to carry objects without sacrificing speed and other opportunities to escape from predators on the forest floor and in the forest fringe areas. According to this theory "living out in the savanna away from trees would have come long after ground living, knuckle-walking and tool making and using" in protohominid phylogeny (Washburn, 1967, p. 24).

Tuttle (1967, 1969a,b,d, 1970), while not subscribing to a knuckle-walking theory of hominid origins, did not deny that such theories might be entertained reasonably in the absence of unambiguous evidence to the contrary. However, hypothetical phylogenies of the African apes which nonparsimoniously posit extremely long parallel histories of knuckle-walking in ancestral gorillas and chimpanzees seemed improbable. Tuttle (1969b, 1970) favored instead a model of pongid radiation that is similar to that of Washburn and Sarich but fitted within a somewhat more extensive temporal framework.

Largely on the basis of inferences from the total morphology, naturalistic behavior, ecology, and biomolecular evidence in extant apes, Tuttle (1969b,d) proposed a model of hominoid radiation that would have gorilla, chimpanzee, orangutan, and man derived from relatively large-bodied pongid apes which possessed rather advanced morphological proclivities for suspensory behaviors in which forelimbs were employed prominently. Ancestral orangutans probably diverged from the other large-bodied hominoids before the development of knuckle-walking in ancestral African apes. Selection for habitation in swamp forests may have been a factor leading to several characteristic features in Pleistocene and Recent orangutans.

Ancestral hominids probably also diverged from the hominoid stem before the development of knuckle-walking adaptations and subsequent furcation of ancestral chimpanzees and gorillas. Tuttle, like Washburn, has been concerned over the excessive bias (probably attributable to overextrapolation from the presumed paleoecology of South African australopithecines) that protohominids and their immediate predecessors must have evolved in response to selection for open country (cf., "savanna") living. Tuttle (1969b) suggested that it might be fruitful for theorists to consider instead that somewhat different selective complexes in the arboreal habitats of prehominids and pretroglodytians may have predisposed their proximal descendants to bipedalism and knuckle-walking, respectively, as they adapted to life on the ground. Tuttle (1969b, p. 960) concluded that

> evidence from hominid upper limbs generally indicates that the ancestors of man probably engaged in some form of suspensory posturing, and that they assumed bipedal posturing very soon after venturing to the ground. Accordingly, man probably did not pass through a knuckle-walking stage; the phylogeny of man and apes probably represents a dichotomous pattern of evolution.

Tuttle (1969a,b,d, 1970) attempted to clarify the manner by which a heritage of suspensory adaptations in the hands of large-bodied apes might have predisposed them to knuckle-walking as they expanded their niches to include terrestrial substrates. He rejected as too simplistic the notion that prototroglodytians possessed inflexible "brachiating" hooks that could only be employed initially for terrestrial posturing and locomotion by topsy-turvy placement in knuckle-walking postures. Instead he proposed that the early prototroglodytians probably possessed relatively long "flexible" hands. Thus, they developed *de novo* many special features of muscles, ligaments, and bones that culminated in stable knuckle-walking postural mechanisms. Alternative pathways to achieve this result are: (1) fist-walking→ incipient knuckle-walking→ consistent knuckle-walking, or (2) modified palmigrade locomotion→ incipient knuckle-walking→ consistent knuckle-walking (Tuttle, 1969b, p. 957).

Some sort of facultative flexed-finger posture would be requisite in order for long-fingered hands to serve as terrestrial supporting props. The subsequent development of knuckle-walking may be associated with the transition from more or less strict supportive functions of the hand to its employment as an augmentary propellant organ in rapid terrestrial locomotion (Tuttle, 1969b, 1970).

Tuttle considers that extant African apes are the products of an excurrent pattern of evolution with the common chimpanzee as a fair approximation of ancestral gorilla. The common prototroglodytian ancestor of gorillas and chimpanzees probably had greater opportunities for terrestrial progression and feeding than ancestral orangutans did and perhaps they more frequently exploited resources on the forest floor, in clearings, and in forest fringe areas. Whereas ancestral gorillas became increasingly adapted to forest floor subsistence patterns, ancestral chimpanzees remained arboreal in feeding and nesting activities (Tuttle, 1970). The large size of gorillas probably developed during the shift to terrestrial habitation and it eventually forced them to nest ever more frequently on the ground. Their enormous size probably developed as part of a terrestrial foraging–feeding and defensive complex which enabled them to exploit foods consistently in the shrub layer of the forest (Tuttle, 1970).

Lewis (1969, 1971, 1972a,b) adopted the viewpoint of certain traditional "brachiationists." He subscribed to a model in which the lesser apes, the great apes, and man are branches of a common hominoid stock, all members of which evidence clearcut morphological adaptations for "brachiation" as defined by Avis (1962). Thus he suggested a very early separation of the Hylobatidae from the emergent Hominoidea and before the achievement of a *Dryopithecus africanus* grade of wrist structure. Ancestral orangutans diverged next with a grade of wrist structure "approximating more closely to, but still inferior to, that of *D. africanus*." (Lewis, 1972b, p. 56.) He suggested that although *Pan, Gorilla,* and

Homo could be derived from the *Dryopithecus* (*Proconsul*) *africanus* grade without violating any basic anatomical or evolutionary principles, the range of variation in the human radiocarpal joint strongly suggests derivation from a structural grade more like that of *Pan* than *Proconsul*. He was especially compelled to favor the *Pan* model because *Pan*, *Gorilla*, and *Homo* evidence early fusion of the os centrale to the scaphoid bone, a feature which he interprets as a hallmark of advanced suspensory locomotion and which distinguishes them from *Dryopithecus africanus*, *Pongo*, and the Hylobatidae (Lewis, 1972b, p. 56). Lewis did not discuss the role that knuckle-walking may have played in hominoid evolution.

In summary, the models of Simons, Pilbeam, Washburn, Sarich, Tuttle, and Lewis are not as dissimilar in most basic features as were the models presented by theorists prior to the 1960's. All authors agree that the African apes and man share particularly close affinities. Yet the precise patterns of hominid, African pongid, and orangutan evolution continue to be subjects of controversy within this shrinking sphere of consensual possibilities.

We provisionally favor the model of Tuttle (1969b), noting that he has avoided commitment to a discrete time scale.

We disagree with aspects of models by Simons and Pilbeam on the phylogeny of African apes, and particularly their notions about long histories of knuckle-walking and extreme parallelism in ancestral chimpanzees and gorillas. Their vacillation on the exact nature of chimpanzee ancestry among available East African dryopithecine apes highlights the ambiguous nature of the fossil evidence when considered more or less apart from other sources of evidence. The search for superlatives among East African dryopithecine apes may have led Simons and Pilbeam into a pit of nonparsimony when they suggested that ancestral chimpanzees and gorillas diverged before the emergence of ancestral hominids. All available evidence from extant species indicates persuasively that among the Hominoidea, the African Pongidae are most closely related to one another and that they diverged relatively late in the hominoid phylogenetic sequence.

We tentatively agree with Lewis' principal conclusion that fundamentally similar morphological complexes in the wrists of great apes and man probably supports a brachiationist hypothesis, i.e., that the extant large-bodied hominoids radiated from a single group of pongids that were fairly advanced in the employment of forelimb suspensory behaviors. Moreover, the remarkable similarities among *Pan, Pongo, Homo,* and *Dryopithecus* (*Proconsul*) *africanus* in certain details of wrist structure provide a substantive base for inferences that the extant Pongidae and

Hominidae diverged from one another probably no earlier than early Miocene. But for reasons outlined before (pp. 326–332) we believe that the nature of Lewis' data on hominoid wrist joints, the biomechanical explanations of them that he proffers, and his employment of perspectives from behavioral and other morphological studies are not adequate to support substantially his conclusions on specific phylogenetic relationships of the great apes and man. In particular, we fail to comprehend how the information on *Pongo* that he stressed supports inferences that (a) *Pongo* possessed wrists less advanced for "brachiation" than those of *Pan* and (b) *Pongo* must have diverged from a structural grade inferior to that of *Dryopithecus* (*Proconsul*) *africanus* and thus before the radiation of *Pan, Gorilla,* and *Homo*. Finally, we should note that this, of course, does not deny that a persuasive case could be made for the phylogenetic model to which Lewis subscribes. We simply need more complete information on the comparative anatomy and functional capacities of hominoid wrists in order to test evidence from this vital region against models based on other aspects of hominoid morphology and naturalistic behavior.

Subsequent to 1967, many authors were compelled to discuss whether knuckle-walking may have constituted a significant phase not only in African pongid phylogeny but also in prehominid, and even proto-orangutan, evolution. We agree with Tuttle (1969b) that definitive evidence for a stage of knuckle-walking in hominid evolution has not been extracted from the fossil record or from available studies of extant hominoids. Those who favor a knuckle-walking ancestry for man and orangutans may elect to view the facultative "knuckle-walking" hand postures of some men and especially that of "Felix" as evidence for the primacy of knuckle-walking and terrestriality in pongid and hominid evolution. Per contra, however, we subscribe to the following conclusions of Tuttle and Beck (1972, p. 34) regarding the relevance of facultative "knuckle-walking" hand postures in orangutan and man:

> Because a highly advanced arboreal climber and arm swinger like an orangutan is able to place his hands in knuckle walking postures suggests that the ancestors of the African apes might have been similarly, or to a greater extent, predisposed to knuckle-walking by their own special arboreal heritage.
>
> Although the thumb is generally used prominently as a supporting strut during human knuckle walking postures, the fact that man has a predisposition for such placement of manual digits II–V may be at least as provocative for evolutionary inferences as the facultative knuckle walking of an orangutan.
>
> It may be thus argued that man passed through a phase of arboreal climbing and suspensory posturing somewhat more advanced than the anti-brachiationists and prebrachiationists have admitted into their models.

In closing, we challenge that it is incumbent upon those who would posit that knuckle-walking was precedent to "brachiation" to publish their own logically consistent, systematic models on the mechanisms that might be involved in such hypothetical evolutionary developments.

Acknowledgments

This investigation was supported mainly by NSF Grant No. GS-3209 and by a Public Health Service Research Career Development Award No. 1-K04-GM16347-01 from the National Institutes of Health. Supplementary support was provided by NIH Grant No. RR-00165 to the Yerkes Regional Primate Research Center.

We thank Mrs. Eleanor Regenos and Mr. Glenn Shine for their many contributions to the methods employed in the EMG project. We are especially indebted to Mr. Robert Pollard for his assistance with the gorillas. We also thank J. Perry, J. Malone, G. Super, and C. Clayton at the Rehabilitation Research and Training Center of Emory University; Dr. G. H. Bourne (Director), Gen. G. Duncan (Assistant Director), Dr. D. Rumbaugh (former Associate Director), Dr. M. Keeling, J. Roberts, F. Keirnan, and E. van Ormer of the Yerkes Regional Primate Research Center of Emory University; and, S. Toibin and K. Barnes at the University of Chicago for their assistance and cooperation in making this adventure successful.

References

Avis, V. (1962). Brachiation: the crucial issue for man's ancestry. *Southwest. J. Anthropol.* **18**, 119–148.

Basmajian, J. V. (1967). "Muscles Alive: Their Functions Revealed by Electromyography," 2nd ed. Williams & Wilkins, Baltimore, Maryland.

Basmajian, J. V., and Stecko, G. (1962). A new bipolar indwelling electrode for electromyography. *J. Appl. Physiol.* **17**, 849.

Biegart, J. (1971). Dermatoglyphics in the chimpanzee. *In* "The Chimpanzee" (G. H. Bourne, ed.), Vol. 4, pp. 273–324. Karger, Basel.

Broca, P. (1869). L'ordre des primates parallèle anatomique de l'homme et des singes. *Bull. Soc. Anthropol. (Paris)* **4**(2), 228–401.

Day, M. H., and Wood, B. A. (1969). Hominoid tali from East Africa. *Nature (London)* **222**, 591–592.

Ellis, R. A., and Montagna, W. (1962). The skin of primates. VI: The skin of the gorilla (*Gorilla gorilla*). *Amer. J. Phys. Anthropol.* **20**, 79–93.

Erikson, G. E. (1963). Brachiation in New World monkeys and in anthropoid apes. *Symp. Zool. Soc. London* **10**, 135–164.

Fossey, D., and Campbell, R. M. (1970). Making friends with mountain gorillas. *Nat. Geographic* **137**(1), 48–67.

Gratiolet, L. P., and Alix, P. H. E. (1866). Recherches sur l'anatomie du *Troglodytes Aubryi. Nouv. Arch. Mus. Hist. Natur. (Paris)* **2**, 1–264.

Groves, C. P. (1970). "Gorillas." Arco, New York.

Kallner, M. (1957). Die Muskulatur und die Funktion des Schultergürtels und der Vorderextremität des Orang-Utans. *Morphol. Jahrb.* **97**, 554–665.

Lewis, O. J. (1965). Evolutionary change in the primate wrist and inferior radio-ulnar joints. *Anat. Rec.* **151**, 275–286.

Lewis, O. J. (1969). The hominoid wrist joint. *Amer. J. Phys. Anthropol.* **30**, 251–268.

Lewis, O. J. (1971). Brachiation and the early evolution of the Hominoidea. *Nature (London)* **230**, 577–579.

Lewis, O. J. (1972a). Evolution of the hominoid wrist. *In* "The Functional and Evolutionary Biology of Primates" (R. Tuttle, ed.), pp. 207–222. Aldine-Atherton, Chicago, Illinois.

Lewis, O. J. (1972b). Osteological features characterizing the wrist of monkeys and apes, with a reconsideration of this region in *Dryopithecus (Proconsul) africanus. Amer. J. Phys. Anthropol.* **36**, 45–58.

Mayer, Prof. (1856). Zur Anatomie des Orang-Utang und des Chimpanse. *Arch. Naturgesch.* **22**, 281–304.

Montagna, W., and Yun, J. S. (1963). The skin of primates. XV. The skin of the chimpanzee (*Pan satyrus*). *Amer. J. Phys. Anthropol.* **21**, 189–204.

Napier, J. R. (1959). Fossil metacarpals from Swartkrans. *Fossil Mammals of Africa No.* **17**. Brit. Mus. Natur. Hist., London.

Owen, R. (1859). "On the Classification and Geographic Distribution of the Mammalia." Parker, London.

Oxnard, C. E. (1967). The functional morphology of the primate shoulder as revealed by comparative anatomical, osteometric, and discriminant function techniques. *Amer. J. Phys. Anthropol.* **26**, 219–240.

Pilbeam, D. R. (1967). Man's earliest ancestors. *Sci. J.*, **3**, 47–53.

Pilbeam, D. R. (1968). The earliest hominids. *Nature (London)* **219**, 1335–1338.

Pilbeam, D. R. (1969). Tertiary Pongidae of East Africa: evolutionary relationships and taxonomy. *Bull. Peabody Mus. No.* **31**. Yale University, New Haven, Connecticut.

Pilbeam, D. R. (1970). "The Evolution of Man." Thames and Hudson, London.

Pilbeam, D. R. (1972). "The Ascent of Man: An Introduction to Human Evolution." Macmillian, New York.

Sarich, V. (1971). A molecular approach to the question of human origins: a critical appraisal. *In* "Background for Man. Readings in Physical Anthropology" (P. Dolhinow and V. M. Sarich, eds.), pp. 60–81. Little, Brown, Boston, Massachusetts.

Sarich, V., and Wilson, A. C. (1967). Immunological time scale for hominid evolution. *Science* **158**, 1200–1203.

Schaller, G. B. (1963). "The Mountain Gorilla: Ecology and Behavior." Univ. Chicago Press, Chicago, Illinois.

Schreiber, H. (1936). Die Extrembewegungen der Schimpansenhand. *Morphol. Jahrb.* **77**, 22–60.

Schultz, A. H. (1933). Notes on the fetus of an orang-utan. *Rep. Lab. Mus. Comp. Pathol. Zool. Soc.* **18**, 61–79.

Schultz, A. H., (1956). Postembryonic age changes. *Primatologia* **1**, 887–964.

Schultz, A. H. (1961). Vertebral column and thorax. *Primatologia* **4** (no. 5), 1–66.

Simons, E. L., and Pilbeam, D. R. (1965). Preliminary revision of early Dryopithecinae (Pongidae Anthropoidea). *Folia Primatol.* **3**, 81–152.

Simons, E. L., and Pilbeam, D. R. (1972). Hominoid paleoprimatology. *In* "The Functional and Evolutionary Biology of Primates" (R. Tuttle, ed.), pp. 36–62. Aldine-Atherton, Chicago, Illinois.

Straus, W. L., Jr. (1940). The posture of the great ape hand in locomotion, and its phylogenetic implications. *Amer. J. Phys. Anthropol.* **27**, 199–207.

Susman, R. (1973). Facultative terrestrial hand postures in an orangutan (*Pongo pygmaeus*) and pongid evolution. *Am. J. Phys. Anthropol.* (in press).

Tuttle, R. H. (1965). "The Anatomy of the Chimpanzee Hand, with Comments on Hominoid Evolution." University Microfilms, Ann Arbor, Michigan.

Tuttle. R. H. (1967). Knuckle-walking and the evolution of hominoid hands. *Amer. J. Phys. Anthropol.* **26**, 171–206

Tuttle, R. H. (1968). Some swamp forest adaptations of the orangutan. *Bull. Amer. Anthropol. Ass.* **1**, 141.

Tuttle, R. H. (1969a). Quantitative and functional studies on the hands of the Anthropoidea. I. the Hominoidea. *J. Moıphol.* **128**, 309–364.

Tuttle, R. H. (1969b). Knuckle-walking and the problem of human origins. *Science* **166**, 953–961.

Tuttle, R. H. (1969c). Terrestrial trends in the hands of the Anthropoidea: a preliminary report. *Proc. 2nd Intern. Congr. Primatol., Atlanta, Georgia,* Vol. 2, pp. 192–200, Karger, Basel.

Tuttle, R. H. (1969d). The way apes walk. *Sci. J.* **5**A(5), 66–72.

Tuttle, R. H. (1970). Postural, propulsive, and prehensile capabilities in the cheiridia of chimpanzees and other great apes. *In* "The Chimpanzee" (G. H. Bourne, ed.), Vol. 2, pp. 167–253. Karger, Basel.

Tuttle, R. H. (1972a). Introduction. *In* "The Functional and Evolutionary Biology of Primates" (R. Tuttle, ed.), pp. vii–xx. Aldine-Atherton, Chicago, Illinois.

Tuttle, R. H. (1972b). Functional and evolutionary biology of hylobatid hands and feet. *In* "Gibbon and Siamang" (D. M. Rumbaugh, ed.), Vol. 1, pp. 136–206. Karger, Basel.

Tuttle, R. H., and Beck, B. B. (1972). Knuckle walking hand postures in an orangutan (*Pongo pygmaeus*). *Nature (London)* **236**, 33–34.

Uzzell, T., and Pilbeam, D. (1971). Phyletic divergence dates of hominoid primates: a comparison of fossil and molecular data. *Evolution* **25**, 615–635.

Virchow, H. (1929). Das Os centrale carpi des Menschen. *Morphol. Jahrb.* **63**, 480–530.

Walker, A. C., and Rose, M. D. (1968). Fossil hominoid vertebra from the Miocene of Uganda. *Nature (London)* **217**, 980–981.

Washburn, S. L. (1950). The analysis of primate evolution with particular reference to the origin of man. *Cold Spring Harbor Symp. Quant. Biol.* **15**, 67–78.

Washburn, S. L. (1967). Behavior and the origin of man. *Proc. Roy. Anthropol. Soc.*, pp. 21–27.

Washburn, S. L. (1968a). "The Study of Human Evolution," Condon Lectures, Oregon State System of Higher Education, Eugene, Oregon.

Washburn, S. L. (1968b). Speculations on the problem of man's coming to the ground. *In* "Changing Perspectives on Man" (B. Rothblatt, ed.), pp. 193–206. Univ. Chicago Press, Chicago, Illinois.

Washburn, S. L. (1971). Analysis of human evolution. *Amer. J. Phys. Anthropol.* **35**, 299.

11

Locomotor Adaptations in Past and Present Prosimian Primates

ALAN WALKER

Introduction

The types of observation that a researcher makes, together with the way in which those observations are analyzed, are to a large extent governed by his viewpoint of the particular problem. My viewpoint is that of a paleontologist and the principal problems discussed here are the types of conclusions that can be reached concerning the possible locomotor activities of extinct prosimians. Like most workers, I have not spent much time trying to answer questions that are probably unanswerable given the present evidence. I have not concerned myself, for instance, with gait analysis since there seems to be no obvious way in which skeletal morphology and gait patterns can be correlated; even body proportions have not given themselves to correlation with gait patterns (Hildebrand, 1967). The observations that can be made on extinct species are limited to skeletal morphology and proportions and interest in living prosimian species, as far as locomotion goes, may initially be confined to matching observations made on the fossils with those made on the same skeletal parts of living species. However, the next step, that of correlating the skeletal anatomy of these species with their known locomotion proves difficult in that very little is known about their locomotion in the wild and what little was known often proved to be inadequate or erroneous. I have made, therefore, a personal survey of the locomotor behavior of as many prosimian primates as possible. These observations have been qualitative but, where possible, have been made in the field and supplemented where necessary with observations on captive animals and the use of cine film. In the case of one species, *Galago demidovii,* I have attempted a quantitative study using captive specimens. This study, to be published elsewhere, will include other *Galago* species so that any specific locomotor differences might be found. These observations of locomotor behavior form the basis for a locomotor classification that has been used when dealing with the fossil remains.

I am certain that it would be possible to discriminate between all living species by their locomotion alone if enough careful field and laboratory research were carried out. However, even from the superficial and mostly unquantified data so far available it is possible to identify the resemblance between the overall locomotor patterns of some species and to conclude that other species seem to have a more individual locomotor style. If this were not possible then practically nothing could be said about the possible locomotor habits of extinct species; in any case, the sort of information required to distinguish locomotor differences at the specific level is not likely to be obtained for most living species and would be impossible, given the nature of the fossil record, to gather for extinct species. The use of other aspects of the fossil assemblage, such as paleobotanical remains, sedimentary evidence, associated fauna, and so on, should not be used as evidence of a species' locomotor preferences without the utmost caution. So many factors are involved in the process of fossilization that only the most careful taphonomical studies are likely to sort out even the major ones. To my knowledge there are very few primate-bearing fossil sites where this type of study has been made.

The Locomotion of Living Prosimians

CHEIROGALEINAE

Cheirogaleus major and C. medius, Microcebus murinus, and Phaner furcifer (Martin, 1972; Petter, 1962; Rand, 1935; Shaw, 1879; Walker, this paper)

Quadrupedal locomotion over a range of speeds seems to be the general arboreal pattern. The body is held parallel to the support surface and the tail trails behind, with the exception of *Phaner* where it may occasionally be held vertically. Leaping proclivities vary, ranging from little in *C. major* to fairly marked, at least in some situations, in *M. murinus.* On the ground, quadrupedal walking and running is the dominant mode but hopping and landing on all fours is seen in *Microcebus*. *Microcebus* also uses a "cantilever" posture from time to time in which the body is held away from the support and the hind feet alone maintain the posture. When resting and feeding these species adopt a quadrupedal stance with the limbs slightly flexed.

INDRIIDAE

Avahi laniger, Propithecus diadema, and P. verreauxi, and Indri indri (Jolly, 1966; Petter, 1962; Rand, 1935; Walker, this paper).

Active arboreal locomotion consists of leaps from one vertical trunk to another, the body being held upright. The forelimbs are held in a variety of positions during the leap, varying from being held at the sides for small jumps to being held above the shoulders at the start of large ones. Progression along the larger horizontal branches is by bipedal hopping, the arms held close to the chest. Resting postures are typically clinging to an upright support or sitting upright in a fork of a tree. On the ground, slow quadrupedal or bent knee, slow bipedal steps may be taken but where speed is required the dominant mode is always bipedal hopping. Feeding postures vary, including hanging by all fours and standing upright with three or four limbs maintaining the grip; direct mouth feeding is common.

DAUBENTONIIDAE

Daubentonia madagascariensis (Lauvedon, 1933; Petter, 1962; Petter and Peyrieras, 1970; Walker, this paper)

Slow and active quadrupedal progression in trees is the noted pattern. Climbing slowly up and down large vertical trunks is common as is walking upside down under large horizontal branches with all four extremities supporting the body. Leaps of quite large distances are often made. When in search of food the head is held close to the branch being inspected. The use of the attenuated middle fingers in feeding and drinking gives rise to forearm gestures of unique type, such as the extreme supination with extended wrist when extracting larvae from bark. Resting postures range from being rolled up in a ball in the large nest to sitting on all four flexed limbs. When on the ground, slow to fast quadrupedal locomotion is employed with the weight on the forelimbs taken on the thenar and hypothenar eminences and with the wrists strongly dorsiflexed.

Lemurinae

Hapalemur griseus, Lemur catta, L. fulvus, L. mongoz, and Varecia variegatus (Jolly, 1966; Petter, 1962; Petter and Peyrieras, 1971; Rand, 1935; Shaw, 1879; Walker, this paper)

Active to slow quadrupedal locomotion in the trees, almost invariably on the upper surface of supports is the basic pattern found with these species. The tail is actively used, being either held arched over the back or swung from side to side. Leaps of considerable distance are frequently made, particularly by *H. griseus* and *L. catta,* these species also using vertical stems more than the others. On the ground, slow to fast quadrupedal locomotion is the rule, but the *H. griseus* from Lac Alaotra has a tendency to break out in short hops, landing on all four limbs. Feeding and resting positions are extremely varied, but resting on vertical trunks that have no fork is seen for long periods only in *H. griseus.*

Lepilemur mustelinus (Charles-Dominique and Hladik, 1971; Petter, 1962; Rand, 1935; Walker, this paper)

Normal progression in the trees is by leaps from one vertical trunk to another or by hops up one vertical trunk using both fore- and hindlimbs together. Resting postures typically involve clinging to a vertical stem. On the ground, bipedal hopping is the rule but quadrupedalism may be used for slow progression.

Galaginae

Galago alleni, G. crassicaudatus, G. demidovii, G. elegantulus, G. inustus, and G. senegalensis (Bartlett, 1863; Bishop, 1964; Charles-Dominique, 1971; Haddow and Ellice, 1964; Hall-Craggs, 1964, 1965; Jouffroy, this volume; Vincent, 1969; Walker, this paper)

Active leaping locomotion is the rule for these species when in the trees, although some move more quadrupedally than others, especially under certain habitat conditions. Slow quadrupedal movements are made by all species. Feeding and resting postures vary from "cantilevering" to sitting upright. On the ground, slow quadrupedal movements change to bipedal hopping as soon as the animal is pressed.

Lorisinae

Arctocebus calabarensis, Loris tardigradus, Nycticebus coucang, and Perodicticus potto (Bishop, 1964; Charles-Dominique, 1971; Hill, 1953; Petter and Hladik, 1970; Walker, this paper)

These species show slow climbing quadrupedal locomotion in the trees. No leaping is seen but individuals can roll themselves up and drop off a branch to avoid a predator. Usually the animals travel on the upper side of the support, but movements underneath supports, especially when moving from a vertical to a horizontal branch, are fairly common. During normal quadrupedal climbing a strong grasp, with at least three extremities, is used. The hindfeet are each placed immediately behind the forefoot of the same side before the forefoot is released for the next stride and the spine is thus involved in sinuous side-to-side movements. On the ground, quadrupedal progression is the only reported mode and here the hands and feet are laterally deviated. Feeding postures vary, ranging from a quadrupedal crouch to hanging by the feet alone. Resting postures usually involve the grasp of a support by all four limbs with the body above the support and the head tucked between the forelimbs.

TARSIIDAE

Tarsius spp. (LeGros Clark, 1924; Hill *et al.*, 1952; Montagna and Machida, 1966; Thomas, 1896)

Active leaping locomotion using vertical supports seems far the most common locomotor activity. The usual resting posture is clinging to a vertical stem in which position the tail may or may not be involved in support. During a leap the tail trails inertly. On the ground, progression is by bipedal hops.

SUMMARY AND SUGGESTED LOCOMOTOR CLASSIFICATION

There seem to be three main locomotor types within the prosimians. Most species habitually use one of these three types but there are a few that have a total pattern intermediate in character between two major types. The three main types are discussed below.

Vertical Clinging and Leaping (Napier and Walker, 1967)

This pattern is characterized by a leaping mode of progression through the trees in which the two hindlimbs, used together, provide the propulsive force of locomotion. The trunk is held in a vertical position before and after each leap and vertical supports are preferred. On the ground, this locomotor habit manifests itself in bipedal hopping with the trunk held erect and the arms almost never involved in support. Tail movements are often up and down but during a leap the tail usually trails inertly. Species using this pattern type include: *Avahi, Galago, Indri, Lepilemur, Propithecus,* and *Tarsius.*

TABLE I

OBSERVATIONS FORMING THE BASIS OF THE AUTHOR'S LOCOMOTOR CLASSIFICATION

Species	In the wild	In captivity	On cine film
Cheirogaleus major	−	+	−
Cheirogaleus medius	−	+	−
Microcebus murinus	+	+	+
Phaner furcifer	−	+	−
Hapalemur griseus	+	+	+
Lemur catta	+	+	+
Lemur fulvus	+	+	+
Lemur mongoz	+	+	+
Lepilemur mustelinus	−	+	+
Varecia variegatus	+	+	+
Avahi laniger	+	−	−
Indri indri	+	−	+
Propithecus verreauxi	+	+	+
Daubentonia madagascariensis	−	+	+
Galago alleni	−	+	−
Galago crassicaudatus	+	+	+
Galago demidovii	+	+	+
Galago inustus	+	−	−
Galago senegalensis	+	+	+
Arctocebus calabarensis	−	+	+
Loris tardigradus	−	+	+
Nycticebus coucang	−	+	+
Perodicticus potto	+	+	+
Tarsius syrichta	−	−	+

Active Quadrupedalism

This is a form of locomotion involving the use of all four limbs in climbing, walking, and running. Leaping propensities vary from group to group. Ground locomotion is essentially the same as arboreal locomotion except that the hands and feet are placed in a plantigrade rather than a flexed position. The hindlimbs provide most of the propulsive effort and the forelimbs act as supporting struts. *Cheirogaleus, Daubentonia, Hapalemur, Lemur, Microcebus,* and *Phaner* use this basic pattern of locomotion.

Under certain habitat conditions some species of *Galago, Hapalemur, Lemur,* and *Microcebus* use a locomotion intermediate between these two main types already described above. This type of locomotion is typified by an intermediate amount of leaping and by hopping on the ground where the forelimbs are used for support.

Slow Climbing Quadrupedalism

An arboreal slow climbing form of locomotion in which no leaping is involved is the last main mode of locomotion. The foot is placed directly behind the hand of the same side before the hand is released to make the next step. At any one time three extremities are involved in grasping the support. Species following this type of locomotion include *Arctocebus, Loris, Nycticebus,* and *Perodicticus.*

The Morphological Correlates of Locomotion in Living Prosimians

The postcranial morphology of the Lorisinae is quite distinctive and characteristic. Increased fore- and hindlimb mobility, increased vertebral column mobility and extremely powerful grasping hands and feet of a distinctive type (Grand, 1967) are all dependent upon a highly modified skeleton. Thus far no fossil remains of this type have been found, but they should be easily recognized if and when they are. The differences between species that use the other two main types of locomotion are in many ways more subtle. The major emphasis will be placed here on distinguishing between active arboreal quadrupeds and vertical clingers and leapers.

The principal problem of finding discriminant features between the skeletal parts of the active quadrupeds and vertical clingers and leapers is one of allometry. Both locomotor groups have small to large members and the rule of allometric difference (Reeve and Huxley, 1945) obscures some discriminant features. There are also those species that do not fit comfortably into one locomotor category or the other. It is difficult, therefore, to assign many single features to one locomotor group, but the use of combined features gives better results than any single feature used alone.

Limb Proportions (Table II)

The brachial index is usually higher (i.e., the radius is relatively longer) in vertical clingers and leapers. The index is smaller for slow climbing quadrupeds and the radius is just slightly longer than the humerus in active arboreal quadrupeds. The crural index seems to have little discriminant value, the tibia being slightly shorter than the femur in most species. The intermembral index is the most useful discriminant of all and distinguishes clearly all three locomotor types.

TABLE II
LIMB PROPORTIONS OF SOME LIVING PROSIMIANS

	Brachial index	Crural index	Intermembral index
Active arboreal quadrupeds (25 specimens of 9 species)	95.8–117.9	87.1–123.3	67.7–74.2
Slow climbing quadrupeds (19 specimens of 4 species)	94.5–112.7	89.5–103.8	83.5–94.3
Vertical clingers and leapers (29 specimens of 10 species)	100.0–121.3	83.1–101.6	49.2–67.4

Scapula

The scapula is the postcranial element most affected by allometry. For example, the scapulae of *Tarsius* and *Indri* are very different in shape, the smaller one is very elongated transversely and the larger one is an equilateral triangle in outline. The forelimb functions of the two are, however, very similar. In large species the infraspinous fossa is large and is probably a requirement of greater mobility, the increased size of the infraspinatus fossa giving a greater couple-arm length to the m. trapezius/m. serratus anterior couple (Ashton and Oxnard, 1964).

Clavicle

The clavicle presents no easily identifiable differentiating feature and as this bone is frequently missing or cut in most osteological collections it has proved difficult to assess any features that can be related to posture or locomotion.

Humerus

In general the humerus of vertical clingers and leapers can be distinguished from those of quadrupeds by the following features:

1. Height of tuberosities. In quadrupeds the tuberosities (especially the greater) are more developed than in vertical clingers and leapers; the greater tuberosity nearly reaches, and sometimes surpasses, the summit of the head which is never the case in vertical clingers. The presence of a large greater tuberosity is possibly related to the extensor role of m. supraspinatus where the increased lever arm length would afford a more powerful recovery of the forelimb in walking.
2. Depth of bicipital groove. In quadrupeds the bicipital groove is usually much deeper and more well defined than in vertical clingers and leapers. This is possibly due to the more active role played by the long

head of m. biceps as a flexor of the elbow in the recovery phase of walking as well as its role in shoulder joint stability.

3. Size of brachialis flange. The brachialis flange is relatively larger in the vertical clingers and leapers than in quadrupeds. This may be associated with the need for flexion of the elbow during climbing and in maintaining the erect sitting posture on vertical stems. Flexion at the elbow joint prevents the trunk from swinging backward away from a vertical support.

4. Size of medial epicondyle. This is relatively larger in vertical clingers and leapers than in quadrupeds, and possibly reflects the increased importance of the digital flexors in the former. In climbing and in clinging to vertical supports the hands of vertical clingers and leapers are actively involved in gripping where the forelimb is subject to tensile forces, whereas the flexion of the digits in a branch walking quadruped is mostly a stabilizing factor and the forelimb is subjected to compressive forces.

5. Shape of the capitulum. The capitulum is more spherical and more isolated from the trochlea in vertical clingers and leapers. This is possibly related to the increased powers of pronation and supination in these forms.

Ulna

1. Direction of the olecranon process. In vertical clingers and leapers the olecranon process is directed more anteriorly, i.e., flexed toward the humerus, than in quadrupeds. As almost all the propulsive effort is provided by the hindlimbs in vertical clinging and leaping, the lever arm of the olecranon is relatively poorly positioned as a lever arm at high angles of elbow extension. In quadrupeds, on the other hand, the lever arm is more mechanically efficient at high angles of powered extension of the elbow and is used in the propulsive phase of walking and running. The m. triceps mass is also used in controlled flexion of the elbow when the forelimb is subjected to extremes of compression, as for instance, at the end of a leap.

2. Depth of sigmoid notch and height of coronoid process. In quadrupedal forms, where the weight of the body is carried partly by the forelimb, the greater sigmoid notch is deep and the coronoid process is high. This configuration resists anterior dislocation of the humerus and the weight is passed to the ulna via the base of the coronoid process. In vertical clingers and leapers, the forelimb is subjected mainly to tensile forces and the greater sigmoid notch is shallow and the coronoid process small.

3. Length of styloid process. The styloid process is relatively longer and more conical in vertical clingers and leapers than in quadrupeds. This is

possibly associated, once again, with the tensile forces to be met by the forelimb in the first group and the compressive forces to be met by the forelimb in the second. In terrestrial quadrupeds the styloid process is more flattened distally than in arboreal quadrupeds.

Radius

The radius of most vertical clingers and leapers is much less straight than that of quadrupeds. Except in *Tarsius* spp., the smallest vertical clinging forms, the radius is strongly curved laterally. The neck is better defined in vertical clingers and leapers than in quadrupeds. The increased power of pronation and supination in the vertical clingers and leapers is probably the main factor in this difference of curvature. The supinator and pronator muscles have a greater mechanical advantage acting on a strongly curved radius than on an almost straight one.

Pelvis

Allometric factors strongly influence the shape of the pelvis (Fig. 1). In *Tarsius*, the smallest vertical clinging and leaping form, the ilium is rodlike. In *Indri*, the largest vertical clinging and leaping form, the ilium is expanded, the anterior superior iliac spine is huge and laterally hooked, and the gluteal surface is extensive. The anterior inferior iliac spine is present in the pelves of all vertical clingers and leapers, but some of the quadrupeds also have a large spine. It may not be possible, even with species of the same body size, to distinguish the two locomotor groups by pelvic morphology and this is even more difficult with the smaller species. In general, though, the ilia are more expanded and the anterior superior

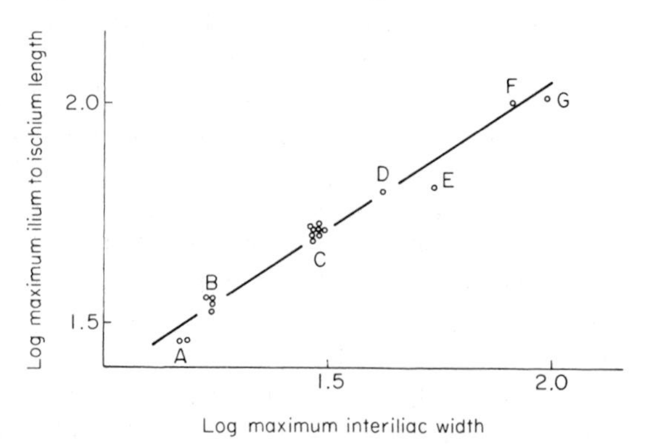

Fig. 1. Allometry in the pelvis of various vertical clinging and leaping species. A, *Tarsius;* B, *Galago senegalensis;* C, *Galago crassicaudatus;* D, *Avahi;* E, *Lepilemur;* F, *Propithecus diadema;* G, *Indri.*

and anterior inferior iliac spines are more prominent in vertical clingers and leapers than in quadrupeds. The ratio between iliac length and ischial length shows that the ilium is long relative to the ischium in most leaping forms. The ratio is highest, paradoxically, in the slow climbing quadrupeds. The expansion of the gluteal surfaces in leaping forms and the shortness of the ischium lever arm for the adductors shows that, at least near full extension, m. gluteus maximus and m. medius act as hip extensors and, with the short trochanteric lever arm, fast extensors. The greater relative length of the ilium in leaping prosimians is at variance with the condition met with in hopping mammals of other orders, in which the ischium is relatively longer (Maynard-Smith and Savage, 1956).

Femur

1. Shape of head and greater trochanter. The femur in vertical clingers and leapers is a long, straight bone with a spherical head (cylindrical in smaller species) set on a short neck. There is usually a posterior extension of the articular surface of the head on to the neck. The greater trochanter overhangs the anterior aspect of the shaft in vertical clingers, probably correlated with the extensor action of the glutei and the greatly developed vasti. In quadrupeds, the femur is relatively shorter, the head always spherical, and the trochanter does not usually overhang the anterior aspect of the shaft.

2. Patellar groove and the direction of the femoral condyles. In vertical clingers and leapers the patellar groove is deeper and narrower than in quadrupeds. The lateral patellar tubercle is more prominent than the medial. The floor of the groove in its anterior part is usually raised above the anterior surface of the shaft, having the effect of increasing the diameter of the pulley surface around which the extensors act through the patella. This, in turn, gives the joint greater powers of flexion than a joint with a smaller pulley diameter as is seen in the quadrupeds. The femoral condyles are oriented posteriorly in vertical clingers and leapers and posteriorly and distally in quadrupeds. This difference in orientation is presumably related to the high degree of knee flexion in the vertical clinging forms in the habitual resting posture.

Tibia

1. Compression of shaft. The tibia of the vertical clinging and leaping forms is more laterally compressed than in quadrupedal forms. This has the effect of affording more attachment for the powerful plantar flexors of the foot that are actively involved in leaping; the mediolaterally compressed bone is suited to accommodate forces acting in a sagittal plane,

such as those incurred in leaping. In quadrupeds, where the accent on powerful plantar flexion of the foot is less and the bone is subjected to stresses from different directions, as in walking along uneven surfaces, the bone is more circular in section.

2. Plane of the condylar surfaces. In accord with the direction of the femoral condyles, the plane of the tibial condylar surfaces is tilted to face more posteriorly in vertical clinging and leaping forms. In quadrupeds the surfaces are set more at right angles to the long axis of the bone.

Fibula

The fibula offers no reliable features that discriminate between the two main locomotor groups.

Hand and Foot

Differences in the hand and foot between the two main locomotor groups are also confused by allometry and by the intermediate forms. In general, however, there is a tendency for the hands and feet of the vertical clinging and leaping species to be dominated by the elements of the ulnar and peroneal borders. This can be correlated with the typical vertical clinging posture, in which turning forces due to gravity and friction act upon the outside borders of the hand and foot. In arboreal quadrupeds, the dominance is central in the hand and foot and the functional axis tends to coincide with rays 3 or 4, instead of between 4 and 5. In slow climbing quadrupeds, the ulnar and peroneal dominance is also seen, but this is seemingly a function of the reduction of digits II and III to form a vice-like hand and foot. Osteologically this tendency to ulnar and peroneal dominance in the vertical clinging species is seen in the relatively increased size of the lateral digital bones and of the hamate, triquetral, and cuboid. The phalanges of the vertical clinging and leaping species are longer, relative to the rest of the hand, than in the quadrupeds. However, the phalanges of the larger vertical clinging species are relatively shorter than those of the smaller vertical clinging species, presumably because of weight factors. The hands and feet of arboreal quadrupeds seem to be efficient organs for gripping branches whereas those of vertical clinging and leaping species appear to be better adapted for wrapping around vertical trunks. The thumbs of the smaller vertical clinging and leaping species (e.g., *Tarsius* spp.) are poorly differentiated from the rest of the hand; those of the larger species (e.g., *Propithecus* spp. and *Indri indri*) are well differentiated. This is again probably related to factors involving the supporting of a proportionately much larger mass.

In the foot, the development of an elongated tarsus is seen in *both* vertical clingers and leapers and arboreal quadrupeds. This development appears to be governed by allometric factors, but the observed regression lines for the two locomotor groups are of different gradients. For a given size, a vertical clinging and leaping form has a relatively longer calcaneal load arm than an active quadruped. This elongation of the calcaneum and navicular is associated, as Hall-Craggs (1965) has shown, with the need for great acceleration in the push-off phase of leaping. Maynard-Smith and Savage (1956) have demonstrated the same point in hopping mammals of other orders, where there is a need to raise the center of mass as quickly as possible at the beginning of each hop.

The hallux of the vertical clingers and leapers is relatively large and bears a prominent peroneal tubercle, presumably an indication of the power of the grasping hallux and in response to the greater component of gravitational forces acting on the foot when grasping vertical stems. The peroneal tubercle is less well developed in quadrupeds and even less so in digitigrade quadrupeds among the Old World Monkeys, demonstrating a loss of powered adduction of the hallux and a decrease in the power of the peroneus longus as an hallucal adductor. As in the hand of vertical clingers and leapers, there is a tendency for the axis of the foot to be toward the peroneal border.

The Presumed Locomotion of Extinct Prosimians

The Nature of the Fossil Record

The chances of the bones of prosimian primates becoming fossilized in a sedimentary environment, uncovered again by tectonic and/or erosional factors, and then discovered, are fairly remote. Very many factors are involved, ranging from ancient local geomorphological conditions to correct recognition of specimens in the field and in collections. Consequent upon these low chances is a scattered and biased sample in the fossil record where populations from whole continents have not been sampled, for us in millions of years and yet samples from succeeding populations in a very limited area from a very short time span are sometimes surprisingly complete. To take the example of the African prosimians, only two very narrow time ranges have been sampled: one, about 20 to 18 million years BP yields only remains of local populations from about six small areas in East Africa; the other, about 2.0 million years BP has a small sample from one tiny area of East Africa that is now part of Olduvai Gorge. With the Madagascan prosimians we have no knowledge of any prosimians older than about 5000 years BP, but of species living after that time and those still extant we have a fairly reasonable record. Even with this erratic sampling, the lack of positive identification of limb bones by association with cranial and dental parts is frustrating. For reasons of frequency of preservation and the nature of neontological taxonomy, cranial parts are the best material upon which to base nomenclatorial types. Even the lack of association of different parts of the postcranial skeleton causes severe problems.

The number of complete fossil skeletons of prosimians is very small and again for a number of reasons some parts of disarticulated skeletons have more chance of being preserved or found than others. Despite all these limitations it is possible to deduce some information from the existing remains of fossil prosimians. These are most appropriately dealt with region by region. There are postcranial remains of Palaeocene to Eocene species from North America and Europe, of Miocene and Pleistocene species from East Africa and of very recently extinct Madagascan species.

The Palaeocene and Eocene Prosimians of North America and Europe

Many fragmentary specimens of Early Tertiary primate limb bones are known. Because they are usually fragmentary and often doubtfully associated, they have received little serious attention. The best-known form, *Notharctus,* has been the subject of a classic monograph by Gregory (1920). All these early forms can be said to be prosimian in grade, but Simons (1961) has pointed out the dangers of assigning them too readily to higher taxa using arguments based on living species. Whatever links may prove to exist between these early primates and modern prosimians, their limb bones are all that is available to give an indication of the locomotor adaptations at that time. Since I have not yet had the opportunity to study the originals of many of the North American fossils, I will restrict my comments to the better-known specimens.

Limb bones of *Plesiadapis* species are known from the United States and France. Simpson (1935) has given an account of the former and a published account of the new specimens from Europe is eagerly awaited. The specimens so far published remind one of a robustly built tree shrew, with rather short limbs and clawed extremities. Most statements have implied that the locomotion of *Plesiadapis* species was squirrel-like, implying arboreal locomotor habits, but as yet there is no convincing evidence. It is not beyond the bounds of possibility that they were flying species with a patagial apparatus and it would seem to be possible to test this hypothesis fairly easily. One can safely say that the limb bones of *Plesiadapis* do not look like those of any living prosimian primate.

The adaptive radiation within the prosimians during the Eocene involved different families on the two major land masses. In North America, the omomyids and notharctids were involved, while in Eurasia (with the possible exception of *Pelycodus*) it was the tarsiids and the adapids. *Hemiacodon,* a North American omomyid, is well known from cranial remains and some limb bones have been referred to the genus and described by Simpson (1940). The most striking feature of the hind limb skeleton is the overall resemblance with that of *Galago* species, with some similarities to *Tarsius* species. For example, the femur is straight and slender, has posteriorly directed condyles and a raised patellar groove. The fibula was unfused, at least proximally. Parts of the foot are known and show distal elongation, but not as marked as in modern vertical clinging and leaping forms. One point of difference, however, is that all the distal tarsal elements are elongated, not only the calcaneum and navicular. This would have the disadvantage of placing the midtarsal joint in the middle of the load arm. The first metatarsal has a very well-developed peroneal tubercle as in living vertical clinging and leaping forms. It would seem that *Hemiacodon* had a locomotor style similar to that of modern *Galago* species, but that the extreme elongation of the tarsus as seen in the living species was not present.

Gregory's account of the anatomy of *Notharctus* still remains a classic of its kind. Gregory was hampered by the fact that in 1920 very little was recorded of the locomotion of living lemurs, but when reading his descriptions one is struck by the continual references to the postcranial morphology being near to *Lepilemur* or *Propithecus,* and Gregory's specimens mostly matched species of these genera rather than *Lemur.* The intermembral index of 64.59 for a specimen of *N. osborni* is within the range of the indices for modern vertical clinging and leaping forms. The broad, and in some parts very detailed, resemblances to *Lepilemur* and *Propithecus* make it fairly certain that in general the locomotion of species of *Notharctus* was of the vertical clinging and leaping type. The closely related genus *Smilodectes* is also very similar in postcranial morphology and proportions to *Notharctus.*

Of the European tarsiids, limb bones of *Necrolemur,* and *Nannopithex* are known. Weigelt (1933) pointed out the marked resemblances between the hind limb of *Nannopithex* and *Tarsius* and noted that the elongation of the tarsus in the fossil form was not so extreme. He also claimed that the fibula of *N. raabi* was fused distally to the tibial shaft. Simons (1961) casts doubt on this assumption and points out that the original surface of the tibia is lost and that it is impossible to determine whether or not the two elements were fused. There is some doubt as to the allocation of the femur and tibiofibula that Schlosser (1907) referred to *Necrolemur edwardsi* and *Necrolemur antiquus,* respectively. Schlosser also referred a calcaneum to *Necrolemur,* assigning it to *N. antiquus* in the text and *N. edwardsi* in the caption to the plate. This fits well in size with the tibiofibula, but the femur is much too large for both the other bones. The exact assignments remain in doubt but it is likely that they belong to two species of *Necrolemur,* the femur to *N. antiquus* and the tibiofibula and calcaneum to *N. zitteli.* The femur has all the characteristics of those of vertical clinging and leaping species and the tibiofibula is closely similar to that of *Tarsius* spp. The calcaneal elongation is not as great as would be expected in a modern vertical clinging and leaping form of that size but is greater than the expected elongation in an active quadrupedal form.

Limb bones of *Adapis* species from the Eocene and Oligocene of Europe are only poorly known and Le Gros Clark (1959) doubted whether they have, in all cases, been correctly attributed to the genus. Certainly some of the bones attributed to *Adapis parisiensis* by Filhol (1883) are not primate, but others probably are. A fragment of humerus* and a pedal phalange of ?*Adapis magnus* from the English Oligocene have been described by Day and Walker (1969). Humeri of both *A. parisiensis* and *A. magnus* show features of the vertical clinging and leaping type (Day and Walker, 1969). A femur and a tibia described by Filhol are peculiar for primates and may not belong to the order. In any case they are of such different sizes that they must belong to different species at least. Gregory (1920) described foot bones of *Adapis* that seem closely similar to those of *Notharctus,* but the material is fragmentary and perhaps the most that can be said is that *Adapis* possessed large, robust, and grasping feet.

The Miocene and Pleistocene Prosimians of East Africa

Of the Miocene lorisids of Kenya and Uganda, parts of humeri, femora, tibiae, calcanea and phalanges have been described (Walker, 1970). More remains are now available for study including further specimens of humeri, femora, tibiae, a talus, and phalanges as well as vertebrae. There are five currently recognized lorisid species of the genera *Progalago, Komba,* and *Mioeuoticus* present in the faunal assemblage. Only a few of the limb bones are associated with identifiable teeth. The unassociated

* Since this writing I have discovered, during a study of otter humeri, that this fragment is in fact, that of an aquatic mammal. This error highlights the problem of dealing with fragmentary material without an adequate comparative background.

limb bones have a size range that matches that of the dental and cranial remains and tentative species assignments have been made according to size (Walker, 1970). Whether or not these assignments are correct in all cases does not seem to matter much for the present purposes since the limb bones of all the size groups show the same morphological features. In almost all respects, including fine detail, the fossils resemble the same bones of living species of the genus *Galago*. The only skeletal element in which the modern *Galago* condition is not as closely matched is the calcaneum. Of three different sized calcanea, the distal elongation that is seen in living vertical clinging and leaping species of that size is not reached. The elongation is, however, greater than would be expected for an active quadruped of the same size. The synovial joint between the navicular and calcaneum that Hall-Craggs (1966) reported in *Galago* was also apparently present in the Miocene species. The postcranial skeletons of modern galagos seemingly differ little from ancestral forms 20 million years ago.

Fragments of limb bones were found associated with dental parts of *Galago senegalensis* in Bed I, Olduvai Gorge, Tanzania. Simpson (1965) could find no meaningful differences between the fossil dentitions and those of modern *G. senegalensis* and I find that the same is true of the limb bones. There is no aspect of the morphology or proportions that is not entirely consistent with these specimens being of *G. senegalensis*.

THE EXTINCT LEMUROIDS OF MADAGASCAR

Large collections of the limb bones of recently extinct lemuroids of Madagascar have been made. None of these extinct species is known from specimens older than about 5000 years BP and there is considerable direct and indirect evidence to suggest that the arrival of man on the island was the prime cause of their extinction (Walker, 1967a). In a study of the limb bones the main problem is that of correctly attributing the postcranial material (Walker, 1967b). Confusion has been caused in the past by incorrect species assignments of limb bones. The steps taken in attributing the limb bones are listed below.

1. When complete, or partially complete, skeletons of a species are known that include recognizable cranial parts, then limb bones of those skeletons are used as the basis for descriptions. Species for which this applies are *Megaladapis edwardsi* (von Lorenz, 1905; Lamberton, 1934c), *Archaeolemur majori* (British Museum Collection), *Archaeolemur edwardsi* (British Museum Collection), and *Daubentonia robusta* (Lamberton, 1934b).

2. When one species of a genus is well known, the limb bones of

another species of the genus can easily be recognized by inference, taking into account the differences in size of the cranial remains. Species for which this applies are *Megaladapis grandidieri, Megaladapis madagascariensis, Varecia insignis,* and *Varecia jullyi.*

3. Where fragmentary cranial remains are associated with limb bones and when other methods of reasoning do not invalidate the association, it is considered reasonable to take the association as genuine. This is the case for *Hadropithecus* (von Lorenz, 1902) and *Mesopropithecus* (Lamberton, 1938).

4. Where two genera are closely similar in cranial characters and where the limb bones are well known for one genus, it is thought reasonable to look for similar postcranial bones for the second genus. This reasoning can be applied to *Hadropithecus stenognathus* in conjunction with premise (3) and using inferred similarities with the closely related genus *Archaeolemur.*

5. If, after checking the collections by the reasoning given above, limb bones and cranial parts of compatible size remain, then a tentative association can be made. This method must be checked against the faunal lists from other sites. This is the only way, at present, of assigning limb bones to *Palaeopropithecus ingens* and *Archaeoindris fontoynonti,* and in a survey of all the specimens in London and Tananarive, it holds good. The final check, of course, is the finding of an articulated skeleton.

Insofar as the limb bones of most extinct Madagascan species are isolated finds, limb proportions for the species have been calculated by basing the usual proportional indices on the means of as many bones as possible (see Table III). When checked against articulated material, this method seems reasonably accurate.

Several authors have, in the course of mainly descriptive studies, briefly given their opinions as to the mode of life, including locomotion, of these forms. A summary of the opinions of early authors can be made for each genus: *Varecia insignis*—a bent-limbed arboreal quadruped leaping about in the tops of the branches (Carleton, 1936). *Archaeoindris fontoynonti*—arboreal (Piveteau, 1961). *Palaeopropithecus*—aquatic (Standing, 1908); arboreal locomotion as in *Bradypus* and *Choleopus,* the sloths (Carleton, 1936); aquatic, climbing trees and plunging into lakes (Sera, 1938, 1950); possibly arboreal and aquatic phases alternated with the seasons, a swamp existence being the equivalent of aestivation (Lamberton, 1944–1945); arboreal (Piveteau, 1961). *Archaeolemur*—arboreal (Carleton, 1936); arboreal, had begun to experiment with brachiation but failed to reach the level attained in higher primates (Lamberton, 1937); aquatic (Sera,

1950). *Megaladapis*—arboreal but perhaps with more of the habits of the "cave-bear" (von Lorenz, 1905); arboreal locomotion as in *Pan* and *Gorilla* (Standing, 1908); ground living (Lamberton, 1934c); aquatic (Sera, 1950). "Presumably aquatic" (Hill, 1953); arboreal (Zapfe, 1963).

The "aquatic" theory was originated by Standing (1908) to explain how the skulls of *Palaeopropithecus* from Ampasambazimba could have been bitten by crocodiles. The limb bones assigned by him to this genus were, unfortunately, those of *Megaladapis*. His conclusions on the limb morphology, that there were clear indications of swimming adaptations, are now seen to be invalid. His assessment of the limbs now assigned to *Palaeopropithecus* as that of a "brachiating" form is more or less correct. Sera, in two papers (1938, 1950) continued to use Standing's attributions and added an astragalonavicular of a crocodile as a capitate, a talus of *Megaladapis* as a talus, and a fibula, also of *Megaladapis*, as a clavicle. Using these bones, he reconstructed the mode of climbing and diving of *Palaeopropithecus*.

Sera found several features of the skull of *Palaeopropithecus* that he took as indicating a complicated glottal apparatus for use in diving, saying that the skull of the howler monkey also closely approximated to this condition. The most likely explanation of Sera's "thyreohyaloid pit" is that the boxlike hyoid of the howler monkey was also developed in *Palaeopropithecus*. In a later paper, Sera (1950) developed his "aquatic" theory which included almost all the fossil primates known and even postulated an aquatic origin for man. In view of the nature of his evidence, all stemming from the fact that Standing did not recognize a death assemblage at Ampasambazimba, the aquatic mode of life of any extinct lemuroid is discounted. Lamberton (1956) gives a complete account of the dismissal of Sera's theory.

Carleton (1936) was far from correct in supposing that *Archaeolemur* had many traces of brachiating habits in its forelimb skeleton. Not one of the characteristics of brachiating forms enumerated by Napier and Davis (1959) is present in *Archaeolemur*. Carleton was much closer to the truth in her assessment of *Palaeopropithecus* as a slothlike form, for in Lamberton (1944–1945) and in this study, it is seen that this was an hanging form, much like the modern orangutan.

The recognition of *Megaladapis* as an arboreal form has been slow to gain favor, probably because of a misconception of the size of even the largest species (*M. edwardsi* was only the size of a large dog). Zapfe (1963), in a recent communication, has pointed out that the hands and feet of *Megaladapis* give a clear indication of an arboreal existence. It is seen from this summary that attention has been concentrated on *Megaladapis* and *Palaeopropithecus,* two of the largest forms.

Varecia insignis and V. jullyi

These two extinct species of *Varecia* were placed in the genus *Lemur* by Standing and the subgenus *Pachylemur* by Lamberton, but enough morphological and behavioral differences from the other lemurs warrant the use of Gray's generic name for the ruffed lemurs. The skeletal remains of these two species are quite close in form to those of *V. variegatus.* They differ, however, in their greater robusticity and their calculated limb proportions. The fore- and hindlimbs are of more equal length than those of the modern species. It may be that this indicates a certain amount of ground living activity but apart from the rather robust bones, the intermembral and brachial indices, and the shortness of the cuboid, there is little else in the skeleton to suggest terrestrialism. The relative shortening of the hind limbs would presumably have limited the amount of leaping in the total locomotor pattern and the general indications are of more stealthy locomotor habits than living species.

Archaeolemur and Hadropithecus

Much of the evidence from the limb bones of these two forms points to their having had a quadrupedal form of locomotion—a locomotion of the kind found today in terrestrial cercopithecoid monkeys. Interestingly, it is the cranial anatomy of these two forms that is closest to the cranial anatomy of the monkeys, so much so that Standing (1908) considered them to represent retrogressive anthropoids, not lemuroids. In only one element, the scapula, does the known skeleton of *Archaeolemur* species not show the features typical of terrestrial cercopithecoid monkeys. Since articulated skeletons of these species are known and there is no doubt that the assigned scapulae are correctly allocated, *Archaeolemur* is an example of the difficulties inherent in making predictions of locomotor behavior on single parts of an integrated whole.

Taking *Archaeolemur* species first, the following are some of the features that indicate a terrestrial quadrupedal digitigrade locomotion similar to that of modern species of *Papio.* The greater tuberosity of the humerus surpasses the height of the head and is very rugose. This feature is considered by Jolly (1964) to be related to the extensor action of m. supraspinatus during the recovery phase of walking and is well developed in terrestrial monkeys. The m. brachialis flange is poorly developed in *Archaeolemur.* This is also characteristic of terrestrial monkeys for, as

Jolly points out, this muscle would only normally be active in the recovery phase of the limb. In arboreal forms, powered flexion of the elbow during climbing leads to a developed m. brachialis and its associated bony flange. The medial epicondyle is poorly developed and is directed backward and medially in *Archaeolemur* and in the terrestrial monkeys. This is probably related to less important digital flexors and, as Jolly suggests, may increase the moment arm of m. pronator teres thereby leading to a decrease in forelimb mobility. The need for rigidity of the elbow joint in terrestrial forms probably results in the heightening of the coronoid process of the ulna and this affects the shape of the trochlea. In arboreal forms, the elbow joint stability is sacrificed and a more mobile radius, with a clearly defined and separated capitulum, is found. Napier and Davis (1959) asserted that the form of the keel between the trochlea and capitulum provides a stabilizing element for the ulna in both flexion and extension. This may be taken to be an attempt at elbow joint stabilization in forms with well separated radius and ulna. *Archaeolemur* has a poorly rounded capitulum and the trochlear and capitular surfaces are not well differentiated, as in terrestrial cercopithecoids. A wide and well-defined olecranon fossa with deep margins is seen in both *Archaeolemur* and terrestrial cercopithecoids and is to some extent related to the posteriorly displaced medial epicondyle. However, in this type of elbow joint the posteromedially directed olecranon process is accepted tightly into the deep fossa and provides great joint stability in positions of extension.

The ulna of terrestrial cercopithecoids displays some distinctive features that are found in *Archaeolemur*. A posteromedially directed olecranon process is considered by Jolly (1964) to be related to greater mechanical advantage of m. triceps in positions of extension. The straighter olecranon process of arboreal forms is most efficient at small angles of flexion at the elbow joint and least efficient at high angles of flexion. Jolly also pointed out that a depressed radial notch on the ulna is related to forearm stability in terrestrial forms. *Archaeolemur* also has a well excavated radial notch which contrasts the type found in arboreal primates where the notch is usually set at quite a high angle to the long axis of the shaft. Again the contrast between forearm mobility in the arboreal forms and forearm stability in the terrestrial forms is seen and again *Archaeolemur* matches the terrestrial condition.

The radius of *Archaeolemur* matches those of the larger digitigrade monkeys in that it is stout and pillarlike and only slightly curved, in order, presumably, to transmit mainly compressive forces. The neck is poorly defined, contrasting with the condition found in arboreal forms.

In the hindlimb, the development of a large greater trochanter of the femur, a strong anterior bowing of the femoral shaft, and a shallow patellar groove are features matched in terrestrial monkeys. As in the forelimb, the hindlimb bones are robust, with relatively large joint surfaces.

In the hand and foot, features indicating digitigrade locomotion are found. The foot is better known than the hand and the reduced length of the tarsal bones, especially the short, broad cuboid, matches the baboon condition. The metatarsal heads have a distinct "step" on the dorsal aspect that is possibly a limiting factor in hyperextension of the metatarsophalangeal joints, as in terrestrial monkeys. The fifth metatarsal of *Archaeolemur* has a very well developed styloid process for m. peroneus brevis, again a feature associated with the feet of terrestrial forms. The calculated limb proportions for *Archaeolemur* are close to those of ground-living Old World Monkeys. The brachial and intermembral indices are closest to those of *Erythrocebus* and the crural index to that of *Macaca*. Apart from *Theropithecus, Erythrocebus* is the most terrestrially committed of all the monkeys.

It seems, then, that in only one major feature (the scapula) does the known skeleton of *Archaeolemur* not resemble that of a terrestrially quadrupedal species. What bones we have of *Hadropithecus* are rather similar to but more gracile than those of *Archaeolemur*. Recently, Jolly (1970) has attempted to bring evidence from the limb bones to help support his thesis that *Hadropithecus* was to *Archaeolemur* as *Theropithecus* is to *Papio*. The sample that we have for *Hadropithecus* is unfortunately too small to attempt to give mean limb proportions and some of the elements may not belong to the genus. Jolly (1970) unfortunately calculated a humerofemoral index of *Archaeolemur* using my figures for radius and tibia (Walker, 1967b), and it would have suited his case better, probably, if he had omitted the limb bone evidence.

Palaeopropithecus

Based on the results of several lines of inquiry, *Palaeopropithecus ingens* (I do not believe that there is a separate species *P. maximus*) must have moved in very much the same way as an orangutan. The calculated intermembral index is just within the upper limits of the range of variation seen in gibbons and *Pongo*. The brachial index, being less that 100, is closer to those of *Pongo* than the gibbons. Many features of the limb bones are matched by those of arm-swinging or arm-hanging primates and the forelimb shows clear evidence of being a well-developed suspensory organ.

Scapula. The glenoid is directed cranially and the glenoid cavity angle of Oxnard (1963) is 106°, clearly falling in that author's "brachiating" range of 101 to 124°. Unfortunately, other angles used by Oxnard (1963) and Ashton and Oxnard (1964) are either difficult to record or depend upon the exact insertion of muscle fibers that in fossil specimens is impossible to see. For instance, the lower angle is very rounded in *Palaeopropithecus* and angles of between 60 and 90° can be taken equally plausibly. Between prosimians and anthropoids muscle insertions may be strikingly different and lead to confusion when taking measuring points. As an example, the insertion of m. trapezius in lemuroids is along the whole length of the spine (Jouffroy, 1962) whereas in anthropoids this is usually along the lateral part only. Where was m. trapezius inserted in *Palaeopropithecus*? In only one reliable case, then, can the measurements of the scapula of *Palaeopropithecus* be compared with those taken by Ashton and Oxnard. This case, however, does suggest an habitually suspended thorax.

Humerus. Medial torsion of the head of the humerus in *Palaeopropithecus* is from 105 to 130° and three out of five values fall within observed ranges for gibbons and *Ateles.* Both greater and lesser tuberosities are low, an essential feature of an extremely mobile shoulder joint. The deltoid index (an estimate of the relative length of the deltoid crest) is high in "brachiating" forms (Napier and Davis, 1959) and values for *Palaeopropithecus* fall close to those of *Pan* with a range of 54.5 to 57.9. The medial epicondyle was well developed in *Palaeopropithecus,* as it is in "brachiating" forms, but as all arboreal lemurines have a large medial epicondyle the significance of this feature relates to the strong development of the flexor muscles of the hand. The distal condyle is well differentiated into a ball-like capitulum and clearly defined trochlea with a strong keel between the two parts. Evidently movements of pronation and supination were considerable. The range of the trochlear index (as defined by Napier and Davis, 1959) is very wide indeed and covers the range of species of several locomotor types given by Napier and Davis.

Ulna. The olecranon process of *Palaeopropithecus* is very short indeed and is associated with a relatively small m. triceps mass (Oxnard, 1963), typical of forms in which the flexors are concerned with weight transmission and the extensors deal mainly with extending the arm for the next reach. The condition seen in gibbons where the coronoid process curves backward to make a keeled hook (Le Gros Clark and Thomas, 1951) is also seen in *Palaeopropithecus.* This is, presumably, a stabilizing mechanism for the elbow joint, preventing lateral or medial deviation and making the joint more of a simple hinge. The shaft of the ulna is considerably attenuated and strongly convex posteriorly. The styloid process is a discrete spherical, or slightly pointed, knob with a clearly defined neck.

Radius. The shaft of the radius is markedly bowed laterally and has a strongly developed interosseous membrane crest. The relative length of the neck in *Palaeopropithecus* is greater than that found in cercopithecoid monkeys and approaches, but never reaches, the proportions found in gibbons. The carpal surface is tilted anteriorly, relative to the shaft. Napier and Davis (1959) considered that this feature facilitates flexion of the wrist and is important in "brachiating" forms, but this could equally apply to knuckle-walking species as well.

The Hand. The metacarpals and phalanges of *Palaeopropithecus* are noteworthy for their extreme length and narrowness. The distal articular surfaces of the metacarpals, as well as the phalanges, are grooved. All elements are strongly curved toward the palm. The maximum lengths of metacarpals and proximal and middle phalanges are estimated by adding the maximum lengths of these strongly curved bones, yielding a combined chord length of 236 mm (no terminal phalanges are known). The hand in life would have had the appearance of a very long thin hook. We have no bones of the first digit or carpus.

Pelvis. The pelvis of *Palaeopropithecus* is most remarkable. In many ways its structural characteristics relating to upright posture are more advanced than those of gibbons. The ilium is long but the ischium is relatively short. The iliac crest curves anteriorly and the anterior superior iliac spine faces anterolaterally. The ventral surfaces of the ilia are concave, to produce a basinlike pelvis. The curving forward of the ilia, as in man, is taken by Le Gros Clark (1959) to be associated with stronger abdominal wall musculature to support the viscera in an upright form. The acetabula point downward and laterally, recalling the human condition. The pubis is wide and flat and hints at a strongly developed m. rectus abdominis. There are six fused vertebrae in the sacrum. Schultz (1964) records four as the maximum number in lemuroids but from two to nine are found in lorisines. In the absence of other features reminiscent of the lorisine skeleton, it is fairly safe to assume that in this case the resemblance lies with the "brachiators" in which high numbers of sacral vertebrae (the majority have from five to six; Schultz, 1964) are found.

Femur. In this bone the closest resemblance is with the sloths, not other primates. Of other primates, however, the orangutan matches the *Palaeopropithecus* condition most closely. The high angulation of the neck, the small greater trochanter, the anteroposteriorly flattened shaft and the wide, shallow patellar groove are all features that the femora of *Pongo* and *Palaeopropithecus* share. The emphasis seems to be on hip joint mobility rather than stability and powerful adduction rather than powerful extension of the thigh (Appleton, 1921, presents a discussion of the diameters of femoral shafts in relation to function). *Palaeopropithecus* femora also exhibit a "carrying angle" of the femoral condyles ranging from 4 to 13°.

Tibia and Fibula. These are rather thin, relatively long bones with no malleoli. This presumably means that ankle joint mobility is greatly increased with stability sacrificed as a consequence.

Foot. The foot, like the hand, is a long thin, rather rigid, hook. As in the hand, grooves for the flexor tendons are well developed. It is difficult to think of a foot posture that would allow *Palaeopropithecus* to engage in terrestrial locomotion, since the long, hooklike foot and the mobile ankle joint are so completely unsuitable for weight bearing.

To summarize, it seems that *Palaeopropithecus* was adapted for an almost completely arboreal existence with the body mainly suspended from the arms but with all four, hooklike, extremities involved in climbing. The closest living locomotor model would be the orangutan, which uses steady, rather slow, four-limbed climbing, rather than the gibbons that use a fast brachiating locomotion.

Archaeoindris fontoynonti

This species, closely related to *Palaeopropithecus* (it perhaps should be included in that genus), is known from the type skull and a few limb bones. Fragments described by Lamberton (1934a) are probably from one immature individual, and two fibulae in Tananarive, together with a femur called "*Lemuridotherium*" by Standing (1910), are of an adult specimen. In some major respects these bones resemble *Megaladapis* and until this species is better known it seems best to assume that *Archaeoindris* had a presumed locomotion habit similar to *Megaladapis.*

Megaladapis

The osteology of the two larger species (*M. edwardsi, M. grandidieri*) are quite well known and, although the evidence is poor, what little we have of *M. madagascariensis* seems to suggest that, apart from its smaller size, it was essentially similar in limb morphology. No primate species, past or present, seems to have the peculiar skeletal morphology or proportions of *Megaladapis.* The nearest living analog is not even a eutherian mammal, but is the koala, *Phascolarctos cinereus.* In both general skull morphology and in some details of the limb bones, the koala and *Megaladapis* have clear resemblances. The major differences lie in the hands and feet (marsupials have clawed extremities) but even so the koala does have a large grasping hallux that bears a flattened nail.

The limb bones of *Megaladapis* and the koala resemble each other in the following respects. The limb proportions of *Megaladapis* are given in Table III. The means of five *P. cinereus* indices are, brachial index 111.1, crural index 78.9, and intermembral index 95.4. The differences can be explained as allometric. A linear allometric regression is obtained when log forelimb length is plotted against log hindlimb length for species of both genera and hence a projected species of *Megaladapis* smaller than *M. madagascariensis* would be expected to have limb proportions close to those of the living koala.

TABLE III
CALCULATED LIMB PROPORTIONS OF SUBFOSSIL LEMUROIDS[a]

	Numbers of limb bones						
	Humerus	Radius	Femur	Tibia	Brachial index	Crural index	Intermembral index
Varecia							
V. insignis } *V. jullyi*	8	6	10	10	113.06 (S.E. 1.95)	85.16 (S.E. 2.99)	95.96 (S.E. 3.63)
Varecia (Lamberton, 1944)	51	15	52	39	113.90	89.57	98.20
Palaeopropithecus	23	11	17	9	95.33 (S.E. 9.01)	101.02 (S.E. 4.87)	147.04 (S.E. 4.16)
Archaeolemur	8	8	11	9	107.01 (S.E. 3.62)	91.07 (S.E. 3.22)	88.73 (S.E. 2.10)
British Museum skeleton					106.20	91.08	88.67
Megaladapis edwardsi	5	6	7	10	88.34 (S.E. 2.23)	72.71 (S.E. 1.34)	121.42 (S.E. 1.84)
Megaladapis grandidieri	6	3	6	3	88.57 (S.E. 2.94)	83.14 (S.E. 2.40)	115.74 (S.E. 3.34)
Megaladapis madagascariensis	—	—	2	1	—	88.6	—

[a] Proportional indices and their standard errors (S.E.) of the mean are given.

Scapulae of both *P. cinereus* and *Megaladapis* are elongated transversely with strong, high spines, and well-developed acromial and coracoid processes. In the humerus, shared characteristics are the backwardly facing head, well-developed greater tuberosity, moderately developed deltopectoral crest, greatly expanded m. brachialis flange, large medial epicondyle, and rounded and separated capitulum. The ulnae have stout, straight olecranon processes, very high coronoid processes with shallow greater sigmoid notches, stout shafts, and long styloid processes. The radii are alike in their curvatures, rounded heads, and details of the m. pronator teres insertions. The hand bones of *Megaladapis* indicate a strong grasping hand with the pollex set against the other digits. Long phalanges as seen in vertical clinging and leaping species were present. In both the koala and *Megaladapis,* the pelvis has a well-developed ilium with conspicuous beaklike anterior superior iliac spine, large anterior inferior iliac spine, and short but stout ischium and pubis. The femur of *P. cinereus* is more slender and cylindrical than those of *Megaladapis* but strong points of similarity are the short neck, well-rounded head, large greater trochanter, constricted digital fossa, wide patellar grooves, and backwardly facing condyles. The resemblances between the tibiae of the two forms is quite remarkable and include the morphology of the condyles, the low attachment of the m. sartorius complex on the shaft, and short stout malleoli. The foot of *Megaladapis* is typically lemuroid and was obviously a grasping organ, with a strong abducted hallux and long digits.

The skeletal resemblances are so strong between this extinct lemuroid and the marsupial that, in the absence of any suitable primate as a living model, something very like the locomotion of the koala should be taken as the presumed locomotion of *Megaladapis.* Koalas are leaf-eating animals that feed exclusively on certain species of *Eucalyptus.* The resting postures are essentially the same as the vertical clinging seen in some prosimians, with a vertical trunk strongly grasped by the feet and the forelimbs wrapped round the support. Short leaps are recorded between adjacent vertical stems. Analysis of cine film shows that the leaps of the koala are essentially the same as those of *Propithecus.* When climbing vertical stems, the koala uses a series of upward hops and when on the ground uses either a very slow quadrupedal locomotion or a froglike hopping very similar to that seen in the Alaotran *Hapalemur griseus.* The presumed locomotion of *Megaladapis* could, therefore, easily be a modified vertical clinging and leaping locomotion in which leaping powers were reduced in favor of large body size.

Mesopropithecus and Neopropithecus

Species of these genera (if they are generically distinct) have only poorly known postcranial remains, but these resemble small *Megaladapis* bones. Until more evidence is available it seems wisest to leave open the question of presumed locomotor habits.

Daubentonia robusta

The limb bones of this species are virtually identical with modern *D. madagascariensis* except that they are roughly one-third longer.

Locomotor Developments in the Prosimians

From the available remains of Eocene to Oligocene prosimians from America and Europe it seems that the dominant locomotor habit, if not the only one in those times, was the vertical clinging and leaping locomotion like that seen today in indrisines and tarsiers. There is no strong evidence of quadrupedal adaptations of the *Lemur* type. Since vertical clinging and leaping locomotion is hindlimb dominated, the early theory of Gidley (1919), that the grasping foot developed before an efficient grasping hand, seems to hold good. In some living species that use this type of locomotion there is evidence that a basic leaping locomotion has been further developed along the same trend (in the even greater elongation of the tarsus in galagines and *Tarsius* spp., for instance), but others have limb skeletons so close in proportions and morphology to the fossil species that it seems simpler to think of their having retained an early successful locomotor habit.

In Africa, the galagines have developed the leaping locomotion and are still a successful group, being widespread geographically and also occupying a number of varying habitats. The fossil evidence suggests that the Miocene ancestors differed little in locomotor ability from their modern descendants. The lorisines of Asia and Africa have developed a peculiar locomotion of their own that is considered to be related to their diet and their method of catching prey (Walker, 1969). This is in essence an arboreal locomotion and these animals are restricted to areas of fairly dense forest. Other African radiations, first of hominoid then of cercopithecoid forms, probably contributed to the restriction of prosimians to nocturnal and mainly insectivorous and gum-eating habits. Why the lorisines and not galagines should be present in Asia remains a mystery and all attempts to explain this lean heavily on speculation about the nature of the Afro-Arabian and Eurasian link in the mid-Tertiary. *Indraloris* is now known from further specimens and is probably not a lorisine (Tattersall, 1969), so the fossil record of lorisine evolution in Asia seems to be completely lacking.

In Madagascar the prosimians seem to have produced, without competition from higher primates, a remarkable variety of locomotor types. The recent extinction of many species of lemur has deprived us of a living laboratory of primate locomotor adaptation. Only two basic types of locomotion are seen today in the remaining lemurs. All the slower, larger, and terrestrial species have vanished together with the elephant birds within the last millenium (Walker, 1967a). It must be stressed that these extinct lemurs are so recently extinct that they cannot be considered as ancestors of living species; in fact the subfossil lemurs are found lying side-by-side with the bones of identifiable Recent species which, because they were not extinct, excited no comments from the paleontologists. It is also worth recording that it was the indrisine lemurs and their relations that were mainly involved in this adaptive radiation and not the lemurine lemurs. The genus *Lemur,* often used as an example of a typical Madagascan form, has probably undergone fairly rapid speciation in the last few thousand years and one of the indrisine lemurs would probably represent the more typical stock.

The development of large species (e.g., *Megaladapis*) with locomotor habits rather like those of the living koala and others (e.g., *Palaeopropithecus*) that resembled the orangutans means that the Madagascan lemurs have met the adaptive problems of increase of body size and answered it in at least two different ways. One adaptive solution, that of using forelimb suspension has been used independently by several other primate groups—the gibbons, certain great apes, certain South American monkeys, and oreopithecines. The second answer, that seemingly retains more of the primitive locomotor pattern, has not been seen in any other group of primates. Ground-living adaptations are seen in two clearly related genera, *Archaeolemur* and *Hadropithecus,* and as in cercopithecoid monkeys that are terrestrially adapted, digitigrade quadrupedal locomotion seems to have been of selective advantage. Even details of teeth and facial structure parallel the condition seen in terrestrial cercopithecoids. In some respects the correspondence is not so exact and, for example, further studies of the shoulder girdle of *Archaeolemur* would be well worthwhile in order to find an explanation for the rather expanded scapula that contrasts with the baboon condition. The locomotion of the insectivorous aye-aye is unique in many ways and a quantitative study of this species' locomotion contrasted with that of another quadrupedal lemur would probably reveal differences due to dietary habits.

The smaller lemurs, many of which are still extant, have locomotor patterns of basically two types, vertical clinging and leaping and active quadrupedalism, and there are several species in which the distinction is not clear cut, suggesting the derivation of one type from the other.

The prosimians, taken as a whole, give indications of locomotor developments from a primitive hindlimb dominated type through to most types of arboreal and ground locomotion seen in other primate groups elsewhere in the world. Some of the developments on the isolated block of Madagascar enable the testing of functional interpretations based on one or two independent developments elsewhere. The Madagascan radiation demonstrates the adaptive potential of the early primate stocks and helps us to distinguish certain recurring trends of locomotor evolution within the order.

References

Appleton, A. B. (1921). The influence of function on the conformation of long bones. *Proc. Cambridge Phil. Soc.* **20**, 374–387.

Ashton, E. H., and Oxnard, C. E. (1964). Functional adaptation in the primate shoulder girdle. *Proc. Zool. Soc. London* **142**(1), 49–66.

Bartlett, A. D. (1863). Description of a new species of *Galago*. *Proc. Zool. Soc. London*, pp. 231–233.

Bishop, A. (1964). Use of the hand in lower primates. *In* "Evolutionary and Genetic Biology of Primates" (J. Buettner-Janusch, ed.), Vol. 1, pp. 133–225. Academic Press, New York.

Carleton, A. (1936). The limb bones and vertebrae of the extinct lemurs of Madagascar. *Proc. Zool. Soc. London*, pp. 281–307.

Charles-Dominique, P. (1971). Eco-éthologie des prosimiens du Gabon. *Biol. Gabon.* **7**, 121–228.

Charles-Dominique, P., and Hladik, C. M. (1971). Le *Lepilemur* du sud de Madagascar: ecologie, alimentation et vie sociale. *Terre Vie* **25**, 3–66.

Day, M. H., and Walker, A. (1969). New prosimian remains from Early Tertiary deposits of Southern England. *Folia Primatol.* **10**, 139–145.

Filhol, H. (1883). Observations relatives au Mémoire de M. Cope intitulé: Relation des horizons rendement des Débris d'Animaux vertèbres fossiles en Amérique. *Ann. Sci. Geol.* **14**, 1–51.

Gidley, J. W. (1919). Significance of the divergence of the first digit in the primitive mammalian foot. *J. Wash. Acad. Sci.* **9**, 273–280.

Grand, T. I. (1967). The functional anatomy of the ankle and foot of the slow loris. *Amer. J. Phys. Anthropol.* **26**, 207–218.

Gregory, W. K. (1920). On the structure and relations of *Notharctus*, an American Eocene primate. *Mem. Amer. Mus. Natur. Hist.* 3, 49–243.

Haddow, A. J., and Ellice, J. M. (1964). Studies on bush-babies (*Galago* spp.) with special reference to yellow fever epidemiology. *Trans. Roy. Soc. Trop. Med. Hyg.* **58**, 521–538.

Hall-Craggs, E. C. B. (1964). The jump of the bush-baby—Photographic analysis. *Med. Biol. Illust.* **14**, 170–174.

Hall-Craggs, E. C. B. (1965). An analysis of the jump of the lesser Galago (*Galago senegalensis*). *J. Zool.* **147**, 20–29.

Hall-Craggs, E. C. B. (1966). Rotational movements in the foot of *Galago senegalensis*. *Anat. Rec.* **154**, 287–293.

Hildebrand, M. (1967). Symmetrical gaits of primates. *Amer. J. Phys. Anthropol.* **26**, 119–130.

Hill, W. C. O. (1953). Primates, vol. 1, Edinburgh University Press, Edinburgh.

Hill, W. C. O., Porter, A., and Southwick, M. D. (1952). The natural history, endoparasites and pseudo-parasites of the tarsiers (*Tarsius carbonarius*) recently living in the Society's gardens. *Proc. Zool. Soc. London* **122**(1), 79–119.

Jolly, A. (1966). "Lemur Behaviour." Chicago University Press, Chicago, Illinois.

Jolly, C. J. (1964). The origins and specialization of the long-faced Cercopithecoidea. Ph.D. Thesis, University of London, London.

Jolly, C. J. (1970). *Hadropithecus:* a lemuroid small-object feeder. *Man* **5**, 619–626.

Jouffroy, K. K. (1962). La musculature des membres chez les lémuriens de Madagascar. *Mus. Nat. Hist. Natur.* **26**(2), 1–322.

Lamberton, C. (1934a). *L'Archaeoindris fontoynonti* Standing 1908. *Mem. Acad. Malgache* **17**, 1–39.

Lamberton, C. (1934b). *Cheiromys robustus* sp. nov. Lamb. *Mem. Acad. Malgache* **17**, 40–46.

Lamberton, C. (1934c). Les *Megaladapis. Mem. Acad. Malgache* **17**, 47–105.

Lamberton, C. (1937). Les hadropitheques. *Bull. Acad. Malgache* **20**, 127–170.

Lamberton, C. (1938). Nouveaux lémuriens fossiles du groupes des propitheques. *Mem. Acad. Malgache* **27**, 1–54.

Lamberton, C. (1944–1945). Bradytherium ou palaeopropitheque? *Bull. Acad. Malgache* **26**, 89–140.

Lamberton, C. (1946). Les pachylemurs. *Bull. Acad. Malgache* **27**, 7–22.

Lamberton, C. (1956). Examen de quelques hypothèses de Sera concernant les lémuriens fossiles et actuels. *Bull. Acad. Malgache* **34**, 51–65.

Lauvedon, L. (1933). Le Aye-aye. *Terre Vie* **3**, 142–152.

Le Gros Clark, W. E. (1924). Notes on the living tarsier (*Tarsius spectrum*). *Proc. Zool. Soc. London,* pp. 217–233.

Le Gros Clark, W. E. (1959). "The Antecedents of Man." Edinburgh University Press, Edinburgh.

Le Gros Clark, W. E., and Thomas, D. P. (1951). Associated jaws and limb bones of *Limnopithecus macinnesi. Fossil Mammals Africa No.* 3, 1–27. Brit. Mus. Natur. Hist., London.

Martin, R. D. (1972). A preliminary field-study of the lesser mouse lemur (*Microcebus murinus* J. F. Miller 1777). *Z. Tierpsychol.* **9**, 43–89.

Maynard-Smith, J., and Savage, R. J. G. (1956). Some locomotor adaptation in mammals. *J. Linn. Soc. London* Zool. **42**(288), 603–622.

Montagna, W., and Machida, H. (1966). The skin of primates, 32. The Phillipine tarsier (*Tarsius syrichta*). *Amer. J. Phys. Anthropol.* **25**, 71–84.

Napier, J. R., and Davis, P. R. (1959). The fore-limb skeleton and associated remains of *Proconsul africanus. Fossil Mammals Africa* **16**, 1–69. Brit. Mus. Natur. Hist., London.

Napier, J. R., and Walker, A. C. (1967). Vertical clinging and leaping—a newly recognized category of primate locomotion. *Folia Primatol.* **6**, 204–219.

Oxnard, C. E. (1963). Locomotor adaptations of the primate forelimbs. *Symp. Zool. Soc. London* **10**, 165–182.

Petter, J.-J. (1962). Recherches sur l'écologie et l'éthologie des lémuriens malgaches. *Mem. Mus. Nat. Hist. Natur.* **A27**, 1–146.

Petter, J.-J., and Hladik, C. M. (1970). Observations sur le domaine vital et la densité de population de *Loris tardigradus* dans les forêts de Ceylon. *Mammalia* **34**, 394–409.

Petter, J.-J., and Peyrieras, A. (1970). Nouvelle contribution a l'étude d'un lémurien malgache, le Aye-Aye (*Daubentonia madagascariensis* E. Geoffroy). *Mammalia* **34**, 167–193.

Petter, J.-J., and Peyrieras, A. (1971). Observations éco-éthologiques sur les lémuriens Malgaches du genre *Hapalemur*. *Terre Vie* **24**, 356–382.

Piveteau, J. (1961). Behaviour and ways of life of the fossil primates. *In* "Origin and Evolution of Man" (S. L. Washburn, ed.). Wenner-Gren Foundation, New York.

Rand, A. L. (1935). On the habits of some Madagascar mammals. *J. Mammal.* **16**, 89–104.

Reeve, E. C. R., and Huxley, J. S. (1945). Some problems in the study of allometric growth. *In* "Essays on Growth and Form" (W. E. Le Gros Clark and P. Medawar, eds.). Oxford Univ. Press, Oxford. pp. 121–156.

Schlosser, M. (1907). Beitrag zur osteologie und systematischen stellun der Gattung *Necrolemur,* sowie zur Stammesgeschiche der Primaten uberhaupt. *N. Jahrb. Min. Geol. Palaeontol.,* pp. 197–226.

Schultz, A. H. (1964). Age changes, sex differences and variability as factors in the classification of primates. *In* "Classification and Human Evolution" (S. L. Washburn, ed.). Wenner-Gren Foundation, New York.

Sera, G. L. (1938). Alcuni caratteri scheletrici di importanza ecologica e filletica nei Lemuri fossili ed attuali. Studi sulla paleobiologia e sulla filogenesi dei primati. *Paleontol. Ital.* **38**, 1–113.

Sera, G. L. (1950). Ulteriori osservazioni sui Lemuri fossili ed attuali. *Paleontol. Ital.,* p. 47.

Shaw, G. A. (1879). A few notes upon four species of lemurs, specimens of which were brought alive to England in 1878. *Proc. Zool. Soc. London,* pp. 132–136.

Simons, E. L. (1961). Notes on Eocene tarsioids and a revision of some Necrolemurines. *Bull. Brit. Mus. Natur. Hist.* **5**, 45–69.

Simpson, G. G. (1935). The Tiffany fauna, Upper Paleocene. II. Structure and relationships of *Plesiadapis*. *Amer. Mus. Nov.* **816**, 1–30.

Simpson, G. G. (1940). Studies on the earliest primates. *Bull. Amer. Mus Natur. Hist.* **77**, 185–212.

Simpson, G. G. (1965). *In* "Olduvai Gorge" (L. S. B. Leakey, ed.), pp. 15–16. Cambridge Univ. Press, Cambridge.

Standing, H. F. (1908). On recently discovered sub-fossil primates from Madagascar. *Trans. Zool. Soc. London* **18**, 59–162.

Standing, H. F. (1910). Note sur les ossements subfossiles provenant des fouilles d'Ampasambazimba. *Bull. Acad. Malgache* **7**, 62–64.

Tattersall, I. (1969). More on the ecology of *Ramapithecus*. *Nature* (*London*) **224**, 821–822.

Thomas, O. (1896). On the mammals obtained by Mr. John Whitehead during his recent expedition to the Philippines with field notes by the collector. *Trans. Zool. Soc. London* **54**(2), 387–398.

Vincent, F. (1969). Contribution a l'étude des prosimiens africains: le galago de Demidoff. Thèse de doctorat d'Etat des Sciences Naturelles, Paris.

von Lorenz, L. L. (1902). Ueber *Hadropithecus stenognathus*. *Denkschr. Kais. Akad. Wiss. Wien* **27**, 243–254.

von Lorenz, L. L. (1905). *Megaladapis edwardsi* G. Grandidier. *Denkschr. Kais. Akad. Wiss. Wien. Math. Natur.* **77**, 451–490.

Walker, A. (1967a). Patterns of Extinction among the subfossil Madagascan lemuroids. *In* "Pleistocene Extinctions" (H. E. Wright and P. S. Martin, eds.). Yale Univ. Press, New Haven, Connecticut.

Walker, A. (1967b). Locomotor adaptations in recent and fossil Madagascan lemurs. Ph.D. Thesis. pp. 1–535. University of London, London.

Walker, A. (1969). The locomotion of the lorises, with special reference to the potto. *E. Afr. Wildl. J.* **7**, 1–5.

Walker, A. (1970). Post-cranial remains of the Miocene Lorisidae of East Africa. *Amer. J. Phys. Anthropol.* 33, 249–262.

Weigelt, J. (1933). Neue Primaten aus der mitteleozanen (oberlutetischen) Braunkohle der Geiseltals. *Nova Acta Leopold* **1**, 97–156 and 321–323.

Zapfe, H. (1963). Lebensbild von *Megaladapis edwardsi* (Grandidier). *Folia Primatol.* **1**, 178–187.

Subject Index

M

N

U

V

W

Z